A Practical Guide to Industrial Ecology by Input-Output Analysis

Shinichiro Nakamura

A Practical Guide to Industrial Ecology by Input-Output Analysis

Matrix-Based Calculus of Sustainability

Springer

Shinichiro Nakamura
Faculty of Political Science and Economics
Waseda University
Tokyo, Japan

ISBN 978-3-031-43683-3 ISBN 978-3-031-43684-0 (eBook)
https://doi.org/10.1007/978-3-031-43684-0

This Springer imprint is published by the registered company Springer Nature Switzerland AG
The registered company address is: Gewerbestrasse 11, 6330 Cham, Switzerland

Paper in this product is recyclable.

To my wife Miyuki

Foreword

A Practical Guide to Industrial Ecology by Input-Output Analysis is intended as a textbook for students from diverse backgrounds taking a university course in Industrial Ecology (IE). Such courses are likely to be offered in a School of Engineering, compatible with the name "Industrial", but "Ecology" suggests that it might be offered in a different School, probably in a department of ecology—or in a new, interdisciplinary program related to the environment. The book's title focuses on "Input-Output Analysis" (IOA), a domain created by Wassily Leontief, a Nobel Laureate in economics, that is used mainly by economists but also by other researchers and practitioners concerned with a wide range of topics. The book is explicitly addressed not only to undergraduate and graduate students but also to both researchers and policymakers. The seven chapters and the Appendix can be read in any order; each is well structured so one can focus on only the sub-topics of direct interest. At the same time, all the diverse topics and chapters are well integrated. I read the manuscript following my interests and curiosity, and it became clear that the book fully delivers on its ambitious promises.

I first met Shinichiro Nakamura in Europe in the mid-1980s when we were both invited to serve as managing editors for a new economics journal. Shinichiro was a professor of economics at Waseda University in Tokyo at this time, and I had recently succeeded Professor Leontief as director of the Institute for Economic Analysis at New York University. The central focus of this book is not on extending economic theory or empirical analysis based on it, but on matter and energy, the natural laws governing them, and the practice of physics, especially of physical chemistry. In response to my questioning him on this apparent shift in perspective, Shinichiro explained that he became interested in an input-output framework that examines the full scope of waste management challenges, and he clearly chose Industrial Ecology as the most suitable home among other potential options. He prepared for this endeavor by an extensive reading of textbooks in physics and chemistry. He also met regularly with engineering professors in Japan working on a system engineering model of waste treatment and incorporated their model within his Waste Input-Output modeling framework. He subsequently collaborated with metallurgists on a well-funded project on material recycling, enabling him to deepen his knowledge of metallurgy. He was eventually named a Professor of Industrial Ecology in the same Faculty of Political Science and Economics, his title from 2010 until his retirement in 2022.

The International Society for Industrial Ecology (ISIE) was created in 1990 and has its roots in engineering. Professor Leontief developed the basic concepts behind input-output economics in the 1930s and 1940s, and the International Input-Output Association (IIOA) was established in 1988. University degrees have long been offered by departments of engineering and departments of economics around the world, but now many departments correspond to sub-disciplines or include content that crosses over disciplinary borders. While the contemporary mainstream of economics is based on macroeconomics and microeconomics, the book describes in detail how input-output economics covers an entire economy at a mesoeconomic level of detail. This feature makes it possible to capture the interdependence of the production sectors of the economy in terms of products and processes, including their use of resources, as well as the interdependence between sector-level demand and supply. The book explains how the quantification of inputs and outputs, for individual production sectors and for the economy as a whole, avoids double counting in the latter. This is assured by the conceptual structure of the sector-level input-output technological requirements that comprise the core of the database that is discussed in detail in the book.

Input-output analysis was introduced within Industrial Ecology from its earliest years. At the present time a significant portion of the members of ISIE are also members of the Section called Environmentally Extended Input-Output analysis (EEIO). EEIO situates Life-Cycle Assessment (LCA) and Material Flow Analysis (MFA) within an input-output modeling framework, these being three of the central analytic constructs of IE. Shinichiro's fundamental contribution is the creation of what he named Waste Input-Output (WIO), which focuses on the end-of-life stage of products and infrastructure. His WIO model extensions and compilation of engineering data have been used in many contexts to evaluate alternative technological options for reducing and processing wastes and resource recovery. Much of this work has been done in collaboration with Yasushi Kondo, his co-author of the first book on WIO, and with engineer colleagues.

I find two chapters particularly informative. Chapter 2, on chemical processes and physical flows, provides concrete and detailed descriptions of the provision of diverse plant and animal sources of foods. It also contains comparable detail for cement and for use of major metals. This perspective on production is rarely if ever addressed in other books on IOA. The longest chapter of the book is Chap. 5: it provides a thorough description of the central role of hybrid LCA within WIO and EEIO.

Two topics that importantly get special attention in the book are the choice of units of measurement for different variables and the representation of investment in built capital. Input-output data from statistical databases for the past are compiled in money values only, while data quantifying resource requirements, obtained from other sources, are measured in physical units. The text discusses measuring products and process components in mixed units and managing their diversity. The input-output framework could then make it possible to examine the impacts of potential changes in physical quantities on unit prices, and vice versa. Secondly, a single, economywide vector of investment is found in all standard input-output databases, and the book points out the need to extend it into a matrix of sector-specific investment outlays with a column for each production sector. This step establishes an estimate for sectoral wastes associated with the sector's investment in built capital.

The ISIE website describes Industrial Ecology as "Science for sustainability and circular economy". WIO establishes boundaries that cover, and delimit, the circular economy and provides methods for evaluating alternative scenarios for proceeding. The book covers this material. Importantly, WIO has, in addition, already stimulated the emergence of a distinct, substantial, and growing research community that explicitly identifies its work as Waste Input-Output. Building on Shinichiro's initiative, I suggest supplementing WIO by DIO—Development Input-Output. That step could address the early temporal stages of a fully dynamic framework, namely by putting new built capital in place to realize sustainable development objectives, complementing WIO's coverage of the end-of-life stages.

New York, USA
July 2023

Faye Duchin

Preface

This book aims to provide practical guidance on methodological issues in Industrial Ecology (IE) from the perspective of Input-Output Analysis (IO). While excellent textbooks are available for IO and IE individually, there is a lack of resources that integrate the two fields. This book aims to fill that gap by focusing on the practical application of IO to IE, specifically within the context of Life-Cycle Assessment (LCA) and Material Flow Analysis (MFA).

The intended readership of this book comprises students, researchers, and practitioners interested in conducting quantitative analysis in IE while taking into account technological constraints. Unlike my previous book with Yasushi Kondo, titled Waste Input-Output Analysis (Springer, 2009), which was primarily a textbook on IO with applications to IE, this book has a distinct focus on IE with IO as the primary methodology employed. It evolved from my lectures on Industrial Ecology at Waseda University in Tokyo from 2010 to 2022, catering to undergraduate and graduate levels in Japanese and English. Additionally, it draws on various other teaching experiences, including the Erasmus Mundus Master's Program, MIND and CIRCLE.

The main goal of this book is to serve as a practical guide on utilizing IO to conduct quantitative analysis in IE. To achieve this goal, a significant portion of the content will consist of numerical tables and mathematical equations, including stoichiometric equations that represent chemical reactions representing the underlying technological constraints.

Given that IE is a quantitative discipline, effectively addressing issues in IE projects requires processing data using mathematical equations. As a nonnative English speaker, I personally find mathematical expressions particularly helpful in conveying my thoughts effectively. Mathematics serves as a universal language that transcends linguistic barriers.

To ensure a comprehensive understanding of the subject matter, I believe it is crucial for students of IE, regardless of their academic backgrounds and interests, to have a foundational knowledge of physics and chemistry before delving into the analysis. Consequently, the book incorporates numerous stoichiometric equations that are less commonly found in standard IE textbooks and are rarely included in major IO textbooks.

The ultimate objective of IE is to identify effective strategies for making our world more sustainable. In fact, IE can be considered the science of sustainability (Amit Kapur and Thomas Graedel, in *Encyclopedia of Energy*, 2004, Elsevier). By providing practical insights and emphasizing the integration of IO within IE, this book aims to equip readers with the necessary tools to conduct quantitative analysis in the field of sustainable science, leveraging the power of IO. The practical application of IO involves the use of matrix calculus, which is a fundamental aspect of the analysis. To highlight this key aspect, I have chosen as the subtitle for this book matrix based calculus of sustainability.

I would like to express my sincere gratitude to the individuals who have provided invaluable assistance throughout the process of writing this book. Their unwavering support and contributions have been integral to its successful completion. I extend my special thanks to Yosuke Shigetomi for dedicating his time to reading the entire book, and to Anthony Newell, my former colleague at Waseda, for meticulously reviewing the English language used throughout the entirety of the manuscript. I am deeply indebted to Faye Duchin, Tetsuya Nagasaka, Kazuyo Matsubae, Stefan Pauliuk, and Aurup Ratan Darh for their valuable insights and feedback on earlier versions of Chaps. 1 and 2. Additionally, I would like to express my appreciation to Christoph Helbig for providing valuable insights and feedback on Chap. 7. Furthermore, I would like to extend my sincere gratitude to the following individuals for their invaluable help in answering my questions, and or reviewing passages, and or providing constructive comments: Peter Berrill, Ichiro Daigo, Sebastian Dente, Seita Emori, Thomas Gibon, Edgar Hertwich, Shunichi Hienuki, Takehito Hiraki, Hiroki Hondo, Mark Jacobson, Shigemi Kagawa, Arunima Mark, Stefano Merciai, Kenichi Nakajima, Jun Nakatani, Keisuke Nansai, Nuri Onat, Hung Suck Park, Stefan Pauliuk, Burak Sen, Konstantin Stadler, Jan Streeck, Osamu Takeda, Tommy Wiedamann, Paul Wolfram, Richard Wood, Siqin Xiong, Yi Yang, and Ryosuke Yokoi (in alphabetical order of family name). However, it is important to emphasize that I bear sole responsibility for any remaining errors.

Kamakura, Japan
July 2023

Shinichiro Nakamura

Contents

List of Figures

List of Tables

Chapter 1
Introduction

Abstract Industrial Ecology (IE) aims to discover solutions for complex environmental problems and effectively tackle sustainability challenges by minimizing environmental impacts and harmonize human development with environmental stewardship. What sets IE apart is its reliance on a quantitative system approach, which forms the foundation of our discussions. To begin, we explore the necessity of this quantitative approach, using electric vehicles (EVs) as a prominent example. As the name implies, IE draws inspiration from the conceptual framework of ecology, particularly during its early stages of development. In light of this, we delve into relevant ecological principles that hold significant relevance for IE, such as the concept of food chains. Moving forward, we provide a concise explanation of the primary methodologies employed within IE, including Life Cycle Assessment (LCA), Material Flow Analysis (MFA), Industrial Symbiosis (IS), and Environmentally Extended Input-Output Analysis (EEIO). These methodologies play crucial roles in understanding and managing the environmental implications of industrial activities. As we conclude the chapter, we offer an overview of the book's contents, outlining the themes of subsequent chapters and highlighting their potential value for diverse audiences. This serves as a guide for readers, allowing them to navigate the material based on their specific interests and needs.

1.1 Industrial Ecology

The primary objective of Industrial Ecology (IE) is to discover solutions for complex environmental problems and effectively tackle sustainability challenges [1]. IE achieves this by adopting a systems perspective, which involves examining the complete life cycle of products or processes, as well as how society utilizes material and energy resources. By doing so, IE aims to prevent burden shifting between different life cycle stages, such as production, use, and end-of-life (EoL) stages, where efforts to reduce environmental impacts in one stage may inadvertently create larger impacts in other stages.

Furthermore, IE relies on scientific knowledge and models to comprehend the intricate relationships between material and energy use and their resulting

S. Nakamura, *A Practical Guide to Industrial Ecology by Input-Output Analysis*,
https://doi.org/10.1007/978-3-031-43684-0_1

environmental impacts. This comprehensive understanding allows IE to address multiple environmental issues simultaneously and avoid burden shifting between impacts. For example, while reducing one type of impact, such as greenhouse gas emissions, it is important to consider potential increases in other impacts, such as water pollution. This holistic approach ensures that IE considers the broader environmental context and avoids unintentional trade-offs.

In summary, IE focuses on finding sustainable solutions to complex environmental problems through a thorough understanding of the systems involved. By considering the complete life cycle and multiple environmental issues, IE aims to prevent burden shifting and promote a comprehensive approach to sustainability.

The importance of taking a systems approach can be illustrated through the example of replacing internal combustion engine-based vehicles (ICVs) with battery electric vehicles (BEVs). While BEVs have no tailpipe emissions and are generally considered cleaner than ICVs, their overall emissions must be evaluated by considering the emissions associated with generating the electricity needed to run them. If the electricity comes from fossil-fuel power plants, it is unclear whether BEVs emit fewer pollutants and CO_2 than ICVs. Therefore, it is essential to consider a larger system, including electricity generation and consumption, when evaluating the environmental impacts of BEVs compared with ICVs. Simply looking at tailpipe emissions alone can be misleading and not give the full picture of a vehicle's environmental impact.

Of the life cycles of a vehicle, the above discussion focuses solely on the driving or usage phase. However, before a vehicle is ready for use, it must first be produced, and emissions occur during this phase as well. While one might assume that collecting data on emissions from automobile manufacturers' factories provides insight into production-phase emissions, it is important to note that such emissions constitute only a small fraction of the total emissions associated with vehicle production (Chap. 6). Most emissions occur during the production of materials such as steel and aluminum, particularly in the case of ICVs. With BEVs, research indicates that battery production is a significant source of emissions during the production phase, and the emissions associated with the production of an EV are often higher than those of a comparable ICV (Chap. 6). By extending the system under consideration to include the upstream processes of a vehicle, it is no longer clear which vehicle emits fewer emissions.

Furthermore, there are environmental impacts other than CO_2 and air pollutants. One also needs to consider the resource implications of a vehicle. A BEV typically requires more scarce metals, such as neodymium (Nd), lithium (Li), and cobalt (Co), than an ICV (Chap. 6). On the other hand, any product, including a BEV, will eventually be discarded and go through an end-of-life (EoL) process, which includes recycling materials and reusing parts and components. If the scarce metals in an EoL-BEV are recovered and recycled to substitute for their virgin counterparts, the resource impacts of the BEV can be reduced. Therefore, considering the impacts occurring only in the production and use phases of an EV will be incomplete and misleading. The impacts occurring in the EoL phase must also be considered to fully assess the environmental impacts of a vehicle.

Consideration of the whole life cycle of a product, consisting of the production, use, and EoL phases, is necessary to consider the total impacts of a product and avoid problem shifting. Arguing that an EV is environmentally friendly as it has no tailpipe is a typical example of problem shifting, neglecting the impacts of electricity generation and materials production. The above example of a BEV indicates the need to take a system-wide (or holistic) perspective to properly consider a product's environmental impacts. Instead of focusing on a small system representing only a fraction of the product's life cycle and supply chain, one must look at a whole system encompassing the whole life cycle and the entire product supply chain. Taking a system-wide perspective requires quantitative information because of the interdependence of the many factors that make up a system. Relying on qualitative information alone cannot provide a proper understanding of the impacts of a product.

1.2 Roots of IE

Although there are excellent textbooks on IE, such as [2–4], we will briefly explore the historical roots of IE to understand the origins of its name. Reference [5] introduced the term "industrial ecosystem" by drawing an analogy between industrial and biological ecosystems. While it is debatable whether natural ecosystems are an appropriate model for IE, it is true that natural ecology has inspired IE to develop many of its basic ideas and concepts. Therefore, we will discuss the fundamental features of biological ecosystems and their relevance to IE in this context. We will primarily refer to [6] for our discussion on biological ecosystems.

A biological ecosystem is a conception of nature that considers biotic communities of organisms and the biophysical environment as an integrated whole [6]. The essential characteristics of a biological ecosystem that are highly relevant to IE include food chains, which comprise at least three trophic levels and represent the core structure of the biological part of ecosystems. Two types of food chains, live-plant-based and detritus-based, determine the energy and material flow pathway throughout a system. Species in an ecosystem can be grouped into four categories based on their trophic function: producers (plants), primary consumers (herbivores), secondary consumers (carnivores), and decomposers. Plants, the only producers in the ecological sense, can produce food from nonorganic materials. Ecosystems are largely self-contained in the sense that energy and materials bound in living plant biomass and nonliving detritus are mostly recycled within the system rather than diffusing or dissipating from one system to another.

The concept of a food chain, describing the flow of nutrients from plants to herbivores and then to top carnivores, involves two opposing flows of nutrients, one referring to living organisms and the other to dead organisms. The detritus food chain involves the process by which dead organisms are decomposed by animals, insects, and bacteria, eventually becoming inorganic substances that green plants can use as nutrients. Recycling within the system of energy and materials is likely the aspect of a biological ecosystem that initially inspired the development of IE, as seen in early definitions of IE from the 1990s, which focused on identifying opportunities to reduce wastes and pollution in the materials-intensive sectors by utilizing low-value by-products as raw materials for others [3, 4].

The early definitions of IE share similarities but are comparatively narrow compared to a more recent one [2]. They define IE as the study of technological organisms, their use of resources, potential environmental impacts, and how they can be restructured for global sustainability. One notable aspect of this definition is the term "technological organisms", which includes nonliving things. They suggest that industrial ecosystems are similar to biological ecosystems in terms of the flow of materials, energy, goods, and services among different sectors. They also note that the flow of end-of-life products to waste streams is similar to the detritus food chain in biological ecosystems. However, the analogy to biological ecosystems breaks down quantitatively regarding the detritus food chain in industrial ecosystems. Evidence indicates that anthropogenic waste is not being recycled effectively, resulting in serious environmental problems such as marine plastic contamination [7–12]. Still, it is not true that there is no unrecycled waste in nature [13].

1.2.1 *Industrial Ecosystems: An Open Subset of Planetary Ecosystems*

A key factor distinguishing industrial ecosystems from biological ecosystems is the absence of "primary producers" organisms with photosynthetic capabilities [13]. This fact reflects that industrial ecosystems are nothing but an "open" subset of planetary ecosystems. Their very existence depends on the fixation of atmospheric carbon by green plants. A system is "open" if it can exchange both matter and energy with its surroundings, "closed" if it can exchange energy only, and is "isolated" if there is no exchange with its surroundings. Earth is almost a closed system, except for tiny amounts of meteoroids hitting its surface, spacecraft escaping its gravity, and light substances (such as helium) escaping into space. Earth receives electromagnetic energy from the sun at an average rate of 340 W m^{-2} (see Sect. 2.6).

1.2.2 *A Visual Representation of the Basic-Economy-Environment Account*

Figure 1.1 illustrates the basic-economy-environment account proposed by [14]. The symbol $\boldsymbol{\alpha}$ represents the internal flow of goods and services within the economy, while $\boldsymbol{\beta}$ represents the flow of valuable biological goods and services, such as resources and pollination services, from the ecosystem into the economic system. The symbol $\boldsymbol{\gamma}$ represents the flow of unwanted outputs or consequences from the economic system into the ecosystem, such as pollutants, waste, and habitat losses, and $\boldsymbol{\delta}$ represents the

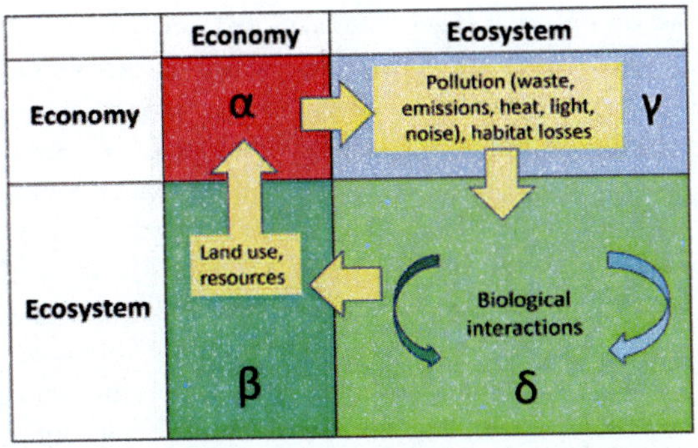

Fig. 1.1 The economy and the biological ecosystem

flows within the ecosystem, which represent the interdependence and repercussions among the living organisms. As long as economic activities remain "small" relative to the size of the supporting ecosystem (Earth), human-induced inflows and outflows do not noticeably impact the ecosystem, and the flows γ and β remain negligible. In this case, IE is not necessary.

However, human activities have become so dominant on Earth that they have necessitated the creation of the concept of the Anthropocene, representing a new geological epoch dominated by humans [15–17]. For instance, a recent extensive study on the biomass distribution on Earth [18] found that humans and livestock outweigh all vertebrates (animals with backbones) combined (including wild birds), except for fish and that the biomass of wild mammals shrank from 0.04 P (10^{15}) g-C before the onset of human civilization, 100,000 years ago, to 0.003 Pg-C. Another example is [19], who used the term "anthropogenic mass" to quantify the human-made mass and compared it to the total living biomass on Earth, which is approximately 1.1 E (10^{18}) g. The study found that Earth reached a crossover point. In 2020 (with a margin of error of 6 years), the anthropogenic mass, which has been doubling roughly every 20 years, exceeded all global living biomass. Thus it is clear that IE is very much needed.

1.3 Methods of IE

IE employs various methods, including LCA, MFA, industrial symbiosis (IS), and EEIO [1]. LCA is a systemic analysis of environmental flows and impacts throughout the life cycle of a product. It has broadened to include life cycle costing (LCC) and social LCA (SLCA) to assess sustainability based on the triple bottom line (TBL) framework, which encompasses the three dimensions of environmental, economic, and social sustainability. MFA quantifies mass and energy flows in industrial systems from individual plants to the global economy. IS studies the exchange of waste as a resource among nearby industrial facilities. Because of its similarity with MFA, we do not treat it separately in this book. EEIO quantifies environmental footprints based on exchanges between economic sectors and the environment.

Of these methods, EEIO is unique in that it uses an established mathematical tool/model that was initially developed for economic analysis. Its format has been used as an essential component of national economic accounting worldwide. Large-scale adoption of EEIO in IE began in the 1990s with LCA and in the 2000s with MFA, although IO itself has a longer history dating back to the 1930s–1940s. In contrast, LCA and MFA are more object-oriented, focusing on specific objectives. LCA assesses life cycle environmental impacts, and MFA quantifies the flow of specific materials/substances. It is important to note that there is no intention to rank these methods by any standards but rather to highlight their specific points.

1.3.1 IO as a Basic Methodological Framework Encompassing the Major Methods of IE, LCA, and MFA

MFA focuses on the physical flows of materials and substances, while IO tables are compiled in monetary units as part of national economic accounting, but they share many similarities. MFA has its origins in IO models ([20], p. 314), as well as in the metabolism tradition that studies process flows such as total city material consumption or waste streams [21, 22]. As we will see in Chap. 7, EEIO is widely applied to MFA. Historically, EEIO focused on the manufacturing and use phases of products, with little attention paid to the end-of-life phase, including waste management and recycling. However, with the development of Waste IO (WIO), EEIO can now accommodate all life stages of products [23, 24].

This book aims to introduce EEIO as a basic methodological framework encompassing two major methods of IE: LCA and MFA. Alternatively stated, we introduce these methods based on EEIO to help readers become familiar with IE's major methodologies more efficiently than studying them separately. We hope that this strategy will aid readers in understanding the interconnectedness of IE methods and their potential to contribute to sustainability.

1.4 Outline of This Book

Chapter 2 focuses on the fundamental concept of a process in IE, which serves as a connecting point in the interdependent flows of inputs and outputs, including waste and emissions. Examples of actual processes, including crop cultivation, fertilizer, livestock, metals, and cement, are presented to illustrate this concept. Additionally, household food consumption is considered a process, where inputs (food) are converted into outputs (human waste and emissions) to provide insight into the environmental impacts of food consumption. The chapter also highlights the significance of the global cycling flow of major elements such as carbon (C), nitrogen (N), phosphorus (P), and sulfur (S) in understanding the environmental impacts related to their flows. Finally, the chapter introduces the basic idea of IO as a quantitative representation of the interdependence among processes that emerge through the flow of inputs and outputs among them, providing a crucial foundation for subsequent chapters. The appendix at the end of this chapter offers essential knowledge on the atmospheric physics necessary for comprehending the science behind greenhouse gases (GHGs) and their impact on climate change.

Chapter 3 introduces IO, including key concepts such as input coefficients, the Leontief inverse, the Leontief quantity model, and productive conditions. It starts with a simple one-sector model based on the production of a single staple and progresses to a two-sector model that includes livestock production. Throughout the chapter, our focus is on models where flows are measured in physical units. The chapter also discusses environmental extensions, such as GHG emissions, water- and land footprints, and waste generation and recycling, including Waste Input-Output analysis (WIO), with numerical examples based on actual data. To keep the modeling as simple as possible, this chapter considers a hypothetical agrarian-based economy where human muscle power serves as the sole energy source for operating production processes with no utilization of fossil fuels. GHG emissions result from the anaerobic decomposition of organic waste and livestock enteric fermentation. The supply of human labor is endogenized by considering the energy required to produce human muscle power, which aligns with the Miyazawa model. The discussion in this chapter starts without using matrix algebra, relying on simple arithmetic only, which may be helpful for readers unfamiliar with the use of matrices. Once the basic concepts are explained for the two-sector model, the chapter uses matrix algebra to reiterate the basics of IO in matrices, which apply to any number of sectors. Additionally, the chapter introduces the IO model of cost and price as the dual to the quantity model, providing a more comprehensive understanding of the IO framework.

The models discussed in Chap. 3 involve well defined processes where all flows are measured in physical units. Chapter 4 extends the basic IO concepts introduced in Chap. 3 to actual IO tables involving secondary products and by-products, measured not in physical but monetary units. We begin by discussing the definition of sectors in the presence of by-products and secondary products, followed by the commodity-by-industry approach to account for them. We then move on to regionally extended models, including global Multi-Regional Input-Output (MRIO) models, with a focus on the latest state of major databases such as EXIOBASE and Gloria, which are widely used in LCA. Since these topics are discussed in detail in standard textbooks such as [25], our discussion about them is brief.

Chapter 4 also discusses the treatment of capital goods, capital stock, and endogenization of fixed capital formation in IO. This includes the capital flow table, capital coefficients, and the extension of the standard IO model to endogenize capital formation. We provide a detailed discussion of Lenzen's model of fixed capital formation [26] and its recent variants [27–30]. These models are increasingly used to assess the environmental impacts of fixed capital formation. Finally, we cover a dynamic model with different age cohorts of capital stock introduced by [31], which is highly relevant to IE.

Chapter 5 covers several topics related to integrating emissions into IO models applicable to LCA. We start by discussing the combustion stoichiometry and greenhouse gas emissions that result from fuel consumption. We explore various approaches to model emissions, including exogenous and endogenous approaches, and discuss direct and embodied emissions, as well as Scopes 1, 2, and 3 of the GHG protocol. We also examine how to avoid double-counting in footprint analysis and the relationship between emissions and international trade.

Moving on to LCA, we provide an overview of its basics and discuss the EEIO model from the point of view of ISO-LCA. We then delve into the hybridization of LCA and EEIO, highlighting the sectoral resolution of an IO table and its implications for LCA and HLCA models. Furthermore, we explore waste input-output analysis (WIO), including two types of by-products, the WIO table and model. The chapter also covers the basics of waste treatment processes, such as incineration. Lastly, we touch on other topics related to LCA, such as environmental life-cycle costing (eLCC), social LCA, distributing environmental responsibility, and attributional and consequential LCA.

Chapter 6 offers a comprehensive overview of GHG emissions and mitigation strategies. It begins by discussing global GHG emissions, including their trends by gas, source, and country. The chapter then delves into mitigation strategies related to electricity generation, covering power technologies and renewable energy sources. The mitigation potential of each technology is evaluated based on process-based LCA (PLCA) and hybrid LCA (HLCA).

The use of EVs to reduce GHG emissions is also explored, comparing their environmental impacts with ICVs and assessing their mitigation potentials based on LCA. Additionally, the implications of large-scale deployment of EVs on resource and power requirements are discussed. The chapter highlights challenges related to the intermittence and unreliability of renewable technologies and the implications of decarbonization for materials. Unlike the preceding chapters, this chapter does not offer any new methodology of LCA/IO. Instead, it demonstrates how the methods presented in preceding chapters can be used to evaluate the effectiveness of various mitigation strategies, using EVs as an example.

Chapter 7 discusses the methods and applications of Material Flow Analysis (MFA) models, both static and dynamic. Static MFA models provide a snapshot of a system in time, while dynamic MFA models describe the system's behavior over a specific time interval. The static MFA models are explained by discussing the transfer matrix, which governs the transfer of materials from one production/use stage to another, followed by the application of the Markov chain. The IO-based static MFA based on WIO-MFA is also presented, followed by extensions to accommodate diverse pathways, including packaging and containers. The dynamic MFA models are introduced with a brief description of the standard model based on the basic model of stock growth and lifetime distributions. The MaTrace models are then discussed, which trace the fate of materials over time and across products in open-loop recycling, with explicit consideration of losses and quality of scrap, along with their extensions to accommodate global multiregional interdependency, alloys, and multi-materials. Many case studies are presented for illustration.

1.5 How to Use This Book

The book is designed as a textbook for IO-based Industrial Ecology (IE) courses at the undergraduate and graduate levels, suitable for students with diverse academic backgrounds. Depending on the course goals and the students' backgrounds, the book can be used in different ways. For example, if the goal is to cover both LCA and MFA, and students have limited prior knowledge of the topics covered in the book, the chapters can be followed sequentially. The sections marked with an asterisk (*) refer to specific or advanced content which can be skipped without losing the book's overall integrity.

For students with limited preknowledge of actual production processes, Chap. 2 provides useful exposure to better understand the contents of the subsequent chapters. Even for those who are knowledgeable about actual processes, taking a brief look at Chap. 2 will be useful because Chap. 3 uses data taken from processes discussed in Chap. 2.

Students or professionals with prior knowledge of LCA/MFA who are interested in applying IO are recommended to start with Chap. 4. Although some topics discussed in this chapter are also addressed in textbooks of IO, such as [25], the chapter is specifically tailored for IE applications, making it more accessible to readers.

Students or professionals who are familiar with IO and wish to explore its applications in IE are recommended to begin with Chap. 5. While this chapter covers standard topics of EEIO that are also addressed in IO textbooks, such as [25], it offers unique insights into the modeling of waste and capital formation for evaluating the environmental impacts thereof that are not covered in standard IO textbooks.

Finally, it is necessary to clarify the usage of the term "IO" in this book. "IO" can refer to either "input-output analysis" or "input-output", depending on the context. When we say "in IO", it means "in input-output analysis". However, when we find phrases like "in the IO model" or "an IO table", it refers specifically to "in the input-output model" or "an input-output table", respectively.

Additionally, it is important to note that we use the term IO to encompass not only the standard input-output analysis but also EEIO and other extensions relevant to IE. In Chaps. 6 and 7, the term IO is used instead of EEIO to encompass these various extensions.

References

1. International Society for Industrial Ecology. 2023. What is industrial ecology? https://is4ie.org/about/what-is-industrial-ecology. Accessed on 2023-05-31.
2. Graedel, Thomas E., and Braden R. Allenby. 2010. *Industrial ecology and sustainable engineering*. Pearson Education Inc., Prentice Hall: Upper Saddle River.
3. Manahan, Stanley E. 1999. *Industrial ecology: Environmental chemistry and hazardous waste*. CRC Press.
4. Ayres, Robert U., and Leslie W. Ayres. 1996. *Industrial ecology*. Edward Elgar Publishing.
5. Frosch, Robert A., and Nicholas E. Gallopoulos. 1989. Strategies for Manufacturing the impact of industry on the environment. *Scientific American* 261(September): 144–153.
6. Schmitz, Oswald J. 2010. The evolutionary ecology of trophic control in ecosystems. In *Resolving Ecosystem Complexity* (Chap. 5). Princeton University Press.
7. Sheavly, S.B., and K.M. Register. 2007. Marine debris & plastics: Environmental concerns, sources, impacts and solutions. *Journal of Polymers and the Environment* 15 (4): 301–305.
8. Engler, Richard E. 2012. The complex interaction between marine debris and toxic chemicals in the ocean. *Environmental Science and Technology* 46 (22): 12302–12315.
9. Taylor, M.L., C. Gwinnett, L.F. Robinson, and L.C. Woodall. 2016. Plastic microfibre ingestion by deep-sea organisms. *Scientific Reports* 6 (May): 1–9.
10. Schneider, Falk, Sophie Parsons, Sally Clift, Andrea Stolte, and Marcelle C. McManus. 2018. Collected marine litter - A growing waste challenge. *Marine Pollution Bulletin*, 128(December 2017): 162–174.
11. Arp, Hans Peter H., Dana Kühnel, Christoph Rummel, Matthew Macleod, Annegret Potthoff, Sophia Reichelt, Elisa Rojo-Nieto, Mechthild Schmitt-Jansen, Johanna Sonnenberg, Erik Toorman, and Annika Jahnke. 2021. Weathering Plastics as a Planetary Boundary Threat: Exposure, Fate, and Hazards. *Environmental Science and Technology* (1).
12. Bank, Michael S., Peter W. Swarzenski, Carlos M. Duarte, Matthias C. Rillig, Albert A. Koelmans, Marc Metian, Stephanie Wright, Jennifer F. Provencher, Monica Sanden, Adrian Jordaan, Martin Wagner, Martin Thiel, and Yong Sik Ok. 2021. Global Plastic Pollution Observation System to Aid Policy. *Environmental Science & Technology*.
13. Ayres, Robert U. 2004. On the life cycle metaphor: Where ecology and economics diverge. *Ecological Economics* 48 (4): 425–438.
14. Daly, Herman E. 1968. On economics as a life science. *The Journal of Political Economy*, 76(3): 392–406.
15. Crutzen, Paul J. 2006. The "anthropocene". In *Earth system science in the anthropocene* (pp. 13–18). Springer.
16. Lewis, Simon L., and Mark A. Maslin. 2015. Defining the anthropocene. *Nature*, 519(7542): 171.
17. Monastersky, Richard. 2015. Anthropocene: The human age. *Nature* 519 (7542): 144–147.
18. Bar-On, Yinon M., Rob Phillips, and Ron Milo. 2018. The biomass distribution on Earth. *Proceedings of the National Academy of Sciences of the United States of America* 115 (25): 6506–6511.
19. Elhacham, Emily, Liad Ben-Uri, Jonathan Grozovski, Yinon M. Bar-On, and Ron Milo. 2020. Global human-made mass exceeds all living biomass. *Nature* 588 (7838): 442–444.
20. Scholz, Roland W., and Claudia R. Binder. 2011. *Environmental literacy in science and society: From knowledge to decisions*. Cambridge University Press.
21. Brunner, Paul H., and Helmut Rechberger. 2004. *Practical handbook of material flow analysis*. CRC Press.
22. Brunner, Paul H., and Helmut Rechberger. 2016. *Handbook of material flow analysis: For environmental, resource, and waste engineers*. CRC Press.
23. Nakamura, Shinichiro, and Yasushi Kondo. 2002. Input-output analysis of waste management. *Journal of Industrial Ecology* 6 (1): 39–63.
24. Suh, Sangwon, and Shinichiro Nakamura. 2007. Five years in the area of input-output and hybrid LCA. *The International Journal of Life Cycle Assessment* 12 (6): 351.

25. Miller, Ronald E., and Peter D. Blair. 2022. *Input-output analysis foundations and extensions* (3rd edn.). Cambridge University Press.
26. Lenzen, Manfred. 1998. Primary energy and greenhouse gases embodied in Australian final consumption: an input-output analysis. *Energy Policy* 26 (6): 495–506.
27. Södersten, Carl Johan H., Richard Wood, and Edgar G. Hertwich. 2018. Endogenizing capital in mrio models: The implications for consumption-based accounting. *Environmental Science and Technology* 52(22): 13250–13259.
28. Berrill, Peter, T. Reed Miller, Yasushi Kondo, and Edgar G. Hertwich. 2020. Capital in the American carbon, energy, and material footprint. *Journal of Industrial Ecology* 24(3): 589–600.
29. Södersten, Carl Johan, Richard Wood, and Thomas Wiedmann. 2020. The capital load of global material footprints. *Resources, Conservation and Recycling* 158(January): 104811.
30. Hertwich, Edgar G. 2021. Increased carbon footprint of materials production driven by rise in investments. *Nature Geoscience* 14(3): 151–155.
31. Pauliuk, Stefan, Richard Wood, and Edgar G. Hertwich. 2015. Dynamic models of fixed capital stocks and their application in industrial ecology. *Journal of Industrial Ecology* 19 (1): 104–116.

Chapter 2
Processes and Elemental Flows

Abstract This chapter deals with the fundamental concept of a "process" in Industrial Ecology (IE), which is a connecting point in the interdependent flows of inputs and outputs, including waste and emissions. A process is featured by mass balances between inputs and outputs and is assumed to exhibit constant returns to scale. Examples of actual processes are shown, from crop cultivation, fertilizer (ammonia), livestock, fish farming, metals (iron (Fe), copper (Cu), and aluminum (Al)), and cement. This chapter also considers household food consumption as a process, where inputs (food) are converted into outputs (human waste and emissions) to gain insight into the environmental impacts of food consumption. The global cycling flow of major elements–carbon (C), nitrogen (N), phosphorus (P), and sulfur (S)–is discussed to highlight their significance in understanding environmental impacts. Finally, the basic idea of Input-Output analysis (IO) is introduced as a quantitative representation of the interdependence among processes that emerge through the flow of inputs and outputs among them.

2.1 Process: The Fundamental Concept of IE

Industrial Ecology (IE) is concerned with how human economic activity impacts the environment over time and space. It takes a comprehensive, systems-level perspective that considers the entire life cycle of products. This includes the interconnected flow of products, by-products, waste, and emissions generated by a variety of producers, consumers, and waste managers, which accumulate in the anthroposphere or are dispersed into the environment. In IE, a process refers to a connecting point, or node, in these interdependent flows. This term is of great importance and has historically been used to describe how industrial manufacturing processes acquire raw materials, generate products for sale, and produce waste for disposal [1].

In IE, particularly in MFA (Material Flow Analysis), a process is commonly depicted as a black box, with inputs flowing in and outputs flowing out. The internal transformation processes within the box are not taken into account, and only the inputs and outputs are considered ([2], p. 38). This definition aligns with how processes

S. Nakamura, *A Practical Guide to Industrial Ecology by Input-Output Analysis*,
https://doi.org/10.1007/978-3-031-43684-0_2

are defined in economics [3]. However, a white box representation, which provides more detailed information about the quantitative relationships between inputs and outputs based on scientific and engineering principles, is preferred whenever available. Processes such as waste management, incineration, shredding, and recycling of scrap metals are examples where a white box representation proves beneficial (see Sect. 5.4). Given that the black box representation is prevalent in IE, the term "process" will henceforth refer to a black box process unless otherwise specified.

2.1.1 Unit Process: Theory and Practice

To conduct effective analyses in IE, it is important that the quantitative relationships between inputs and outputs remain stable over time, or at least not subject to sporadic changes within short periods, such as a quarter or a year. Ideally, these relationships should reflect the underlying technological relationships and remain stable as long as the underlying technologies remain unchanged.

However, when a process is defined broadly at a low level of resolution and encompasses a wide range of heterogeneous technologies, such as agriculture or metals, it becomes challenging to expect the quantitative input-output relationships to remain stable. Changes in the product mix of the numerous products associated with these broad definitions can lead to variations in the quantitative input-output relationships, even if they remain stable at a higher level of resolution, such as paddy rice production, salmon aquaculture, blast furnace-based iron production, or zinc production based on the Imperial Smelting Furnace (ISF).

To address this issue, it is desirable to define a process in as detailed and disaggregated a manner as possible. This gave rise to the concept of a unit process, which is defined as "the smallest element considered in a life cycle inventory model (the compilation and quantification of inputs and outputs for a product throughout its entire life cycle) for which input and output data are quantified" ([4], p. 76). However, it is important to note that the unit process is a theoretical concept. In practice, a unit process can represent a single process, such as the rolling of steel, or it can represent an entire facility containing multiple processes, such as a slaughterhouse, as long as it provides sufficient detail for inventory modeling ([4], p. 77).

2.1.2 Elementary Flows: Another Theoretical Construct

In the theory of Life Cycle Assessment (LCA) as described in [5], flows are divided into two categories: flows exchanged between unit processes (such as products, waste, materials, and energy) and flows not exchanged between unit processes (such as resources and emissions) [4]. The latter group is considered free of human transformation since resources are taken from the natural environment without human alteration, and emissions are released into the environment without subsequent human modification [4].

However, a problem arises with this definition due to the absence of a clear and distinct spatial separation between the technosphere and the ecosphere. The interconnected nature of these two spheres renders them abstract [4]. Consequently, the term "elementary flows" will not be utilized in this book.

2.1.3 *Mass Balances: A Fundamental Feature of a Process*

2.1.3.1 Mass Balances

The law of mass conservation states that mass cannot be created or destroyed: it can only be transformed from one form to another. Therefore, in LCA, it is crucial to account for mass and ensure that everything has a designated destination. In a unit process, the mass of inputs should be equal to the mass of outputs since no accumulation occurs within the process [4, 6]. This principle, known as a mass balance, is a universal approach. However, there are exceptions to this rule [4]. For example, atmospheric constituents such as oxygen (O) and nitrogen (N) are typically not counted as resource inputs. Consequently, the mass of outputs may appear larger than the mass of inputs due to the presence of combustion products such as CO_2, H_2O, and NO_X. A more general case involving the accumulation or depletion of materials in the process is considered by [2], where the mass of all inputs into a process equals the mass of all outputs of this process plus a storage term that accounts for the accumulation or depletion of materials within the process.

When there is no accumulation in the process, mass balances can be represented as

$$\sum \text{all inputs} = \sum \text{all outputs} \tag{2.1}$$

and as

$$\sum \text{all inputs} = \sum \text{all outputs} + \text{accumulation or depletion} \tag{2.2}$$

when there is accumulation or depletion. It is worth noting that (2.2), representing the mass balance with accumulation, can be transformed into (2.1) by redefining outputs. This redefinition is particularly relevant in agricultural or livestock production, where the growth of organisms can lead to the accumulation or depletion of mass. For instance, in pig production, mass balances in the form of (2.1) can be established by considering pigs for slaughter, excreta, and dead pigs as outputs, while piglets and feed serve as inputs.

2.1.3.2 Items with No Mass

The principle of mass balance requires inputs and outputs to have a mass, but there are exceptions where inputs and outputs have no mass. For example, energy is an indispensable input for all processes, but it has no mass. Hours of work provided by humans or animals have no mass. The same applies to services. While electrons have a mass of 9.1×10^{-31} kg, electricity is typically assumed to have no mass in the context of IE.

2.1.3.3 Capital Goods

An industrial process occurs within a production facility, such as a blast furnace in the context of iron production from ore. A question arises regarding whether the mass of the blast furnace should be considered as an input for the iron production process and, if so, how its output should be accounted for. To address this question, it is crucial to distinguish between "capital goods" and "operation" (or "current flows").

"Capital goods" refers to the factories or equipment utilized in the manufacturing of products, whereas " operation" encompasses the inputs necessary for their operation and maintenance, such as materials, fuel, power, lubricants, and repair parts. The flows within a process mainly consist of operation. However, in certain cases, such as boreholes in oil extraction or a prebaked anode in aluminum production, the differentiation between operation and capital goods may not be clear [7].

Conversely, in the production of capital goods like power plants, chemical plants, or metal smelters, the parts, components, and building materials are considered part of the operation, along with fuel, electricity, and lubricants. The process outputs in this case encompass assembled capital goods, scrap, waste, and emissions.

Capital Services and Capital Stock

Closely related to the preceding discussion on capital goods (durable) is the differentiation between the services rendered by the durable (capital services) and the actual stock of the durable itself. To clarify this distinction, let us consider the analogy of renting an apartment. When you rent an apartment, you pay a fee for the services provided by the apartment over a specific period, but you do not own the property itself. This same principle applies to leasing machinery and equipment, where the fee covers the cost of the services they provide.

In the case of self-owned capital goods, the price of capital services becomes implicit, and it serves as an important measure of business performance. The inputs required for the operation and maintenance of capital goods are utilized to produce capital services from the existing stock. We will explore the concepts of capital goods, capital stock, and capital services in greater detail in Sect. 4.4.

2.1.4 *Constant Returns-to-Scale: A Fundamental Assumption About a Process*

To facilitate analysis in both IE and IO, it is generally assumed that each process exhibits constant returns to scale. This assumption implies that the process is capable of continuous expansion or reduction in a proportional manner, which is also known as divisibility [8]. This assumption implies that doubling the inputs leads to doubling the outputs and vice versa. Further details regarding this assumption will be presented in Chap. 3.

2.2 Examples of Processes

To offer a more tangible understanding of production processes, this section presents illustrative examples from diverse industries. These examples encompass agriculture (crops and livestock), the chemical industry (ammonia (NH_3)), metals (Fe, Cu, and Al), and cement. However, it is important to note that these examples aim to provide a broad overview of the production processes and should not be regarded as detailed technical information. Interested readers are advised to consult the referenced literature for more in-depth technical details.

2.2.1 *Crop Production Processes*

We will commence by examining a crop production process that relies on biological growth, driven by photosynthesis, to produce crops. This process is supported by inputs such as fertilizer, pesticides, and energy. Taking glucose ($m = 6$) as an example of a carbohydrate, $C_mH_{2m}O_m$, photosynthesis can be represented as ([9], Chap. 15)

$$6CO_2 + 6H_2O \xrightarrow{\text{light}} C_6H_{12}O_6 + 6O_2 \tag{2.3}$$

Since the production process involves growth and accumulation, the mass balance Eq. (2.2) is applicable, and it can be reformulated as (2.1) by incorporating the accumulation in the output. It is important to note that although CO_2 is essential for photosynthesis, we do not consider it an input because atmospheric air is typically not counted as a resource input in unit processes ([4], p. 151).

Table 2.1 An example of the wheat production process (Chile)

Inputs		
Seeds	kg	7.8
Fertilizers	kg	20.8
Pesticides and herbicide	kg	3.8
Diesel	kg	10.2
Land	m^2	2,208
Outputs		
Wheat grain	kg	1,000
Wheat residues[a]	kg	1,500
Emissions to air[b]		
SO_2e	kg	12.5
CO_2e	kg	175.1
Emissions to water[b]		
$PO_4^{3-}e$	kg	2.7

Source [10]. a: The amount of crop residue was estimated based on the residue coefficient taken from [11]. b: The term "e" refers to the equivalent value obtained by aggregating gases with different impact potentials. For more details, please refer to Sect. 2.6.4

2.2.1.1 Wheat Production

Table 2.1 presents a simplified and normalized numerical example of wheat production processes taken from a study conducted in Chile [10]. The original data were aggregated to provide flows per 10^3 kg of the crop, and estimates were made for crop residues, roots, stems, and other components that are not harvested, using residue coefficients (residue-grain ratio) provided by [11]. The impacts of climate change, acidification, and eutrophication are respectively expressed in terms of CO_2 equivalents (CO_2e), SO_2 equivalents (SO_2e), and PO_4^{3-} equivalents ($PO_4^{3-}e$). In this particular example, the main source of acidification impacts is attributed to the application of N fertilizer, while the primary contributor to eutrophication is P fertilizer. Additionally, the use of N fertilizer also has significant effects on eutrophication.

2.2.1.2 Rice Production

Table 2.2 provides a numerical example of rice production processes extracted from a Japanese study [12]. It is worth noting that while Tables 2.1 and 2.2 show similar requirements for seed and land required per unit of crop, they cannot be considered representative examples as they are influenced by specific local conditions and have different system boundaries.

One notable difference between the two tables is the emission of methane (CH_4) in rice production. This characteristic is a result of rice cultivation in paddies, which creates anaerobic conditions due to limited O_2 availability. Under normal conditions

Table 2.2 An example of the rice production process (Japan)

Inputs		
Light fuel oil	L	42.0
Gasoline	L	8.0
Electricity	kWh	14.5
Fertilizer	kg	70.5
Pesticides	kg	1.6
Rice seed	kg	6.2
Land	m^2	2,296
Outputs		
Grain rice	kg	1,000
Rice residues	kg	1,500
Direct field emissions to air		
CH_4	kg	54.2
N_2O	kg	0.2
NH_3	kg	4.4
Direct field emissions to water		
Total N	kg	6.5
Total P	kg	0.7

Source [12]. The data have been simplified (aggregated individual items of fertilizer and pesticides) and normalized to flows per 10^3 kg of crop. Estimates for unharvested parts of the rice plant, such as crop residues, roots, and stems, were derived from residue coefficients provided by [11]

with abundant O_2, the decomposition or breakdown of carbohydrate (dead organisms, such as roots and stems that were left in paddies) by microorganisms follows the equation of aerobic respiration given, as given by [13]

$$C_6H_{12}O_6 + 6O_2 \longrightarrow 6CO_2 + 6H_2O, \tag{2.4}$$

This reaction is the reverse of the photosynthetic reaction (2.3).

However, in the absence of O_2, decomposition takes an anaerobic route, and the overall reaction (after skipping all the intermediate steps) is given by [14]

$$\frac{1}{3}C_6H_{12}O_6 \longrightarrow CH_4 + CO_2 \tag{2.5}$$

This characteristic of anaerobic decomposition makes paddy-rice production a significant global emitter of CH_4 [15]. The same reaction also applies to the decomposition of organic sediment captured in a hydroelectric dam.

2.2.2 *The Haber-Bosch Process of NH_3 Production*

Nitrogen is the very stuff of life, constituting a major part of the nucleic acids that determine the genetic character of all living things and the enzyme proteins that drive the metabolic machinery of every living cell [16]. The nitrogen molecule, N_2, constitutes nearly 80% of Earth's atmosphere by mass. However, more than 99% of this nitrogen is not available to more than 99% of living organisms [17]. Before nitrogen can be used by most living organisms, the strong triple bond in the N_2 molecule has to be broken through processes such as nitrogen fixation and assimilation. These processes convert nitrogen into reactive nitrogen species, denoted as N_r, which are essential for plant growth and include compounds like ammonia (NH_3), ammonium (NH_4^+), nitrate (NO_3^-), and nitrite (NO_2^-) [16].

Until the beginning of the 20th century, the primary source of N_r was biological, with leguminous plants playing a significant role in biological N fixation [16]. However, the Haber-Bosch process, invented in 1913, revolutionized N_r production by enabling industrial synthesis of NH_3 from atmospheric N_2. This breakthrough marked a turning point in human history by providing an unlimited supply of N_r for the first time, which could be utilized to enhance food production [16]. In fact, Fig. 2.1 shows a high correlation between population growth and the growth of NH_3 based on the Haber-Bosch process. It is noteworthy to observe that humans took their entire history to reach the first one billion around 1800. However, it took only 130 years to reach the next one billion, less than 40 years to reach another billion, and less than 14 years to reach yet another billion.

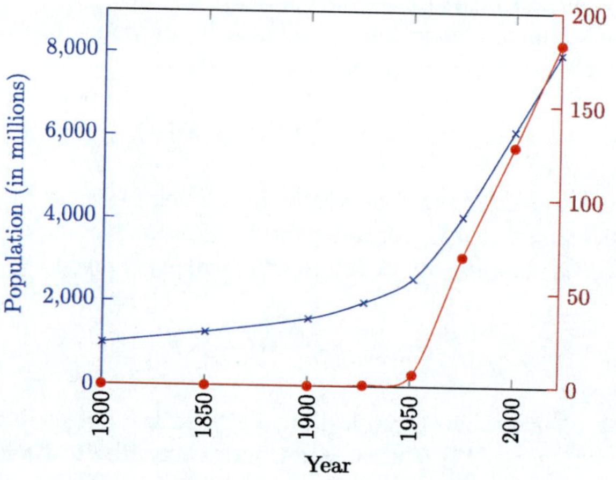

Fig. 2.1 Population growth and ammonia (NH_3) production by the Haber-Bosch process. Tg = million tonnes. *Sources* Population: [18, 19]; NH_3 production [20]

The Haber-Bosch process combines hydrogen and nitrogen on a catalytic iron surface, and the reaction is represented by

$$N_2(g) + 3H_2(g) \longrightarrow 2NH_3(g) \tag{2.6}$$

The process requires high pressures exceeding 100 MPa and high temperatures ranging from 400 to 500 °C. Additionally, hydrogen gas needs to be produced, which is synthesized mainly from natural gas (CH_4) through the steam reforming process

$$CH_4 + 2H_2O \longrightarrow 4H_2 + CO_2 \tag{2.7}$$

It is important to note that the ammonia synthesis process is not dependent on the specific method of hydrogen gas production, but the quality of the gas can impact the design and operational conditions of the process [21]. As there is no accumulation within the NH_3 production process, a simple mass balance Eq. (2.1) is applicable.

Table 2.3 provides an overview of the integrated NH_3 production process with H_2 gas production based on steam reforming. The production of 1 kg of NH_3 requires $23.4 + 0.069 \times 3.6 + 8.10 = 31.74$ MJ of energy input (although modern plants typically require 28 MJ/kg). With a global production of approximately

Table 2.3 NH_3 production process

Inputs		
Water	m^3	1.10E−03
Air	kg	1.50E+00
Natural gas	MJ	2.34E+01
Electricity	kWh	6.94E−02
Fuel	MJ	8.10E+00
Outputs		
NH_3	kg	1.00E+00
Emissions to air		
NO_2	kg	1.00E−03
N_2O	kg	1.48E−05
CH_4 (fossil)	kg	1.20E−05
CO_2 (fossil)	kg	1.46E+00
CO (fossil)	kg	8.40E−05
SO_2	kg	1.00E−05
Particulates	kg	1.20E−06
NMVOC[a], unspecified	kg	1.80E−05
Emissions to water		
NH_3 or NH_4	kg	1.00E−04

Source [21], Tables 11.2, 11.3, and 11.5. Auxiliaries and wastes are neglected. a: NMVOC refers to nonmethane volatile organic compounds. Fossil refers to the emission originating from the combustion of fossil fuel

180×10^9 kg (Fig. 2.1), the global energy requirement for operating the Haber-Bosch process amounts to 5.81 EJ (E denotes 10^{18}), which corresponds to around 1% of global energy production in 2019 (617 EJ) [22]. For more detailed information and recent developments regarding the Haber-Bosch process, please refer to [23–25].

2.2.3 *Animal Production Processes*

We now shift our focus to domesticated animals and consider the production processes related to pigs, cattle, and salmon. Animal production processes share similarities with crop production processes, as they involve accumulation (growth). However, the animal production process stands apart due to the inclusion of external feed input, its subsequent conversion into manure, and the resulting emissions.

2.2.3.1 Pigs and Cattle

Table 2.4 gives an example of the pig production process in the EU [26]. To simplify the illustration, the breeding process of a sow was not considered. To produce 10^3 kg-live weight of pigs, approximately 2,800 kg of feed is needed, of which around 530 kg (dry weight) ends up as manure. For more detailed information on pig production, including sow breeding, please see [26, 27].

Table 2.5 provides an example of the beef production process (suckler cow-calf system) in the EU, taken from [26]. Unlike the pig production process in Table 2.4, the table does not include the mass of manure but displays the associated emissions. To produce 10^3 kg of live weight, cows require around 20,850 kg of feed.

A comparison with the pig production process in Table 2.4 reveals several distinguishing differences. First, the feed composition varies. Cattle, being ruminant animals, can digest fiber-rich biomass like grass, which pigs and humans cannot. Ruminants play a crucial role in converting biomass into high quality protein sources such as meat and milk for human consumption [29]. However, the anaerobic digestion of fiber-rich carbohydrates in the forestomach of ruminants produces methane as a by-product of enteric fermentation [29]. Methane is generated when bacteria use hydrogen to reduce carbon dioxide, as illustrated in the following equation [29]

$$CO_2 + 4H_2 \longrightarrow CH_4 + 2H_2O \tag{2.8}$$

This process makes cattle one of the primary emitters of CH_4, alongside paddy-rice production [30]. In contrast, pigs produce minimal CH_4 emissions, with most of it originating from manure treatment.

Secondly, the meat productivity of pigs is significantly higher than that of cattle, primarily due to the number of offspring produced by female animals and a shorter growth period (6 months for pigs versus 39 months for cattle). A sow typically gives birth to 27 piglets per year, while a cow produces around six calves in her lifetime

Table 2.4 Pig production process per 1000 kg pig live weight

Inputs		
Sow[a]	kg	33
Wheat	kg	1,112
Barley	kg	855
Soybean meal	kg	341
Mineral feed P	kg	497
Electricity	kWh	148
Heat	MJ	541
Land	m^2	4,400
Outputs		
Pig for slaughter (live weight)	kg	1,000
Manure	kg	6,900
Manure: dry matter	kg	528
CH_4 (enteric)	kg	3.7
CH_4 (manure)	kg	16.2
N_2O	g	553
NH_3	kg	7.6
NO_X	g	612

Source [26]. a: This value is derived by the author from the live weight of a sow needed to produce 10^3 kg of fattening pig, accounting for 1/3 of the sow's weight. This derivation is based on certain assumptions: the sow gives birth to 27 piglets per year, with a mortality rate of 1.9%, and a live weight for slaughter of 107 kg, as stated in [26]. It is also assumed that the sow will be slaughtered after three years, with 1/3 of the sow "being slaughtered" each year

(these numbers apply to Japanese cattle farms). The difference in meat productivity aligns with the fact that pigs (pork) are three times more efficient than beef in terms of both protein and heat value when converting feed to meat [31]. However, it is important to note that the calculation of meat productivity relies on simplifying assumptions, and the co-production of milk is not considered. Therefore, these numbers should be interpreted as a rough estimate of the overall tendency. For more details on pig and cattle production, including the breeding of sows and cows, please refer to [26, 27].

2.2.3.2 Salmon Aquaculture

The final example of animal processes is the Norwegian salmon farm presented in Table 2.6, taken from [32]. It requires 1100 kg of feed to produce 1000 kg of salmon from 17 kg of smolts, indicating a high feed conversion efficiency, compared to pigs. However, this calculation does not take into account the amount of feed required to

Table 2.5 Inputs and outputs for a beef-fattening process (suckler cow-calf system)

Inputs		
Cow (slaughter weight)[a]	kg	344
Feed intake in dry matter (DM)		
Home-grown (grass, maize, spring-barley, straw)	kg	20,851
Fertilizer import	kg	
Nitrogen		478
Phosphorous		21.5
Direct on-farm energy used		
Electricity used in stables	MWh	1.07
Electricity used in crop processing	MWh	0.64
Diesel (traction)	GJ	14
Transport	10^3 kg km	
Feed (soy meal)		
By ship		162
By truck		12
Outputs		
Meat (slaughter weight)	kg	10^3
Direct on-farm emissions	kg	
N_2O		26.2
CH_4		476.1
Enteric fermentation		417.6
Manure management		58.5
NH_3		95.6
NO_3		1231
PO_4^{3-}		2.7

Source [28], Tables 2 and 7. a: The slaughter weight of a cow required per 1000 kg of meat (slaughter weight) estimated by the author, assuming that the cow gives birth to 6 calves throughout her life. Among these, three are bull calves, each producing a slaughter weight of 348 kg, while 80% of the remaining heifer calves produce a slaughter weight of 263 kg each. The cow itself has a slaughter weight of 357 kg, as indicated in Table 1 of [28]

produce the fish-derived feed used in salmon farming. It is important to note that the environmental impact of salmon farming is not negligible, especially in terms of nutrient pollution. Nitrogen and phosphorus emissions from salmon farms are primarily linked to the feed and can contribute to the eutrophication of water bodies.

Table 2.6 A salmon farming process (Norway)

Inputs		
Feed	kg	1103
Crop-derived	%	42.4
Fish-derived	%	57.6
Feed transport	10^3 kg-km	290.3
Smolts	kg	17.4
Smolts transport	10^3 kg-km	1.2
Total on-farm energy use	MJ	646.8
Electricity		72.2
Natural gas		0.1
Diesel		570.3
Gasoline		4.2
Outputs (farm-level)		
Salmon	kg	1000
N emissions	kg-N	41.1
P emissions	kg-P	5.2
CO_2 (fuel)	kg-CO_2	39.5

Source [32], Table 1, the Norwegian data. The CO_2 emissions originating from fuels were calculated by the author

2.2.4 *Metals and Cement*

We now shift our focus to nonliving things, beginning with metal production processes. Unlike the case in crop and animal production processes, there is no accumulation involved in metal production processes. The underlying mass follows (2.1).

2.2.4.1 Iron Production Process Based on the Blast Furnace (BF)

The blast furnace is the primary method for producing pig iron. Iron-bearing materials, such as iron ore, sinter, and pellets, are continuously fed into the furnace from the top through a charging system, along with additives such as limestone and reducing agents such as coke [33]. Iron (Fe) typically occurs in iron ore as iron oxide (Fe_2O_3), so to produce iron, the bond between iron and oxygen must be broken. This is facilitated by using coke as a reduction agent in the following reaction

$$Fe_2O_3 + 3CO \longrightarrow 2Fe + 3CO_2 \tag{2.9}$$

CO is generated by the reaction

$$2C + O_2 \longrightarrow 2CO \tag{2.10}$$

Table 2.7 A blast furnace process of iron smelting

Inputs		
Items	Units	Amounts
Coke	kg	380
Sintered ore	kg	1160
Bulky ore	kg	280
Pellets	kg	190
Air	m^3	995
Pulverized coal	kg	120
Outputs		
Pig iron	kg	1000
C	%	4.5
Si	%	0.39
Mn	%	0.27
P	%	0.1
S	%	0.03
Dust	kg	16
BF gases	kg	1800
CO	%	22
CO_2	%	22.8
H_2	%	4.2
Slag	kg	300
SIO_2	%	32.8
CaO	%	42.5
Al_2O_3	%	14.3
MgO	%	6.5
MnO	%	0.26
S	%	0.96

Source JFE steel corporation. http://www.jfe-21st-cf.or.jp/jpn/chapter_2/2d_2.html

The gangue material in iron ore, consisting of Si_2O_3 and Al_2O_3 ([34], Table 4), is separated as slag by adding limestone ($CaCO_3$). The resulting molten slag, composed of Al_2O_3-SiO_2-CaO, is collected in the lower part of the blast furnace. In Table 2.7, $CaCO_3$ is not listed as a separate input because it is contained in the sintered ore and pellets.

The use of coke as a reduction agent makes iron production a major emitter of CO_2. To address this problem, an alternative technological option involves reduction with H_2 [35]. This process requires additional heat and involves the following endothermic (heat absorbing) reactions

$$
\begin{aligned}
3Fe_2O_3 + H_2 &\longrightarrow 2Fe_3O_4 + H_2O \\
Fe_3O_4 + H_2 &\longrightarrow 3FeO + H_2O \\
FeO + H_2 &\longrightarrow Fe + H_2O
\end{aligned}
\tag{2.11}
$$

In addition to requiring extra heat, H_2 must be produced for this iron-making process. For more details on reducing iron oxides with hydrogen, see [35].

2.2.4.2 Aluminum Production

Aluminum (Al), the third most abundant element in the lithosphere after oxygen and silicon (Si) [36], is primarily derived from its oxide known as bauxite. The principal mineral component of bauxite is gibbsite, $Al(OH)_3$ [37, 38]. The production process for aluminum involves the extraction of Al_2O_3 from bauxite using the Bayer process [38, 39]

$$
Al(OH)_3 + NaOH \longrightarrow \frac{1}{2}Al_2O_3 + \frac{3}{2}H_2O + NaOH \tag{2.12}
$$

The extracted Al_2O_3 is then dissolved in molten salt and subjected to electrolysis via the Hall-Héroult process, which utilizes a carbon anode to produce aluminum [38, 40]

$$
2Al_2O_3 + 3C \longrightarrow 4Al + 3CO_2 \tag{2.13}
$$

It is important to note that the C in the above equation does not originate from fuel (it is a solid) but from the anode consumed in the electrochemical reaction. The consumption of the anode is given by $0.334/\epsilon_F$ -kg/kg-Al, where ϵ_F is the current efficiency [40].

Table 2.8 gives an example of an integrated Al production process, including bauxite processing, alumina production, smelting of alumina into primary aluminum, and ingot production, taken from [41]. A remarkable feature of aluminum production is the substantial amount of electricity required for smelting (electrolysis), approximately 15 kWh/kg-Al [42]. Consequently, smelting facilities are typically located in regions with access to cheap electricity, such as Australia, Brazil, Canada, China, Norway, and Russia. In Japan, domestic production was a significant source of aluminum until the mid-1970s, but currently, the country relies entirely on imports.

The high energy cost of smelting implies the importance of recycling aluminum scrap via remelting. Recycling aluminum can be accomplished using less than 5% of the energy required for primary production [43].

Table 2.8 Aluminum production process: from bauxite to aluminum ingot

Items	Units	Amounts
Inputs		
Residual fuel oil	l	2.13E−01
Gasoline	l	1.74E−03
Diesel	l	2.65E−02
Electricity[a]	kWh	1.53E+01
Natural gas	m^3	4.85E−01
Bituminous coal	kg	1.63E−02
Sodium hydroxide (NaOH)	kg	1.40E−01
Quicklime (CaO)	kg	8.68E−02
Bauxite	kg	5.02E+00
Liquefied petroleum gas	l	6.50E−03
Anode	kg	4.55E−01
Outputs		
Aluminum ingot	kg	1.00E+00
CO_2 fossil	kg	1.52E+00
CO	kg	6.00E−02
CH_4	kg	4.15E−04
Nitrogen oxides	kg	3.37E−04
N_2O	kg	2.12E−06

Source [41] Table 3, except electricity, the value of which was taken from [42]. The original data consisting of four disaggregated processes (bauxite, alumina, smelting, and ingot) were integrated into the ingot process by the author. The list of emissions is shortened. a: Taken from [42]

2.2.4.3 Copper Production

Copper (Cu) is primarily found in sulfide minerals, including chalcopyrite ($CuFeS_2$), chalcocite (Cu_2S), bornite (Cu_5FeS_4), and enargite (Cu_3AsS_4), where each tonne of copper is associated with a tonne of sulfur [33]. Copper concentrates undergo roasting to produce blister copper with approximately 98% purity, with the reaction for chalcopyrite represented as given by [44, 45]

$$\begin{aligned} 2CuFeS_2 + 4O_2 &\longrightarrow 2FeO + Cu_2S + 3SO_2 \\ Cu_2S + O_2 &\longrightarrow 2Cu + SO_2 \end{aligned} \tag{2.14}$$

Unlike the iron production process, the roasting process is autothermal, meaning it produces synthesis gas using only the heat from the reaction itself. The sulfides in the feed serve as both fuel and reagents to form sulfates [46]. The resulting blister copper then undergoes refining processes, including pyrometallurgical or fire refining and electro refining, to produce electric copper of up to 99.99% purity [45]. Table 2.9 gives an example of the copper production process.

Fig. 2.2 Ashio copper mine (Japan) circa 1895 The emission of SO_X in copper smelting destroyed the entire vegetation of the surrounding mountains. *Source* Pictorial History of Modern Japan Vol. 5 published by Sanseido. This photograph is in the public domain in Japan

The high concentration of sulfur in copper ore is advantageous for roasting, as it provides the necessary energy. However, this process also results in significant amounts of SO_2, which can cause severe environmental damage if released into the atmosphere. Figure 2.2 illustrates the copper production facilities of the Ashio mine in Japan in the late 19th century, depicting the magnitude of the impact with the complete destruction of vegetation on the surrounding mountains due to emissions. Despite extensive mitigation efforts, the vegetation has still not fully recovered. This incident was one of the first major pollution disasters of its time and led to the emergence of Japan's environmental movement.

To mitigate the emission of SO_2 from flue gases, various flue gas desulfurization processes have been developed [47]. One commonly used processes is wet scrubbing, where limestone ($CaCO_3$) serves as the sorbent. The SO_2 reacts with the sorbent to form synthetic gypsum ($CaSO_4 \cdot 2H_2O$), which has commercial value. The chemical reaction involved in this process can be represented as follows [48]

$$CaCO_3 + SO_2 + 2H_2O + 0.5O_2 \longrightarrow CaSO_4 \cdot 2H_2O + CO_2 \tag{2.15}$$

Table 2.9 The unit process data for 1000 kg of copper: from rock to copper cathode

Ore grade (% Cu)	0.18
Mining type	Open pit
Process	Pyrometallurgical copper production
Feedstock (ore) (kg)	540,000
Energy	
Electricity (MJ)	47,432
Diesel (MJ)	14,409
Heavy fuel oil (MJ)	1,350
Total energy (MJ)	63,192
Auxiliary materials	
Explosives (kg)	375.3
Lime (kg)	126.1
Limestone (kg)	15.9
Sand (kg)	255.8
Oxygen (kg)	705.3
Hydrochloric acid (kg)	0.1
Emissions to air	
CO_2 (kg)	1,050.4[a]
SO_2 (kg)	12.9[a]

Source [45]. a: The value includes direct emissions from diesel and fuel oil burning and the smelter

2.2.4.4 Interconnections Among Nonferrous Metals in Ores, Alloys, Products, and Waste Management

Interconnections Based on Geological Factors

Nonferrous metals, such as Cu, lead (Pb), zinc (Zn), gold (Au), and silver (Ag), are distinguished from Fe and Al by the presence of interconnections based on geological factors, resulting in interdependence in metal production processes. For instance, in the copper refining process, the co-elements Fe, Zn, Au, and Ag are major impurities. During electrolysis, Cu is deposited at the cathode, while Fe and Zn remain ionized in the solution, and Au and Ag settle to the bottom (anode slime) for recovery. The revenue generated from the recovery of Au and Ag is used to offset the cost of the electricity required for electrolysis ([49], p. 617).

Table 2.10 provides additional information on the occurrence of co-elements in ores of carrier, commodity, and bulk metals. It further divides these co-elements into two groups based on the metallurgy of individual commodity metals. This table was produced based on the metal wheel [50]. It is interesting to note that Ag and Au are co-produced in Cu, Pb, and Zn refining processes, although they have their production infrastructure. It is noteworthy that highly valuable, high-tech metals, essential in electronics, are primarily available as co-products only. A typical example of such metals is the platinum (Pt) group metals (PGM), including Pt, palladium

Table 2.10 Metal linkages in metal processing

Carrier metals	Oxidized ores		Sulfide ores	
	Fe	Al	Cu	Pb Zn
Co-elements with considerable own production infrastructure	Pb, Zn		Ag, Au, Mo, Pb, Pd, Pt, Zn	Ag, Au
Co-elements with no or limited dedicated production infrastructure		Ga, V	As, Bi, Co, Ir, Ni, Os, Rh, Ru, Se, Te	As, Bi, Cd, Co, In, Ga, Ge
Co-elements that end up in residues or as emissions.	As, Al, Cr, Cu, Mg, Mn, Sn, Ti, V	Cu, Cr, Fe, Mn, Li, Ti	Ca, Fe, Hg, Sb, Si	Cu, Fe, Hg, Mg, Mn, Sb, Si, Ti

Source [50], Fig. 1, tabulated by the author

(Pd), rhodium (Ro), iridium (Ir), osmium (Os), and ruthenium (Ru). These metals occur in copper ore and are coproduced with copper, making them a by-product of copper production.

Interconnections in Alloys and Products

Base metals, such as Fe and Al, are rarely used in their pure form, but are commonly used in alloys with other metals. Pig iron, produced through a blast furnace, is converted into various types of steel by adjusting for impurities like P and S and adding alloying metals such as chromium (Cr), manganese (Mn), molybdenum (Mo), niobium (Nb), nickel (Ni), vanadium (V), titanium (Ti), and zirconium (Zr), some of which are considered rare. For instance, austenitic stainless steel is an alloy of Fe with Cr and Ni. Similarly, although aluminum is primarily used in its nearly pure form to produce foil materials (and in the case of Japan, the one yen coin), the majority of aluminum is used in alloys with Cu, Fe, Mg, Mn, Ni, Si, and Zn [51]. Zn and Si do not naturally occur as co-elements in aluminum ore or natural mines (Table 2.10), but occur in end-of-life products or "urban mines."

Implications for Waste Management of End-of-Life (EoL) Products

The composition of aluminum products, such as window frames and car engines, can significantly differ from that of natural ores. This implies that the refining processes adopted for natural mines may not be readily applicable to urban mines. EoL products made of metals typically undergo disassembling, shredding, and sorting processes to be transformed into scraps. These scraps are then remelted or refined in order to produce secondary metals. During these processes, unintentional mixing of different metal species is likely to occur, resulting in undesirable combinations of metals, such as Cu and Sn in steel scrap from EoL vehicles, which can cause problems as impurities (tramp elements) during working processes [52]. Poor sorting of electronic components (with cables made of Cu and solder in electronics containing tin (Sn)) from the iron components of an EoL vehicle can result in the intrusion of Cu and Sn into steel scrap.

The fact that metals in products mostly exist not in a pure form isolated from each other but as alloys combined with other metal species has significant implications for the recovery of metals from EoL products. This will be discussed in Chap. 7.

2.2.4.5 Cement

Production Processes

Cement production is the third-largest industrial contributor to global anthropogenic GHG emissions, responsible for about 7% of such emissions [53]. The primary source of CO_2 emissions in cement production is the calcination of limestone, described by the following chemical reaction

$$CaCO_3 \longrightarrow CaO + CO_2 \tag{2.16}$$

Table 2.11 gives an example of the input-output flows associated with cement production in the US [54]. To produce 1000 kg of cement, 1165 kg of limestone is needed, which leads to the emission of 553 kg-CO_2 due to calcination. Furthermore, the consumption of fuel necessary to generate the required heat for the reaction results in the emission of 365 kg-CO_2, bringing the total emissions to 918 kg-CO_2. While the use of alternative heat sources can help reduce emissions from fuel consumption, it is important to note that emissions from calcination are fundamentally constrained by the laws of physics, (2.16).

Interestingly, waste materials, such as bottom ash (from waste incineration plants), fly ash (from flue gas/dust treatment), foundry sand, and blast furnace slag, are used as materials, and residual oil is used as fuel. A large variety of waste can be recycled in cement production as alternative material and fuel sources [55]. Fly ash, a siliceous material obtained from thermal power stations, is an important source of Al_2O_3 and SiO_2, which are essential in the hardening reaction (Pozzolan reaction) of cement [56, 57]. However, reducing thermal power generation as a means of curbing CO_2 emissions could lead to a shortage of fly ash, necessitating the search for alternative sources of these vital materials in cement production. For details of cement production processes, see [53, 54, 58–60].

CO_2 *Uptake by Cement Carbonation*

The process of carbonation, which involves the re-absorption or uptake of CO_2 from the atmosphere, is an important but often overlooked issue in discussions about the environmental impact of cement-based materials. The aforementioned LCA litera-

Table 2.11 Inputs and emissions of the cement production process (US average)

Raw material	$kg/10^3$ kg of cement
Limestone	1,165
Cement rock, marl	207
Shale	52
Clay	60
Bottom ash	10
Fly ash	13
Foundry sand	4
Sand	40
Iron, iron ore	14
Blast furnace slag	20
Other raw material	26
Gypsum, anhydrite	49
Water	88
Fuel and electricity	unit/10^3 kg of cement
Coal, 10^3 kg	0.107
Gasoline, L	0.133
Liquefied petroleum gas, L	0.0143
Middle distillates, L	1.066
Natural gas, m^3	5.569
Petroleum coke, 10^3 kg	0.0223
Residual oil, L	0.0442
Wastes, 10^3 kg	0.0177
Electricity, kWh	144
Emissions	$kg/10^3$ kg of cement
Particulate matter, total	0.201
CO_2	918
SO_2	1.65
NOx	2.42

Source Taken from [54] with items of minor amounts excluded

ture on cement does not consider this issue. Carbonation is an intrinsic property of Portland cement-based concrete, representing the reverse process of (2.16) [61].

Reference [61] proposed the following formula to calculate the amount of CO_2 absorbed per volume of carbonated concrete

$$CO_2 \text{ uptake (kg } CO_2/m^3) = 0.75 \times C \times CaO \times \frac{M_{CO_2}}{M_{CaO}} \tag{2.17}$$

Here, C represents the mass of Portland cement clinker per m^3 concrete, CaO is the mass fraction of CaO in the cement clinker, and M_x is the molar mass of x. Calculations on concrete production in Nordic countries indicate that after 100 years, between 58 and 86% of concrete in Sweden, Norway and Denmark will be carbonated, while the corresponding value for Iceland is 37%. A major factor behind this difference is the treatment of demolished concrete: in Denmark, it is mostly crushed and recycled, while in Iceland, it is disposed of in landfills. For more recent results, including calculations at the global level and more sophisticated models, please refer to [62–65].

2.3 A Process Representation of Food Consumption

Household consumption of goods and services constitutes the largest share of GDP in most economies and is a significant driver of anthropogenic environmental impacts. Consumption of goods and services represents inputs to households, while post-consumption waste represents outputs.

In economics, labor (hours of work) is often treated as an output of the household, with consumption as inputs [66, 67]. In his early work on IO [68], Leontief devoted a section to "Households as an industry" (p. 41), although he later did not use this modeling. In IE, [69] is an example of recent work that has considered human labor as a part of the production system. In Chap. 3, we will discuss the energy (food) requirements to produce human labor in an economy where human muscle power is the sole power source to operate the production process.

Here, our focus is on representing household consumption of food as a process, where inputs (food) are converted into outputs (human waste and emissions), to gain insight into the environmental impacts of food consumption. Representing household consumption as a process does not mean that we are treating the household sector as an endogenous sector. For a process, its inputs and outputs must have mass (Sect. 2.1.3). Since energy has no mass, hours of work cannot be counted as an output of the household consumption process. Unlike the case in economics, where hours of work (or human energy) are the primary output of the household [66, 67], we focus on the physical outputs resulting from food consumption, namely human waste and emissions. Since the conversion of food in our bodies refers to human metabolism, the following can be regarded as a process-based representation of human metabolism. We represent human metabolism as a black box and focus on the flow and stock of major chemical elements entering (as food) and leaving (as human waste and emissions) our body. We begin with the chemical composition of our body and food.

2.3.1 We Are What We Eat

Of the organic elements that make up our body, C has the largest share (21%), followed by N (around 3%), P (0.7%), and S (0.2%) (Table 2.12). These elements come from the food we consume. Table 2.13 gives data on the Spanish diet (food inputs) in terms of food input per person per year, while Table 2.14 gives the elemental composition of food constituents (excluding inorganic ones). For comparison, Table 2.15 gives data on the average food purchase in Japan per year per person. Interestingly, when excluding Beverages, Bottled water, and Other, the total mass of food purchase is nearly the same in both countries, with 562 kg in Spain and 565 kg in Japan. Furthermore, the consumption of meat and fish products is similar in both countries, with 103 kg in Spain and 96 kg in Japan.

For adult humans whose body weight is stable, the amount of food intake is balanced with outputs such as expiration, excreta, flatus, sweat, and odor. On average,

Table 2.12 Elemental composition of body components

	Chemical structures	C	H	N	Ca	O	S	P	Sum
Water	H_2O	0.00	5.56	0.00	0.00	44.44	0.00	0.00	50.00
Lipid	$C_{51}H_{98}O_6$	9.42	1.51	0.00	0.00	1.48	0.00	0.00	12.40
Protein	$C_{100}H_{159}N_{26}O_{32}S_{0.7}$	7.07	0.94	2.14	0.00	3.02	0.13	0.00	13.30
Bone mineral	$Ca_{10}(PO_4)_6(OH)_2$	0.00	0.01	0.00	1.12	1.16	0.00	0.52	2.80
Total		16.49	8.01	2.14	1.12	50.10	0.13	0.52	78.50
Ratio		0.210	0.102	0.027	0.014	0.638	0.002	0.007	1.000

Source An estimate of the elemental composition of the human body for mature men in the US with a bodyweight of 78.5 kg [71]. The chemical structure of bone mineral is from [72]. Elemental decomposition is based on the chemical structures

Table 2.13 Food purchase per capita in Spain in 2005

Product group	kg	%
Eggs	14	1.6
Meat products	66	7.5
Fish and seafood	37	4.2
Dairy products	143	16.2
Bakery products	70	8.0
Vegetable oils and fats	22	2.5
Vegetables	108	12.2
Fruit	103	11.7
Beverages	174	19.8
Bottled water	68	7.7
Other	77	8.7
Total	881	100

Source [73], Table 1

Table 2.14 Elemental composition of organic constituents in food (Spanish diet)

Constituents	% in weight[a]	kg	C	H	O	N	S
Water	75	660.8	0	0.11	0.89	0	0
Protein	3.6	31.7	0.47	0.07	0.29	0.15	0.02
Fat	5.8	51.1	0.77	0.12	0.12	0.00	0.00
Carbohydrate	13	114.5	0.42	0.06	0.52	0.00	0.00
Alcohol	0.78	6.9	0.52	0.13	0.35	0.00	0.00
Fiber	0.061	0.5	0.44	0.06	0.49	0.00	0.00
Total	99	881	109.4	90.1	669.2	4.8	0.6

a: The weight refers to the Spanish diet in Table 2.13. *Source* [74] Table 1 and [73] Table 4. The composition slightly differs from Table 2.12 due to the use of different samples and methods

Table 2.15 Per capita food purchases in Japan in 2018

Items	kg/capita/year
Chicken egg	21
Meat	51
Fish and Seafood	45
Milk and milk products	96
Cereals	104
Potatoes	23
Starch	16
Fat	20
Beans	9
Vegetables	103
Fruits	49
Seaweed	1
Sugars	18
Miso	4
Soy sauce	6
Other food items	5
Total	570

Source The Japanese Ministry of Agriculture and Forestry: Food supply and demand table

humans excrete 80–170 g of feces and 10^3 g of urine per person per day [70]. Therefore, human metabolism can be seen as a process in which inputs of air, food, and water are converted into outputs of CO_2 (expiration), excreta, flatus, sweat, and aromatic compounds.

Table 2.16 C and N flows in the human metabolism process: kg per year per capita

	Consumption	Body[c]	Respiration	Flatus etc.	Excreta[d]	Urine	Feces
C[a]	67	21	59	0.11	6.6	1.9	4.6
N[b]	4.0	2.5	0.0	0.6	3.5	2.2	0.7

a: Estimated for average US body mass of 72.69 kg for both males and females of age 15–39 [70]. b: Taken from [75] on Japanese data, which was adjusted to the US average body mass based on the US-Japan body mass ratio of 72.69–62 kg. c: The amount contained in the body. d: Excreta includes both urine and feces. The decomposition of N in excreta is based on [76]

2.3.2 *C and N Flows in Human Metabolic Processes: Their Origins, and Fate*

Table 2.16 gives an estimate of the C and N flows involved in the human metabolism process in the US [70] and Japan [75].[1] One notable observation from Table 2.16 is the manner in which C and N are emitted from the human body. The majority of C (89%) is emitted as CO_2 through respiration, while solid human waste (feces) accounts for the remaining 7%. In contrast, the majority of N (56%) is emitted as urine, with no gaseous emissions.

2.3.2.1 C Flows Associated with Human Food Consumption

The majority of C in our diet comes from carbohydrates (Table 2.14), which are produced by green plants through the process of photosynthesis (2.3). C moves from the atmosphere to plants through photosynthesis and then from plants to animals, including humans, through the food chain.

Respiration is the metabolic process through which organisms release the energy captured during photosynthesis as heat and perform work. The overall reaction for aerobic respiration (2.4) is the reverse of the photosynthetic reaction (2.3).

In their study on the human C budget in the US, [70] estimated the total C emission originating from the respiration of the US population. It was estimated that the expiration released 15.2 Tg-C (T ,"tera," is 10^{12}) of CO_2, which makes up around 1.2% of the emissions from fossil fuel [70].

The emission of CO_2 from human respiration is not counted the same as the emission from burning fossil fuels as a greenhouse gas contributing to climate change. When comparing the equation for photosynthesis (2.3), which produces carbohydrates by fixing C, with the equation for respiration (2.4), which consumes carbohydrates, we can observe that there is no net change in the amount of CO_2 in the atmosphere. The same quantity of atmospheric CO_2 that green plants fix is

[1] The difference from the Spanish C consumption in Table 2.14 can be attributed to variations in the sample, methods, and data used.

subsequently released into the atmosphere, resulting in no change in the atmospheric concentration of CO_2. Therefore, human respiration is regarded as C-neutral.

2.3.2.2 N Flows Associated with Human Food Consumption

The Haber-Bosch process (Sect. 2.2.2) plays a crucial role in global food production, and its impact cannot be overstated. The production of NH_3 through this process has increased in tandem with the global population (Fig. 2.1). It is estimated that almost half of the world's population depends on NH_3 produced by the Haber-Bosch process to meet their food needs [77]. In 2010, the Haber-Bosch process generated 120 Tg N, which is double the amount of Nr produced by natural terrestrial sources (63 Tg N) [77]. The human impact on the N cycle is much larger than that on the C cycle, where anthropogenic emissions from fuel combustion account for less than 10% of the annual C uptake by terrestrial photosynthesis [78].

The final step in the N cycle is microbial denitrification, which releases N_r back into the atmosphere as N_2, where it originated. Denitrification is an anaerobic process that occurs in the absence of O_2, which converts NO_3^- to N_2 through the following series of reactions [79]

$$NO_3^- \longrightarrow NO_2^- \longrightarrow NO \longrightarrow N_2O \longrightarrow N_2 \tag{2.18}$$

The production of NO_3^- occurs through a series of microbial processes as follows [80]. In the first step, NH_3 is oxidized to NO_2^-

$$NH_3 + H_2O \longrightarrow NO_2^- + 3H^+ + 2e^- \tag{2.19}$$

in the second (last) process, nitrite is oxidized to NO_3^-

$$NO_2^- + H_2O \longrightarrow NO_3^- + 2H^+ + 2e^- \tag{2.20}$$

N_r molecules undergo transformations and move from one environmental system to another before returning to the atmosphere. These transformations result in changes in their chemical states (2.18), which can have impacts on human health and the environment. This process is known as the N cascade [17]. The effects of N_r on human health and ecosystems include respiratory illness, cancer, and cardiac disease caused by NO_X (NO and NO_2) emissions, acidification, eutrophication, biodiversity losses, habitat degradation through NO_2^- and NO_3^-, and climate change through N_2O emissions.

Approximately half of the N_r fertilizer applied to global agroecosystems becomes part of the human food and livestock feed, while the other half enters the N cascade, being released into the atmosphere as NH_3, NO, N_2O, or N_2, or lost to aquatic ecosystems primarily as NO_3^- [17]. The portion of N_r ingested by humans and livestock as food or feed is eventually released as excreta and enters the N cascade.

2.4 A Schematic Representation of Global C, N, P, and S Cycles

We discussed above the C and N flows, which are major elements in our body. Now we will present simplified diagrams that represent the global flows of C, N, and other important elements, namely P and S. Although these diagrams are not highly detailed, they serve to identify common and unique patterns in these flows. Unfortunately, due to differences in data availability, we are unable to utilize a single schematic representation for all the elements. Particularly, our coverage of the P flow will be limited to the anthropogenic flow only, and the flow of S will only be partially quantified.

2.4.1 *The Global Flows of C and S*

The flows of C (Fig. 2.3a) and S (Fig. 2.4a) exhibit qualitative similarities in terms of their exchange among the atmosphere, terrestrial, and water systems (including living organisms, which are not explicitly shown in Fig. 2.4 due to the lack of data). However, they differ significantly in their anthropogenic contributions. The atmospheric emission of C resulting from fuel combustion is less than 10% of the amount fixed by photosynthesis. In contrast, the anthropogenic emission of S surpasses natural emission by approximately a factor of three [82].

Fossil fuel combustion is the primary source of anthropogenic emissions of CO_2, NO_X, and SO_2. By representing fossil fuel as $C_aH_bO_cS_xN_y$, the emissions resulting from its combustion can be expressed as ([86], p. 18)

$$
\begin{aligned}
&C_aH_bO_cS_xN_y + [a + 0.5(b - c) + x]O_2 \\
&\qquad \longrightarrow aCO_2 + bH_2O + xSO_2 + y[\alpha NO + 0.5(1 - \alpha)N_2]
\end{aligned}
\tag{2.21}
$$

Here, $0 < \alpha < 1$ refers to "fuel NO formation," NO originating from the fuel (i.e., the organic nitrogen, such as proteins), while $1 - \alpha$ corresponds to the formation from combustion air. While the original equation in [86] includes Cl as a component of fuel, it is omitted here because of its absence in the discussion so far.

When SO_2 is released into the atmosphere, it reacts with water and oxygen to form sulfuric acids, which contribute to the occurrence of acid rain ([87], Table 3.4)

$$
SO_2 + H_2O + O_3 \longrightarrow 2H^+ + SO_4^{2-} + O_2 \tag{2.22}
$$

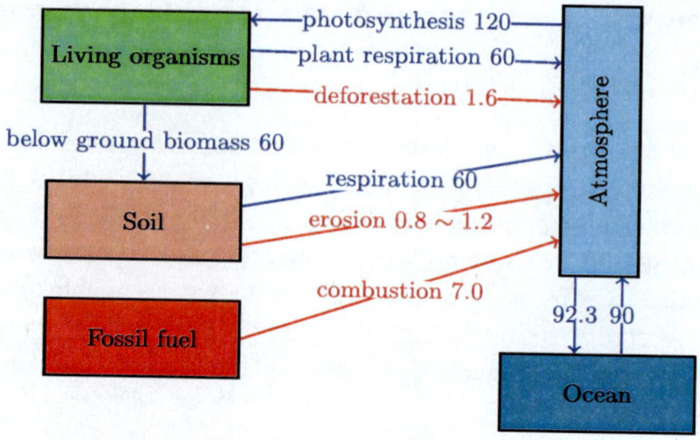

(a) C: P-g-C per year

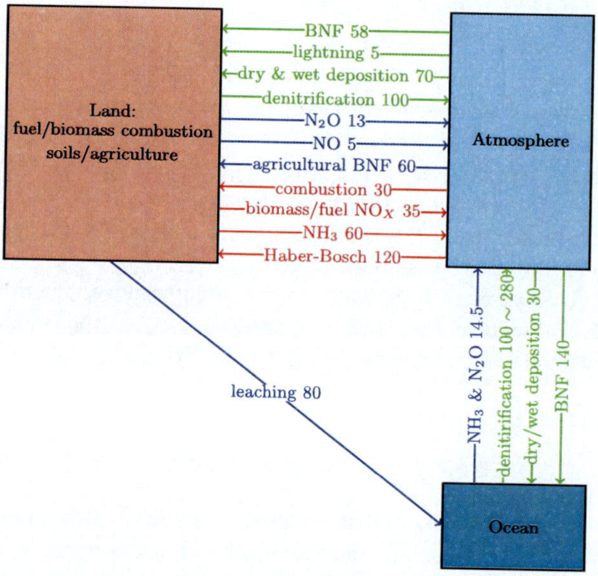

(b) N: T g-N per year

Fig. 2.3 The global flows of C and N BNF: biological nitrogen fixation. In Figure **a**, the flows originating from anthropogenic factors such as fossil fuel combustion and land use are represented in red. In Figure **b**, the flows in red primarily originate from anthropogenic factors, while the flows in green originate from natural factors. The flows in blue represent a combination of both anthropogenic and natural factors, falling somewhere in between. *Source* The data were taken from [81] for the C-flow, and from [77] for the N-flow. The margins of error in the original work were neglected

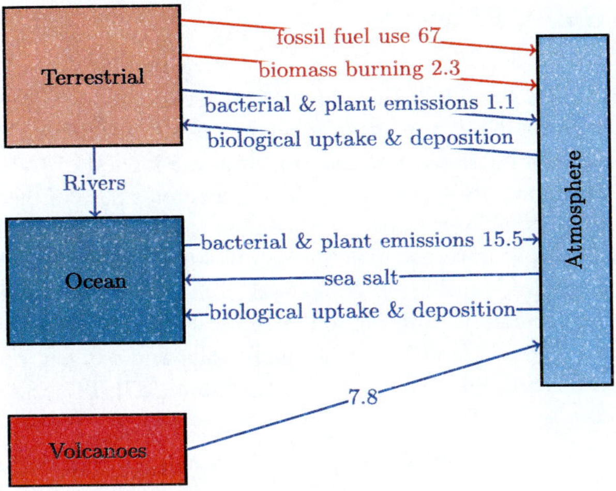

(a) S: T-g-S per year

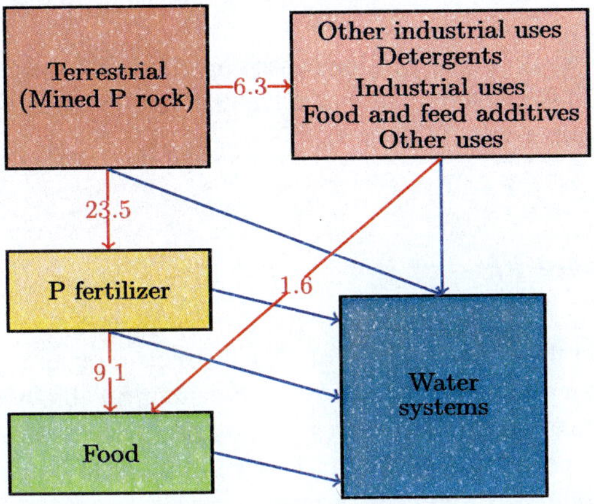

(b) P: T-g-P per year

Fig. 2.4 The global Flows of S and (anthropogenic) P. The P flow consists solely of flows that arise from anthropogenic activities. In Figure **b**, the blue-colored flows represent losses to water (and other sinks). The scheme of P flow is largely based on [84], Fig. 3. *Source* The data were taken from [82, 83] for the S flow, and from [85] for the P flow

2.4.2 The Global Flow of N

The flow of N (Fig. 2.3b) exhibits a unique characteristic in that it needs to be converted to a usable form, N_r, before most living organisms can utilize it. The invention of the Haber-Bosch process enabled the industrial production of N_r (NH_3, ammonia) from the abundant atmospheric N_2 (which constitutes 80% of the atmosphere). This led to anthropogenic dominance in the global N flow, with more N being fixed through the Haber-Bosch process than through biological processes. As mentioned above, N_r species exert various environmental impacts in a process known as the N cascade before ultimately returning to the atmosphere as N_2. When released into the atmosphere, NO_2, NO, and NH_3 react with water and oxygen, resulting in the formation of nitric acids, which contribute to acid rain ([87], Table 3.4)

$$NO + \frac{1}{2}H_2O + O_3 \longrightarrow H^+ + NO_3^- + \frac{3}{4}O_2 \tag{2.23}$$

$$NO_2 + \frac{1}{2}H_2O + \frac{1}{4}O_2 \longrightarrow H^+ + NO_3^- \tag{2.24}$$

$$NH_3 + 2O_2 \longrightarrow H^+ + NO_3^- + H_2O \tag{2.25}$$

In assessing the impacts of acidification caused by different substances, the potential acid deposition onto soil and water is evaluated using SO_2 as the reference substance, resulting in SO_2 equivalents, SO_2e.

2.4.3 The Global Flow of P

Finally, we consider the P flow (Fig. 2.4b). One distinctive feature of the P flow is the absence of atmospheric exchange. Unlike N, P cannot be obtained from the atmosphere and has to be extracted from terrestrial sources. Due to the uneven geological distribution of phosphate rock on Earth, P is considered to be one of the primary elements with the highest supply concern [88].

Despite its scarcity, the release of P into the environment can cause significant environmental impacts. The movement of P is represented by one-way transport from terrestrial sources to ocean or water systems in general (including other natural sinks). Although not shown in Fig. 2.4, the anthropogenic river P flux contributes to a doubling of preanthropogenic flux, indicating the dominant influence of human activities in the global P flow, similar to the S and N flows. The release of P into water systems can result in eutrophication, which destroys aquatic ecosystems by depleting the available oxygen. The explosively increased algae population and the subsequent decomposition of their biomass consume large amounts of oxygen. Additionally, P, N_r, and N also contribute to eutrophication. It is worth noting that the expected global adoption of EVs is likely to significantly raise the demand for P in batteries, particularly lithium-iron-phosphate (LFP) batteries [89, 90].

2.4.4 *The Need to Define a New Geological Epoch Dominated by Humans?*

As mentioned above, human activities have now become the primary influence on the global flows of N, P, and S. These examples represent only a small fraction of the numerous cases that illustrate the overwhelming impact of human actions on Earth's energy and substance cycles [91–95]. The effects of human activities will leave a lasting mark on the geological and stratigraphic records, persisting for millions of years into the future [96, 97]. The global scale of industrial ecosystems has become dominant to the extent that it suggests the advent of a new geological epoch characterized by human dominance. Considering this backdrop, there is an increasing consensus among scientists that Earth is entering a new geological epoch known as the Anthropocene, wherein humans play a central role.

2.5 Input-Output Analysis (IO) and Processes

We will now establish the connection between processes and IO. When considering the interdependence in the economy, there are various aspects to consider, including family/territorial connections, financial interdependence, shareholding, and political orientations. However, the interdependence can also arise through the flow of inputs and outputs among the processes that constitute the economy. For instance, crop cultivation, the Haber-Bosch process, hydrogen production, and power generation are closely interconnected through the flow of NH_3, H_2, and power. As we have seen above, the flow of inputs and outputs in a process is primarily governed by the laws of science, and any process that violates the second law of thermodynamics is not feasible: an eternal machine does not exist.

IO considers the economy as a web of interconnected flows of inputs and outputs among the processes that constitute it. This way of considering connections/interdependence has a significant advantage over the others mentioned earlier because the connections remain stable over time and space, anchored in science and technology. In contrast, family connections can dissolve with the departures or deaths of its members, and financial relationships are highly volatile.

To illustrate this concept, let us consider a simple economy consisting of three processes: A, B, and C. Each process produces one output (product) using two inputs, as depicted in Fig. 2.5 (the upper panel). For simplicity, we will ignore waste and emissions. The interconnections among the processes arise from the flows of inputs and outputs between them. In this example, products A and B are intermediate products used in production, such as steel for manufacturing a vehicle or cement

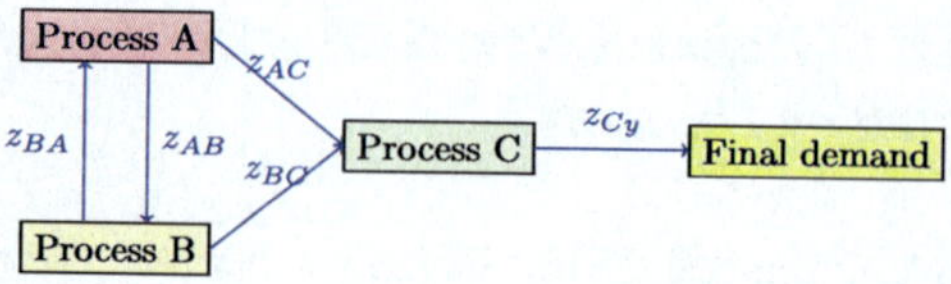

	Process A	Process B	Process C	Final demand	Total output
Process A	0	z_{AB}	z_{AC}	0	x_A
Process B	z_{BA}	0	z_{BC}	0	x_B
Process C	0	0	0	z_{Cy}	x_c

Fig. 2.5 Interconnections among the three processes A, B, and C via the flow of process inputs and outputs. The upper panel illustrates the flow of inputs and outputs among the three processes (A, B, and C). Products A and B are intermediate inputs used in the production of all three products, while C is the final product that is intended for household consumption. It is worth noting that none of the C product is used in the production of A, B, and C. The flow of product i, produced by process i and used in process j to produce product j, is represented by $z_i j$. The lower panel represents the flow depicted in the upper panel as an IO table. The table summarizes the inputs and outputs of the three processes, where the columns indicate the inputs required for each process, and the rows show the destination of the products, with the sum of each row giving the total output of the corresponding product (A, B, or C)

for constructing a building. Product C, on the other hand, is a final product like a vehicle or a building, intended for consumption by the household sector, and none of it is used in the three production processes. By representing these flows in a two-dimensional accounting system (the lower panel of Fig. 2.5), we obtain an IO table. By reading the table column-wise, we can observe the inputs into each process, while reading it row-wise allows us to determine the destination of each product. The sum of each row gives the total output, denoted as x_j, where $j \in$ A, B, C.

The size of an actual IO table depends on the level of resolution used to define processes (sectors), typically ranging from around 100 to 500. In the next chapter, we will delve further into IO and continue our discussion.

2.6 Greenhouse Gases and Global Warming

This supplementary section provides fundamental knowledge of atmospheric physics necessary to understand the science behind greenhouse gases (GHGs) and their impact on climate change. To begin with, it is useful to consider how Earth's surface temperature would be without GHGs.

2.6.1 *What if There Were No GHGs on Earth?*

Understanding the basic principle of global warming requires an understanding of the incoming radiation energy from the sun that warms Earth's surface and the outgoing radiation from Earth and the atmosphere radiated out to space. Solar energy is by far the most important source of energy on Earth's surface.

Solar radiation reaching Earth

Denote by Ω_0 the average intensity of solar energy reaching the top of Earth's atmosphere that directly faces the sun. It is known as the total solar irradiance and has a value of 1366 W per m^{-2} [98]. Since space is a vacuum and does not provide a medium for other types of energy transfer, when the energy enters and leaves the terrestrial system, it must be in the form of electromagnetic radiation [99]. Earth receives solar radiation as a disk with an area of πr^2, where r is the radius of Earth, and it radiates as a sphere with an area of $4\pi r^2$. Therefore, we can establish the following relationship between the average energy falling on one square meter of the top of Earth's atmosphere, Ω_1, and Ω_0 [100]

$$\pi r^2 \Omega_0 = 4\pi r^2 \Omega_1 \tag{2.26}$$

which implies that only one-fourth of Ω_0, 342 Wm^{-2}, reaches Earth's surface.

Blackbody Radiation

What would Earth's surface temperature look like if exposed to the radiation Ω_1 per m^2? Answering this question requires the concept of a blackbody, which is a body that radiates (emits) the same amount of energy as it absorbs. It is important to note that a blackbody does not have to be black. Both the sun and Earth are well approximated by a blackbody (Earth to a lesser extent) [101, 102]. A shiny body that reflects all the energy without absorbing it is not a blackbody.

The intensity of electromagnetic radiation emitted by a blackbody in thermal equilibrium (a state where the body emits the same amount of energy as it absorbs) at temperature T (in Kelvin) and frequency ν is described by Planck's function [102]:

$$B(\lambda, T) = \frac{2hc^2}{\lambda^5}\left(e^{hc/\lambda kT} - 1\right)^{-1} \tag{2.27}$$

where c is the speed of light, h is Planck constant, k is Boltzmann constant, λ is the wavelength and $\nu\lambda = c$. At the top-of Earth's atmosphere, the inflow of solar radiation is balanced by the emission from the surface-atmosphere, which results in thermal equilibrium [103].

Wien's Law on Maximum Wavelength and Temperature

The wavelength of maximum emission is the solution of an extreme problem solved by differentiating $B(\lambda, T)$ with respect to λ and setting the derivative equal to zero

$$\lambda_{max} T = 2898\,\mu\text{m K} \tag{2.28}$$

which constitutes Wien's law. When the temperature of a body increases, the wavelength of its radiation becomes shorter. Since the surface temperature of the sun is about 5770 K, $\lambda_{max} = 0.502\,\mu$m, which falls in the area of visible light.

The Stefan-Boltzmann Law of Radiation

Integration of (2.27) over all wavelengths gives the energy flux of a radiating blackbody as a function of the surface temperature T (in Kelvin) of the body

$$\Omega = \sigma T^4 \tag{2.29}$$

with $\sigma = 2\pi^5 k^4/15c^2 h^3 = 5.67 \times 10^{-8}\,\text{W}^{-2}\,\text{K}^{-4}$.

The Theoretical Temperature of Earth's Surface without GHGs

Here, we mostly refer to [104]. Assuming that Earth radiates all the energy it receives from the sun and is in a state of thermal equilibrium, we can use (2.29) to calculate Earth's surface temperature. However, we must first take into account the fact that a fraction of the incoming solar radiation is reflected back into space by factors such as clouds, ice, snow, vegetation, and deserts, resulting in an albedo, a, of approximately 0.30 [103]. With these factors considered, the planetary heat balance can be approximated as

$$\Omega_0(1 - a)/4 = \sigma T_e^4 \tag{2.30}$$

where T_e is the theoretical surface temperature. Solving this equation for T_e, we get

$$T_e = \sqrt[4]{(1 - \alpha)/4\Omega_0\sigma^{-1}} = 255\,\text{K} \tag{2.31}$$

This temperature would be the prevailing temperature if Earth's atmosphere was completely transparent to radiation and the radiation into space (heat loss) occurred at Earth's surface. It is noteworthy that the value of T_e obtained from terrestrial outgoing long-wave radiation measured by satellites is typically in the range 252–254 K, providing a crude means for validating [104].

The average surface temperature of Earth is 288 K (15 °C), which is about 33 K higher than the theoretical temperature. Earth's atmosphere is not completely transparent to the radiation. This difference is due to the presence of gases in the atmosphere that absorb some of the radiation, making the atmosphere opaque to it. Greenhouse gases (GHGs) are responsible for this effect, without which Earth would be a frozen planet entirely covered by ice. Without the blanketing of the GHGs, known as the *natural greenhouse effect* ([101], p. 16), life as we know it would never have existed on Earth. The actual surface temperature of 288 K indicates that the bulk of the heat loss cannot occur at the ground level, as this would violate the energy balance described in equation (2.30). As the temperature decreases with height in the troposphere, it reaches the theoretical temperature of 254 K at around 6.5 km above the ground.[2] This altitude is where the bulk heat loss of Earth must occur according to the equation.

It is incorrect to regard the presence of GHGs as something harmful to humans: on the contrary, humans owe their very existence to GHGs. It is also true that CO_2 is industrially produced for many applications, including welding, dry ice, soft drinks, neutralizing alkaline wastewater, fire extinguishers, casting, vegetable factories, and livestock anesthetics. Of great concern, however, is the accelerating increase of their concentrations in the atmosphere caused by anthropogenic activities (primarily due to the burning of fossil fuel and land use) and their impacts on Earth's environment, including the climate.

The Wavelengths of the Incoming Radiation from the Sun and the Outgoing Radiation from Earth's Surface

From Wien's law (2.28), it follows that the wavelength of radiation from a blackbody is inverse proportional to its surface temperature. Since the sun has a high surface temperature (about 5770 K K), its electromagnetic radiation is mostly distributed in short wavelengths, visible light, and ultraviolets. A blackbody radiates the energy it absorbs. Accordingly, the solar energy absorbed by Earth is radiated toward space. Since Earth has a much lower temperature (288 K), its radiation is mostly in the infrared region, with $\lambda_{max} = 10.06\,\mu m$. Figure 2.6, top panel, gives the intensity of incoming radiation (on the left) and the outgoing radiation (on the right) based on Planck's function (2.27). The density of outgoing radiation comes in three curves, each referring to different temperature zones on Earth (recall Wien's law).

2.6.2 *Greenhouse Gases (GHGs)*

GHGs refer to gas molecules that absorb (and emit) infrared (IR) radiation. All molecules with three or more atoms are infrared absorbers ([105], p. 459). Accordingly, CO_2, CH_4, N_2O, and H_2O are GHGs. On the other hand, the gases occupying 99.9% of the atmosphere, N_2 (78%), O_2 (21%), and argon (Ar) (0.9%), are not GHGs because they involve fewer than three atoms: two in the case of N_2 and O_2, and one in the case of Ar.

In Fig. 2.6, the colored area below the curves in the top panel shows the transmission of radiation (the passage of electromagnetic radiation through the atmosphere)

[2] In calculating this altitude, the value of $\Omega_0 = 1361\ \mathrm{Wm^{-2}}$ was used [104].

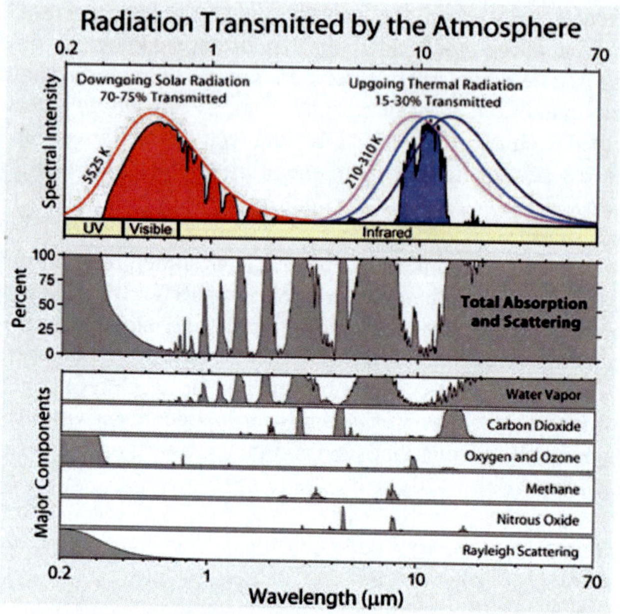

Fig. 2.6 Radiation transmitted by the atmosphere *Source* This image was prepared by Robert A. Rohde for the Global Warming Art project, and is licensed under Creative Commons Attribution-Share Alike 3.0 Unported license https://creativecommons.org/licenses/by-sa/3.0/deed.en

from the sun to Earth (the red curve on the left-hand side) and from Earth to space (the blue curve on the right-hand side), or the area where Earth's atmosphere is transparent to the incoming and outgoing radiation of the relevant wavelengths. On the other hand, the white area under the curves refers to the fraction of absorbed radiation or the area where Earth's atmosphere is opaque to the incoming and outgoing radiation of the relevant wavelengths. Most incoming radiation occurs in visible light regions, while most outgoing radiation occurs in IR. The white area under the red curve on the left refers to the fraction of solar radiation not reaching Earth's surface, mostly absorbed by ozone (O_3) before reaching Earth. It primarily refers to ultraviolet radiation, which is harmful to living organisms. On the other hand, the white area of the right-hand graph refers to the fraction of IR radiation not escaping into space, mostly absorbed by water vapor and CO_2, but also by CH_4, and N_2O, though to a much lesser extent.

The graph in the middle panel of Fig. 2.6 gives the fraction of absorption (IR) and scattering (mostly ultraviolet), with the contribution of each gas given in the bottom panel. We notice that water vapor is the most important GHG (it absorbs the largest share of IR), followed by CO_2. However, water vapor is not counted as an anthropogenic GHG for the reasons mentioned below. It is noteworthy that the largest band of radiation absorbed by CO_2 is close to the highest density of the outgoing IR radiation. Furthermore, the band partly overlaps with the fraction of IR radiation not absorbed by water vapor; the radiation in the band that is not absorbed by water vapor is partly absorbed by CO_2, making CO_2 a good absorber of IR radiation.

Water vapor is not considered an anthropogenic greenhouse gas because human activities have a negligible impact on its long-term greenhouse effect ([101], p. 23). The significance of water vapor's global warming effects should not be underestimated, despite it not being considered an anthropogenic GHG. Any effect that increases the temperature (like the increase in the concentration of CO_2) causes the evaporation of more water into the atmosphere and increases the absorption of IR radiation, setting in motion the mechanism called water vapor feedback [99, 102].

2.6.3 How Do the Greenhouse Effects Work?

Two Terms of the Greenhouse Effects

The greenhouse effects have two terms: the absorptivity of the atmosphere and the difference between the surface temperature and the emission temperature of the atmosphere, referred to as convection [99, 100, 104, 106]. It is remarkable that this fundamental observation was made by Svante Arrhenius in an article published in 1896 [106, 107]. We have discussed the basics of how GHGs absorb IR radiation. When exposed to IR radiation emitted by Earth's surface, GHGs in the atmosphere absorb and re-emit (reradiate) the energy, contributing to the heating of the atmosphere. This energy is then reradiated both upward and downward, with the downward radiation warming Earth's surface. This initiates a sequence of radiation and warming, which repeats multiple times until a thermal equilibrium is established between the incoming solar energy and the outgoing energy escaping from the top of Earth's atmosphere to space, as represented by (2.30) in the planetary heat balance [104].

Convection

Convection plays a crucial role in the greenhouse effects of GHGs, as highlighted by Edward Hulburt in his seminal study [100]. He demonstrated that considering absorption alone leads to a temperature profile for Earth's atmosphere that significantly deviates from the actual one. This emphasizes the importance of accounting for the second term, referred to as convection, which involves the circulation of rising and sinking air masses in the troposphere.[3]

In the troposphere, where most weather phenomena take place, the air temperature typically decreases with altitude at a rate of around 6 K per km. Consequently, the temperature at the height where GHGs emit radiation to space is significantly colder than Earth's surface, usually ranging from 220 to 240 K (–30 to 50 C). GHGs enable Earth to radiate at a temperature lower than its surface, which causes the surface temperature in balance with a given amount of absorbed solar radiation to be higher than would be the case if the atmosphere were transparent to IR radiation.

---

[3] For a test of Hulburt's 1931 model with modern state-of-the-art data, see [104].

When the concentration of GHGs in the atmosphere increases, previously transparent portions of the atmosphere become opaque to IR radiation, thus amplifying the difference between the surface temperature and the radiating temperature at the altitude where most of the radiation occurs. This increase in atmospheric GHGs also elevates the altitude at which the temperature is at 255 K (as shown in (2.31)). Conversely, by raising the altitude where the temperature is 255 K and increasing the distance to Earth's surface, the surface temperature is further elevated. This is because the temperature-altitude relationship mentioned earlier works in both directions, and increasing the distance to the surface enhances the surface temperature.

While discussing GHGs and the greenhouse effect, it is important to note that the analogy between GHGs and a greenhouse made of glass can be somewhat misleading. In a greenhouse with a glass ceiling, warming primarily occurs because the glass prevents the warmed air from escaping into space, which is fundamentally different from how GHGs operate. Earth does not have a physical counterpart to a glass ceiling that traps heat and prevents it from escaping into space.

2.6.4 Global Warming Potentials (GWP)

GHGs differ in their lifetimes and their impact on global warming, also known as radiative forcing. For example, CH_4 has an average lifetime of 12 years and decomposes into CO_2, while N_2O has an average lifetime of about 120 years and breaks down into N_2 and O_2 over time. In contrast, synthetic GHGs like perfluorocarbons (e.g., CF_4, C_2F_6) and sulfur hexafluoride (SF_6) have lifetimes of over 1000 years [101].

To compare the potential climate impact of different GHG emissions, a metric called Global Warming Potential (GWP) takes into account both radiative forcing and lifetime, with CO_2 normalized to a value of 1 due to its long lifespan. Table 2.17 shows the GWPs for eight major GHGs over various time horizons. CO_2 consistently exhibits the smallest GWP across all time horizons, while CH_4 has a GWP of 72, indicating that emitting 1 kg of CH_4 is equivalent to emitting 72 kg of CO_2 in terms of radiative forcing. However, CH_4's GWP decreases rapidly over time, reaching 25 in 100 years and 7.6 in 500 years, due to its shorter lifetime. In contrast, N_2O has a longer lifetime of 114 years, and its GWP declines at a slower rate. The remaining synthetic gases have very long lifetimes and their GWPs increase over time, which raises concerns about their impact on the climate.

GWPs are like the exchange rates of currencies such as the US dollar, euro, Chinese Yuan, British pounds, Swiss franc, and Japanese yen; converted using relevant exchange rates, currencies are made equivalent in their value. Similarly, the emission of several GHGs can be converted to an equivalent value of CO_2 emissions, CO_2e, by adding the values converged by the relevant GWPs. For instance, a simultaneous emission of 1 kg of CO_2, 0.5 kg of CH_4, and 0.1 kg of N_2O amounts to

$$1 \times 1 + 25 \times 0.5 + 298 \times 0.1 = 43.3\,\text{kg-CO}_2\text{e} \tag{2.32}$$

Table 2.17 Lifetime, radiative forcing, and GWPs relative to CO_2

GHGs	Lifetime (years)	Radiative efficiency[a]	Given time horizon		
			20-yr	100-yr	500-yr
CO_2		0.14	1	1	1
CH_4	12	3.7×10^{-4}	72	25	7.6
N_2O	114	3.03×10^{-3}	289	298	153
SF_6	3,200	0.52	16,300	22,800	32,600
NF_3	740	0.21	12,300	17,200	20,700
CF_4	50,000	0.1	5,210	7,390	11,200
C_2F_6	1,0000	0.26	8,630	12,200	18,200

a: W m^{-2} ppb^{-1}. *Source* [108], Table TS.2

for a time horizon of 100 years. The difference to the currency exchange rate is that GWPs are science-based and do not change so far as the scientific knowledge is not updated, whereas a currency exchange rate reflects commerce relations and varies all the time.

References

1. Robert A. Frosch, and Nicholas E. Gallopoulos. 1989. Strategies for manufacturing the impact of industry on the environment. *Scientific American* 261(September): 144–153.
2. Paul H. Brunner, and Helmut Rechberger. 2004. *Practical handbook of material flow analysis*. CRC Press.
3. Kurt Hildenbrand, and Werner Hildenbrand. 1974. *Lineare ökonomische modelle*. Springer.
4. Michael Z. Hauschild, and Mark A.J. Huijbregts. 2015. *Life cycle impact assessment*. Springer.
5. European Commission. 2010. *International reference life cycle data system (ILCD) handbook—General guide for life cycle assessment—Provisions and action steps*. Publications Office of the European Union.
6. S.E. Jorgensen, and I. Johnsen. 1981. *Principles of environmental science and technology*. Elsevier.
7. Frischknecht, Rolf, Hans-Jörg. Althaus, Christian Bauer, Gabor Doka, Thomas Heck, Niels Jungbluth, Daniel Kellenberger, and Thomas Nemecek. 2007. The environmental relevance of capital goods in life cycle assessments of products and services. *The International Journal of Life Cycle Assessment* 12 (1): 7–17.
8. Tialing, C. 1951. An analysis of production as an efficient combination of activities. In *Activity analysis of production and allocation*, ed. Tialing C. Koopmans. Chapter III, 33–97. New York: Wiley.
9. Fromm, Mark, and Herbert Hargrove. 2012. *Essentials of biochemistry*. Berlin Heidelberg, Heidelberg: Springer.
10. Joaquín Herrera Huerta, Edmundo Muñoz Alvear, and René Montalba Navarro. 2012. Evaluation of two production methods of Chilean wheat by life cycle assessment (LCA). *Idesia* 30(2): 101–110.
11. Lal, R. 1995. The role of residues management in sustainable agricultural systems. *Journal of Sustainable Agriculture* 5 (4): 51–78.

12. Hokazono, Shingo, and Kiyotada Hayashi. 2012. Variability in environmental impacts during conversion from conventional to organic farming: a comparison among three rice production systems in Japan. *Journal of Cleaner Production* 28: 101–112.
13. Christopher S Cronan. 2018. *Ecosystem biogeochemistry*, vol. 3. Springer International Publishing.
14. Anukam, Anthony, Ali Mohammadi, Muhammad Naqvi, and Karin Granström. 2019. A review of the chemistry of anaerobic digestion: Methods of accelerating and optimizing process efficiency. *Processes* 7 (8): 1–19.
15. IPCC. 2023. Methane emissions from rice cultivation: Flooded rice fields. https://www.ipcc-nggip.iges.or.jp/public/gl/guidelin/ch4ref5.pdf. Accessed 05 May 2023.
16. James N. Galloway, and Ellis B. Cowling. 2002. Reactive nitrogen and the world: 200 years of change. AMBIO. *A Journal of the Human Environment* 31(2): 64–71.
17. James N. Galloway, John D. Aber, Jan Willem Erisman, Sybil P. Seitzinger, Robert W. Howarth, Ellis B. Cowling, and B. Jack Cosby. 2003. The nitrogen cascade. *BioScience* 53(4): 341–356.
18. UN. 2019. World population prospects 2019. https://www.un.org/development/desa/pd/news/world-population-prospects-2019-0. Accessed 01 July 2023.
19. Johnston, Wm Robert. 2023. Historical world population data. http://www.johnstonsarchive.net/other/worldpop.html. Accessed 01 July 2023.
20. Rouwenhorst, Kevin Hendrik Reindert., Anthony S. Travis, and Leon Lefferts. 2022. 1921–2021: A century of renewable ammonia synthesis. *Sustainable Chemistry* 3 (2): 149–171.
21. Hj Althaus, Mike Chudacoff, Roland Hischier, Niels Jungbluth, Maggie Osses, Alex Primas, and Stefanie Hellweg. *Life cycle inventories of chemicals: Ecoinvent report No. 8, v2.0.* Number 8.
22. IEA. 2023. https://is4ie.org/about/what-is-industrial-ecology. Accessed 31 May 2023.
23. Pattabathula, Venkat, and Jim Richardson. 2016. Introduction to ammonia production. *Chemical Engineering Progress* 112 (9): 69–75.
24. Giddey, S., S.P.S. Badwal, C. Munnings, and M. Dolan. 2017. Ammonia as a renewable energy transportation media. *ACS Sustainable Chemistry and Engineering* 5 (11): 10231–10239.
25. Collin Smith, Alfred K Hill, and Laura Torrente-Murciano. 2020. Current and future role of Haber-Bosch ammonia in a carbon-free energy landscape. *Energy and Environmental Science* 13(2): 331–344.
26. Thu Lan T. Nguyen, John E. Hermansen, and Lisbeth Mogensen. 2011. Environmental Assessment of Danish Pork
27. J.Y. Dourmad, J. Ryschawy, T. Trousson, M. Bonneau, J. Gonzàlez, H.W.J Houwers, M. Hviid, C. Zimmer, T.L.T. Nguyen, and L. Morgensen. 2014. Evaluating environmental impacts of contrasting pig farming systems with life cycle assessment. *Animal* 8(12): 2027–2037.
28. Thu Lan T. Nguyen, John E. Hermansen, and Lisbeth Mogensen. 2010. Environmental consequences of different beef production systems in the EU. *Journal of Cleaner Production* 18(8): 756–766.
29. Angela R. Moss, Jean Pierre Jouany, and John Newbold. 2000. Methane production by ruminants: Its contribution to global warming. *Animal Research* 49(3): 231–253.
30. IPCC. 2018. Agriculture. https://www.ipcc.ch/site/assets/uploads/2018/02/ar4-wg3-chapter8-1.pdf. Accessed 31 May 2023.
31. Shepon, A., G. Eshel, E. Noor, and R. Milo. 2016. Energy and protein feed-to-food conversion efficiencies in the US and potential food security gains from dietary changes. *Environmental Research Letters* 11 (10): 105002.
32. Nathan Pelletier, Peter Tyedmers, Ulf Sonesson, Astrid Scholz, Friederike Ziegler, Anna Flysjo, Sarah Kruse, Beatriz Cancino, and Howard Silverman. 2009. Not all salmon are created equal: Life cycle assessment (LCA) of global salmon farming systems. *Environmental Science & Technology*
33. Mischa Classen, Hans-Jörg Althaus, Silvio Blaser, Matthias Tuchschmid, Niels Jungbluth, G. Doka, Mireille Faist Emmenegger, and W. Scharnhorst. 2009. Life cycle inventories of metals. *Final Report Ecoinvent Data v2* 1(10).

34. Guj, Pietro. 2010. Major iron ore mines in Australia. Technical report, JOGMEC.
35. Daniel Spreitzer, and Johannes Schenk. 2019. Reduction of iron oxides with hydrogen-a review. *Steel Research International* 90(10).
36. Aluiminium Federation. 2020. Aluminium is the third most common metal in the earth's crust. https://alfed.org.uk/aluminium-is-the-third-most-common-element-in-the-earths-crust-2/. Accessed 31 May 2023.
37. Meyer, F.M. 2004. Availability of bauxite reserves. *Natural Resources Research* 13 (3): 161–172.
38. Daniel Brough, and Hussam Jouhara. 2020. The aluminium industry: A review on state-of-the-art technologies, environmental impacts and possibilities for waste heat recovery. *International Journal of Thermofluids* 1–2.
39. Hind, Andrew R., Suresh K. Bhargava, and Stephen C. Grocott. 1999. The surface chemistry of Bayer process solids: A review. *Colloids and Surfaces A: Physicochemical and Engineering Aspects* 146 (1–3): 359–374.
40. Noboru Masuko, and Koichiro Masio. 2015. Present aluminum smelting technology. *Journal of the Japan Institute of Light Metals* 65(2): 66–71.
41. Shahjadi Hisan Farjana, Nazmul Huda, and M.A. Parvez Mahmud. Impacts of aluminum production: A cradle to gate investigation using life-cycle assessment. *Science of the Total Environment* 663: 958–970.
42. Menzie, W.D., J.J. Barry, D.I. Bleiwas, E.L. Bray, T.G. Goonan, Grecia Matos, The Global Flow, and of Aluminum From. 2006. *Through 2025*, 2010. Technical report, U.S: Department of the Interior, U.S. Geological Survey.
43. The Aluminum Assocaition. https://www.aluminum.org/Recycling. Accessed 31 May 2023.
44. Hong, Jinglan, Yilu Chen, Juan Liu, Xiaotian Ma, Congcong Qi, and Liping Ye. 2018. Life cycle assessment of copper production: A case study in China. *International Journal of Life Cycle Assessment* 23 (9): 1814–1824.
45. David Sanjuan-Delmás, RAF Alvarenga, M. Lindblom, Tobias Christoph Kampmann, Lauran van Oers, Jeroen B Guinée, and Jo Dewulf. 2022. Environmental assessment of copper production in Europe: An LCA case study from Sweden conducted using two conventional software-database setups. *International Journal of Life Cycle Assessment* 27(2): 255–266.
46. Güntner, J., and J. Hammerschmidt. 2012. Sulphating roasting of copper-cobalt concentrates. *Journal of the Southern African Institute of Mining and Metallurgy* 112 (6): 455–460.
47. Poullikkas, Andreas. 2015. Review of design, operating, and financial considerations in flue gas desulfurization systems. *Energy Technology & Policy* 2 (1): 92–103.
48. Koralegedara, Nadeesha H., Patricio X. Pinto, Dionysios D. Dionysiou, and Souhail R. Al-Abed. 2019. Recent advances in flue gas desulfurization gypsum processes and applications—A review. *Journal of Environmental Management* 251 (July): 109572.
49. James Brady, and Gerard Humiston. 1986. *General chemistry principles and structure*. Wiley.
50. E. Verhoef, P. Gerard, Markus Dijkema, and P. Reuter. 2004. Process knowledge, system dynamics, and metal ecology. *Journal of Industrial Ecology* 8(1): 23–43.
51. Nakajima, Kenichi, Osamu Takeda, Takahiro Miki, Kazuyo Matsubae, Shinichiro Nakamura, and Tetsuya Nagasaka. 2010. Thermodynamic analysis of contamination by alloying elements in aluminum recycling. *Environmental Science and Technology* 44 (14): 5594–5600.
52. C. Houpert, V. Lanterl, M. Guttmann, and M. Jallon. 1997. Influence of tramp elements in the production of high quality steels using the scrap/electric arc furnace route. *La Revue de Métallurgie-CIT Novembre* 1369–1384.
53. IEA. 2018. Technology roadmap low-carbon transition in the cement industry. Technical report.
54. Medgar L. Marceau, Michael A. Nisbet, and Martha G. Vangeem. 2006. Life cycle inventory of portland cement manufacture. Technical report April, Portland Cement Association.
55. Benhelal, Emad, Gholamreza Zahedi, Ezzatollah Shamsaei, and Alireza Bahadori. 2013. Global strategies and potentials to curb CO_2 emissions in cement industry. *Journal of Cleaner Production* 51: 142–161.

56. Saraswathy, V., S. Muralidharan, K. Thangavel, and S. Srinivasan. 2003. Influence of activated fly ash on corrosion-resistance and strength of concrete. *Cement and Concrete Composites* 25 (7): 673–680.
57. Phuong Trinh Bui, Yuko Ogawa, Kenichiro Nakarai, and Kenji Kawai. 2015. A study on pozzolanic reaction of fly ash cement paste activated by an injection of alkali solution. *Construction and Building Materials* 94: 28–34.
58. Feiz, Roozbeh, Jonas Ammenberg, Leenard Baas, Mats Eklund, Anton Helgstrand, and Richard Marshall. 2015. Improving the CO_2 performance of cement, part I: Utilizing life-cycle assessment and key performance indicators to assess development within the cement industry. *Journal of Cleaner Production* 98: 272–281.
59. Gao, Tianming, Lei Shen, Ming Shen, Litao Liu, and Fengnan Chen. 2016. Analysis of material flow and consumption in cement production process. *Journal of Cleaner Production* 112: 553–565.
60. Pablo Busch, Alissa Kendall, Colin W. Murphy, and Sabbie A. Miller. 2022. Literature review on policies to mitigate GHG emissions for cement and concrete. *Resources, Conservation and Recycling* 182(November 2021): 106278.
61. Pade, Claus, and Maria Guimaraes. 2007. The CO_2 uptake of concrete in a 100 year perspective. *Cement and Concrete Research* 37 (9): 1348–1356.
62. Andersson, Ronny, Katja Fridh, Hakan Stripple, and Martin Häglund. 2013. Calculating CO_2 uptake for existing concrete structures during and after service life. *Environmental Science & Technology* 47 (20): 11625–11633.
63. Fengming Xi, Steven J. Davis, Philippe Ciais, Douglas Crawford-Brown, Dabo Guan, Claus Pade, Tiemao Shi, Mark Syddall, Jie Lv, Lanzhu Ji, Longfei Bing, Jiaoyue Wang, Wei Wei, Keun Hyeok Yang, Björn Lagerblad, Isabel Galan, Carmen Andrade, Ying Zhang, and Zhu Liu. 2016. Substantial global carbon uptake by cement carbonation. *Nature Geoscience* 9(12): 880–883.
64. Zhi Cao, Rupert J. Myers, Richard C. Lupton, Huabo Duan, Romain Sacchi, Nan Zhou, T Reed Miller, Jonathan M. Cullen, Quansheng Ge, and Gang Liu. 2020. The sponge effect and carbon emission mitigation potentials of the global cement cycle. *Nature Communications* 11(1): 1–9.
65. Watari, Takuma, Zhi Cao, Sho Hata, and Keisuke Nansai. 2022. Efficient use of cement and concrete to reduce reliance on supply-side technologies for net-zero emissions. *Nature Communications* 13 (1): 1–9.
66. von Neumann, John. 1945. A model of general economic equilibrium. *Review of Economic Studies* 13 (1): 1–9.
67. Angus Deaton, and John Muellbauer. *Economics and consumer behavior*. Cambridge University Press.
68. Leontief, Wassily. 1951. *The structure of American economy, 1919–1939: An empirical application of equilibrium analysis*. New York, second edi edition: Oxford University Press.
69. Rugani, Benedetto, Daryna Panasiuk, and Enrico Benetto. 2012. An input-output based framework to evaluate human labour in life cycle assessment. *The International Journal of Life Cycle Assessment* 17 (6): 795–812.
70. Tristram O. West, Gregg Marland, Nagendra Singh, Budhendra L Bhaduri, and Adam B. Roddy. 2009. The human carbon budget: An estimate of the spatial distribution of metabolic carbon consumption and release in the United States. *Biogeochemistry* 94(1): 29–41.
71. Jack Wang, F. Avraham Dilmanian, John Thornton, Mary Russell, Santiago Burastero, Manolo Mazariegos, Steven B. Heymsfield, and Richard N. Pierson. 1993. Five-level model: Reconstruction of body weight at atomic molecular, cellular, and tissue-system levels from neutron activation analysis. In J. Ellis, K. Eastman (eds.). *Human Body Composition*, 125–128. Springer.
72. Adele L. Boskey. 2013. Bone composition: Relationship to bone fragility and anti-osteoporotic drug effects. *BoneKEy Reports* 2.
73. Ivan Muñoz, Llorenç Milà i Canals, and Amadeo R. Fernández-Alba. 2010. Life cycle assessment of the average Spanish diet including human excretion. *The International Journal of Life Cycle Assessment* 15(8): 794–805.

74. Ivan Muñoz, Llorenc i Canals, and Clift Roland. 2008. Consider a spherical man a simple model to include human excretion in life cycle assessment of food products. *Journal of Industrial Ecology* 12: 521–538.
75. Kentaro Hayashi, Azusa Oita, Luis Lassaletta, Junko Shindo, Hideaki Shibata, Gen Sakurai, and Sadao Eguchi. 2018. Reducing nitrogen footprints of consumer-level food loss and protein overconsumption in Japan, considering gender and age differences. *Environmental Research Letters* 13(12).
76. Tessari, Paulo. 2007. Nitrogen balance and protein requirements: Definition and measurements. In *Cachexia and Wasting: A Modern Approach*, Chapter 3.2, 73–79.
77. David Fowler, Mhairi Coyle, Ute Skiba, Mark A. Sutton, J Neil Cape, Stefan Reis, Lucy J. Sheppard, Alan Jenkins, Bruna Grizzetti, and James N. Galloway. 2013. The global nitrogen cycle in the twenty-first century. *Philosophical Transactions of the Royal Society B: Biological Sciences* 368(1621): 20130164.
78. Vaclav Smil. 2011. Nitrogen cycle and world food production. *World Agriculture* (Smil): 9–13.
79. J.S. Liou, and E.L. Madsen. 2008. Microbial ecological processes: Aerobic/anaerobic. In *Encyclopedia of ecology*, 2348–2357. Academic Press Location Oxford.
80. EPA. 2002. Nitrification. Technical report, EPA.
81. John Theodore Houghton, Y.D.J.G. Ding, David J. Griggs, Maria Noguer, Paul J. van der Linden, Xiaosu Dai, Kathy Maskell, and C.A. Johnson. 2001. *Climate change 2001: The scientific basis*. Cambridge University Press.
82. Lelieveld, J., G.J. Roelofs, L. Ganzeveld, J. Feichter, and H. Rodhe. 1997. Terrestrial sources and distribution of atmospheric sulphur. *Philosophical Transactions of the Royal Society B: Biological Sciences* 352 (1350): 149–158.
83. Kellog, W., R. Cadle, E. Allenm, A. Lazrus, and E. Martell. 1972. The sulfur cycle. *Science* 175 (4022): 12–26.
84. Dana Cordell, Jan Ol Drangert, and Stuart White. 2009. The story of phosphorus: Global food security and food for thought. *Global Environmental Change* 19(2): 292–305.
85. Chen, Minpeng, and T.E. Graedel. 2016. A half-century of global phosphorus flows, stocks, production, consumption, recycling, and environmental impacts. *Global Environmental Change* 36: 139–152.
86. Alfons Buekens. 2013. *Incineration technologies*. Springer Science \& Business Media.
87. Reinout Heijungs, Jeroen B. Guinée, Gjalt Huppes, Raymond M. Lankreijer, H.A. Udo de Haes, A. Wegener Sleeswijk, A.M.M. Ansems, P.G. Eggels, R. van Duin, and H.P. De Goede. 1992. Environmental life cycle assessment of products: Guide and backgrounds (Part 2). Technical report October, CML, Leiden.
88. Ulrich, Andrea E., and Emmanuel Frossard. 2014. On the history of a reoccurring concept: Phosphorus scarcity. *Science of the Total Environment* 490: 694–707.
89. Bryan M. Spears, Will J. Brownlie, Dana Cordell, Ludwig Hermann, and José M Mogollón. 2022. Concerns about global phosphorus demand for lithium-iron-phosphate batteries in the light electric vehicle sector. *Communications Materials* 3(1): 14.
90. Chengjian, Xu., Qiang Dai, Linda Gaines, Hu. Mingming, Arnold Tukker, and Bernhard Steubing. 2020. Future material demand for automotive lithium-based batteries. *Communications Materials* 1 (1): 99.
91. Paul J. Crutzen. 2006. The "anthropocene". In *Earth system science in the anthropocene*, 13–18. Springer.
92. FAO. 2006. *Livestock's long shadow*. LEAD/FAO.
93. J.N. Rauch, and J.M. Pacyna. Earth's global Ag, Al, Cr, Cu, Fe, Ni, Pb, and Zn cycles. *Global Biogeochemical Cycles* 23(2): GB2001.
94. Bar-On, Yinon M., Rob Phillips, and Ron Milo. 2018. The biomass distribution on Earth. *Proceedings of the National Academy of Sciences of the United States of America* 115 (25): 6506–6511.
95. Mette Bendixen, Jim Best, Chris Hackney, and Lars Lønsmann Iversen. Time is running out for sand. *Nature* 571(7763): 29–31.

96. Monastersky, Richard. 2015. Anthropocene: The human age. *Nature* 519 (7542): 144–147.
97. Simon L. Lewis. and Mark A. Maslin. 2015. Defining the anthropocene. *Nature* 519(7542): 171.
98. NASA. Solar irradiance. https://www.nasa.gov/. Accessed 31 May 2023.
99. Raymond Pierrehumbert. 2011. Infrared radiation and planetary temperature. *Physics Today* (January): 33–38.
100. Hulburt, E.O. 1931. The temperature of the lower atmosphere of the earth. *Physical Review* 38 (10): 1876–1890.
101. John Houghton. 2004. *Global warming*. Cambridge University Press.
102. Guido Visconti. 2016. *Fundamentals of physics and chemistry of the atmosphere Guido Visconti*. Springer International Publishing.
103. Kiehl, Jeffery T., and Kevin E. Trenberth. 1997. Earth's annual global mean energy budget. *Bulletin of the American Meteorological Society* 78 (2): 197–208.
104. Rasmus E. Benestad. 2017. A mental picture of the greenhouse effect: A pedagogic explanation. *Theoretical and Applied Climatology* 128(3–4): 679–688.
105. Peter Atkins, and Julio de Paula. 2009. *Elements of Physical Chemistry*. Oxford University Press.
106. Jean Louis Dufresne, Vincent Eymet, Cyril Crevoisier, and Jean Yves Grandpeix. 2020. Greenhouse effect: The relative contributions of emission height and total absorption. *Journal of Climate* 33(9): 3827–3844.
107. Arrhenius, Svante. 1896. On the influence of carbonic acid in the air upon the temperature of the ground. *The London, Edinburgh, and Dublin Philosophical Magazine and Journal of Science* 41 (251): 237–276.
108. Susan Solomon, Martin Manning, Melinda Marquis, and Dahe Qin. 2007. *Climate change 2007-the physical science basis: Working group I contribution to the fourth assessment report of the IPCC*, vol. 4. Cambridge University Press.

Chapter 3
The IO Model of a Simple Economy Without Fossil Fuel

Abstract This chapter delves into the assumptions and concepts of input-output analysis (IO), including input coefficients, the Leontief inverse, the Leontief quantity model, and productive conditions. It begins with a simple one-sector model centered around the production of a single staple and gradually expands to a two-sector model that incorporates livestock production. Environmental extensions of IO analysis are explored, such as GHG emissions, water and land footprints, waste generation, and recycling, including Waste Input-Output analysis (WIO), with numerical examples based on real data. The chapter focuses on an economy that operates without fossil fuels, relying solely on human muscle power as the energy source for production. GHG emissions primarily arise from anaerobic decomposition in organic waste and enteric fermentation in livestock. The chapter endogenizes the supply of human labor by considering the energy required to produce human muscle power, aligning with the Miyazawa model. The initial discussion in the chapter avoids matrix algebra, using simple arithmetic instead to aid readers unfamiliar with matrices. Once the basic concepts are covered, matrix algebra is introduced to reinforce the fundamentals of IO, which can be applied to any number of sectors. Additionally, the chapter introduces the IO model of cost and price as the dual to the quantity model, enhancing the understanding of the IO framework.

3.1 The One-Sector IO Model

3.1.1 The Basics

We introduce the one-sector IO model by considering a one product economy where the population is solely supported by the production of a single staple food like rice, maize, wheat, or potatoes. We focus on rice production as an example, such as in Japan in the 17th century, where rice was the basis of national income. Rice production requires inputs such as seeds, fertilizer, pesticides, land, water, and human labor (Sect. 2.2.1.2). In our one-sector economy, humans are the sole source of power, and no machines or equipment are involved, meaning no fuel is needed.

S. Nakamura, *A Practical Guide to Industrial Ecology by Input-Output Analysis*,
https://doi.org/10.1007/978-3-031-43684-0_3

Table 3.1 A numerical example of the paddy rice production system

Items	Per ha		Whole region[c]	
	Amount	Units	Amount	Units
Seeds	82.5	kg/ha	2.201E+06	kg
Manure (dry)[a]	700	kg/ha	1.867E+07	kg
Compost (rice straw)	2000	kg/ha	5.335E+07	kg
Labor	2330	h/ha	6.215E+07	h
Land[b]	10000	m^2	2.667E+08	m^2
Rice yield	3900	kg/ha	1.040E+08	kg
Harvest residue	3900	kg/ha	1.040E+08	kg

The data refer to the 17th century traditional organic agriculture in the Jiaxing region of eastern China, taken from [1, 2]

[a] The manure originates from local people and animals, of which the share of the former is around 1/3

[b] Obtained by multiplying the values per ha by the size of the rice farmland

[c] The total farmland of 32603 ha allocated to rice, assuming the same land requirement per yield of rice (3900 kg/ha) and wheat (867 kg/ha)

We will use Table 3.1 as an example of the rice production process in our hypothetical one product economy. Around 2300 hours of labor were involved in producing a hectare of rice and similar food crops per growing season, which is equivalent to a full-time laborer for one year. Yields were favorable, even compared to some mechanized systems of the 20th century. It is important to note that this is only for illustration purposes.

Inputs produced by the production process are termed endogenous inputs, while primary inputs like labor and land are termed exogenous inputs. Compost can be considered an endogenous input if made from rice output; otherwise, it is exogenous. Water and pollination services are not considered for now, but we assume ecological services are abundant.

3.1.1.1 Input Coefficients and Leontief Inverse Coefficients

In IO, input coefficients are essential as they indicate the relationship between a process's inputs and outputs. We denote the amount of product j by x_j and the amount of input i used to produce product j by z_{ij}. In the current one product case, rice grain is the only product, which we denote by "1." From Table 3.1, x_1 is 104 Gg, z_{11} is 2.2 Gg, and $z_{\text{manure }1}$ is 18.7 Gg.

The law of mass conservation requires that producing any amount of output requires some positive input, even if it is small. Rice production is a biological process that involves the growth of living organisms and thus requires rice seeds to initiate the process. Hence, for rice production, we must have

$$z_{11} > 0, \quad \text{for } x_1 > 0 \tag{3.1}$$

We assume that the rice production process is capable of continuous proportional expansion or reduction, which is known as divisibility [3]. This assumption implies that the amount of inputs is proportional to the amount of primary output, resulting in constant returns to scale (CRTS). To describe this relationship, we use input coefficients, which represent the proportionality factor between the input and output of a process. Denoting the input coefficient for rice seeds (input) by a_{11}, we have

$$z_{11} = a_{11}x_1 \tag{3.2}$$

The value of a_{11} represents the amount of rice seeds needed as input to produce a unit of rice. Similarly, we can define input coefficients for other exogenous inputs, such as labor, compost, and land. Denoting the input coefficient for input i by a_{i1}, we have

$$z_{i,1} = a_{i1}x_1, \quad i \in \{\text{labor, compost, land}\} \tag{3.3}$$

To calculate the input coefficients, we can use data on the amount of input and output for a given process. For example, from Table 3.1, we obtain

$$a_{11} = \frac{z_{11}}{x_1} = \frac{82.5}{3900} = 0.021 \tag{3.4}$$

This implies that per unit of rice output, 0.021 units of seed input are needed. Similarly, the input coefficients for compost, labor, and land are calculated as

$$\begin{aligned} a_{\text{compost}1} &= \frac{z_{\text{compost},1}}{x_1} = \frac{2000}{3900} = 0.513 \text{ kg/kg-rice} \\ a_{\text{labor}1} &= \frac{z_{\text{labor hours},1}}{x_1} = \frac{2330}{3900} = 0.597 \text{ h/kg-rice} \\ a_{\text{land}1} &= \frac{z_{\text{land},1}}{x_1} = \frac{10000}{3900} = 2.564 \text{ m}^2\text{/kg-rice} \end{aligned} \tag{3.5}$$

These values indicate that per kg of rice output, one needs 0.51 kg of compost, 0.60 hours of work, and 2.56 m^2 of land.

The assumption of constant returns to scale is implicit in the literature on life cycle assessment (LCA) and is fundamental to our analysis. It is worth noting that even a highly mathematical exposition of LCA [4] takes this assumption for granted and does not mention it explicitly. Similarly, the comprehensive book on LCA [5] does not discuss the possibility of diseconomies in LCA.

Diseconomies of Scale

Diseconomies of scale occur when the supply of "essential inputs," without which production is not feasible, is limited [6, 7]. For example, in rice production, land is an essential input. If the supply of land is limited, increasing rice output requires either expanding production to less suitable areas or increasing the density of rice crops, which can result in lower yields per unit of land, unless there are improvements in the technology of rice production. Another example of diseconomies of scale is the relationship between ore grade and energy intensity in metallic mining, where a decrease in ore grade leads to an increase in the energy required for metal extraction [8].

The Input-Output Relationship without CRTS

If CRTS is not assumed, an alternative formulation can be used, which includes a scale parameter, β

$$z_{i1} = a_{i1} x_1^{\beta_{i1}} \quad \text{for input } i \tag{3.6}$$

Here, CRTS holds for $\beta = 1.0$, while economies of scale occur for $\beta < 1$, and diseconomies of scale for $\beta > 1$ (the subscripts are omitted for simplicity). However, estimating the value of β can be challenging because it depends on specific inputs, processes, and products. In engineering literature, the 0.6 rule is often used with $\beta = 0.6$ [9], but its general applicability is uncertain.

CRTS as a First-Order Approximation to the Actual Input-Output Relationships

Relaxing the CRTS assumption can result in many difficult-to-solve problems. One such problem is the coexistence of processes with different efficiency. A change in demand for a product can then result in a change in the " average" input coefficients, although the individual process-level coefficients are constant. This is a topic related to the aggregation of production functions in economics [10, 11]. Thus, unless we have sufficient knowledge about the process, including the scale parameter β, it is practical and realistic to assume CRTS, which is widely accepted in the LCA community.

However, it is important to acknowledge that the assumption of CRTS has limitations that must be recognized. In IE, the focus is on solving real-world problems, and it is essential to ensure that theoretical assumptions remain relevant to practical situations. For example, assuming that input coefficients are valid across vastly different production levels is unrealistic, and it would be more appropriate to consider these processes as distinct ones. In other words, CRTS should be seen as a convenient first-order approximation to the actual nonlinear relationships that provide a good approximation in the neighborhoods of the data from which the coefficients were obtained. It is worth remembering that all models are "flawed" since they represent an idealized description of reality, but they can still provide valuable insights and useful predictions [12].

3.1.1.2 The Leontief Quantity Model: How Much to Produce to Meet the Final Demand?

When rice is harvested, a portion of it, z_{11}, needs to be kept for production (sowing) in the next period. If there is not enough rice left for production in the next period, the population will be at risk of starvation. The rest $y_1 = x_1 - z_{11}$ is available for consumption, export, or storage. y_1 is referred to as the final demand, and z_{11} is referred to as the intermediate demand. The balance between the supply and demand of rice is then given by

$$\underbrace{x_1}_{\text{total supply}} = \underbrace{z_{11}}_{\text{intermediate demand}} + \underbrace{y_1}_{\text{final demand}} \tag{3.7}$$

From the data in Table 3.1, we have, for the whole region:

$$x_1 = 104.0\,\text{Tg}, \quad z_{11} = 2.2\,\text{Tg}, \quad y_1 = x_1 - z_{11} = 101.8\,\text{Tg} \tag{3.8}$$

We apply the input coefficient to the supply-balance equation to derive the Leontief quantity model, which indicates the amount of product needed to satisfy a given final demand. Using a_{11}, we rewrite (3.7) as

$$x_1 = a_{11}x_1 + y_1 \tag{3.9}$$

We can solve this equation for x_1, given a_{11} and y_1, but first, it is important to consider what we know about a_{11}.

From (3.1), we know that $a_{11} > 0$. Rewriting the balance equation, we get

$$(1 - a_{11})x_1 = y_1 \tag{3.10}$$

which means that, per unit output, $(1 - a_{11})$ units are available for final demand. For a positive amount of output to be available for the final demand, a_{11} needs to satisfy the following condition

$$1 - a_{11} > 0 \tag{3.11}$$

This condition is known as the *Hawkins–Simon condition* [13]. Combining this with $a_{11} > 0$, we get the following *productive condition*

$$0 < a_{11} < 1 \tag{3.12}$$

This condition indicates that a single product economy satisfying this condition can produce the amount of product needed to accommodate the final demand for the product. An increase in a_{11} implies a decline in the yield per seed or a decline in the amount of final demand that a given harvest can accommodate. In contrast, a decrease in a_{11} implies an increase in the yield per seed or an increase in the amount of final demand that a given harvest can accommodate.

When (3.11) is satisfied, we can solve (3.10) for x_1 and obtain

$$x_1 = (1 - a_{11})^{-1}y_1 \tag{3.13}$$

From (3.11), it follows that $x_1 > 0$ for any $y_1 > 0$. Since a negative output makes no sense, this property, the positiveness of the solution, is of fundamental importance in IO. While it is obvious in the current simplest case that the positiveness of the solution results from (3.11), it becomes less obvious in a more general case with multiple processes (sectors), which has been a subject of advanced mathematical inquiry [14, 15].

In (3.13), the term $(1 - a_{11})^{-1}$ gives the amount of rice that is "directly and indirectly" required (detailed below) to deliver one unit of rice to meet the final demand. This term is known as the *Leontief inverse coefficient*. It is a scalar version of the Leontief inverse matrix. Equation (3.13) gives the amount of output needed to satisfy a given final demand, considering a specific technology represented by a_{11}. It represents the *Leontief quantity model* for the case of a single product. In the literature, it has become customary to denote the Leontief inverse matrix as L, presumably as a tribute to its developer. Following this practice, we henceforth denote $(1 - a_{11})^{-1}$ by l_{11}, and write (3.13) as

$$x_1 = l_{11} y_1 \tag{3.14}$$

From (3.11), it follows

$$l_{11} = (1 - a_{11})^{-1} > 1 \tag{3.15}$$

In our numerical example, $l_{11} = (1 - 0.021)^{-1} = 1.0215$. Delivery of 1000 kg of rice to the final demand requires 1021.5 kg of rice production, with the balance of 21.5 kg referring to the rice production necessary to produce seeds. A positive balance (of the amount of production over the amount of final demand) results from the fact that inputs are required to produce outputs and that the inputs have to be produced.

Direct and Indirect Requirements

The amount of output needed to meet a given final demand can be decomposed into "direct and indirect" requirements based on the following expansion of l_{11} as the sum of an infinite series of a_{11}^n for n from 0 to ∞

$$\begin{aligned} 1 + a_{11} + a_{11}^2 + a_{11}^3 + \cdots = l_{11}, \text{ or} \\ (1 + a_{11} + a_{11}^2 + a_{11}^3 + \cdots) y_1 = x_1 \end{aligned} \tag{3.16}$$

Reference [16] refers to the term $(1 + a_{11}) y_1$ as the "direct requirements" and the remaining terms as the "indirect requirements":

$$\text{direct requirements}: \quad (1 + a_{11}) y_1 \tag{3.17}$$

$$\text{indirect requirements}: \quad \left(\sum_{i=2}^{\infty} a_{11}^i \right) y_1 \tag{3.18}$$

Natural Disasters and a_{11}

Natural disasters such as flooding, hurricanes/typhoons, earthquakes, drought, and volcanic eruptions can lead to an increase in a_{11} or a decline in productivity/yield. These events can destroy harvests, resulting in a smaller grain output for the same amount of seed input. An example of the impact of such disasters is the Tambora volcanic eruption in 1815, which occurred in Indonesia and is considered the largest known historic eruption. This event caused devastating crop failures and even led to a year without a summer in Europe [17].[1] In an isolated economy, an increase in a_{11} towards unity could have catastrophic consequences, potentially leading to the collapse of the entire civilization. A productive economy, capable of supporting its population, is characterized by a_{11} being reasonably smaller than unity.

Human history is filled with examples of civilizations experiencing a " collapse," where once-flourishing societies vanished, leaving behind only monuments and ruins. While numerous factors, both common and specific, likely contributed to this phenomenon, a decline in food production productivity is often cited as a common element. For instance, according to [18], human-caused deforestation is attributed to a decline in agricultural productivity, leading to population starvation. However, this perspective is not without criticism from the scientific community [19, 20]. Regardless of the ongoing controversy surrounding the factors that contributed to these collapses, a simple mathematical representation closely associated with the concept of collapse is the scenario in which a_{11} approaches unity.

3.1.2 *Human Labor as the Power Source and Its Energy Requirements

In our simple one-sector economy, human labor (muscle power) is the only power source for production. Humans require food to provide muscle power, just as machines need fuel to operate. We will consider the energy requirements of human labor as the sole power source and derive the total quantity of food needed to satisfy a given final demand for food.

3.1.2.1 The Amount of Applied Power of Human Labor

We calculate the applied power of human labor per capita per year (AP) by multiplying the level of human power per capita by the hours of labor per capita per year. Assuming a standard age structure (55% working population) [21], sex ratio

---

[1] An important cultural by-product of this event was the creation of the novel Frankenstein by Mary Shelley, regarded by many as the best science fiction work ever written.

(50% male/females), 40 h/week for 50 weeks, the applied power per capita per year is given by

$$AP = 75\ \text{W} \times 0.55 \times 2000\ \text{h} \times 60^2\ \text{s/h} = 297.0\ \text{MJ} \tag{3.19}$$

where 75 W refers to the gender average level of human power per capita [21].

3.1.2.2 The Energy Cost of Human Labor

The energy cost of human labor is the energy required to produce human labor. The measure of the energy cost of human labor based on physiological energy requirements is typically in the range of 1.8–2.5 MJ per hour of labor [21]. For instance, [1] assume 2.16 MJ/h. However, this measure ignores the energy needed for basic needs like cooking and shelter, as well as populational and societal structures. We follow the methodology developed by [21] to fully address these points.

The level of energy consumption per capita in this economy, denoted by E in watts, can vary from 370 W (or 7641 kcal/day) for a poor rural economy to 740 W (or 15281 kcal/day) for a fairly developed rural economy [21]. The average year's energy consumption per capita, denoted by EI, is given by

$$EI = E \times 60^2 \times 24 \times 365 \tag{3.20}$$

The energy cost of human power is the amount of energy embodied in a joule of applied power, given by the ratio EI/AP. For an average worker with a power level of 75 W, we have

$$\frac{EI}{AP} = \frac{E \times 60^2 \times 24 \times 365}{75 \times 0.55 \times 60^2 \times \text{Hours of labor}} = \frac{E}{\text{Hours of labor}} \times 212.36 \tag{3.21}$$

The value of EI/AP can range from 40 (for very poor societies with $E = 370$ W) to 80 (for fairly developed rural societies with $E = 740$ W). For a fairly developed rural economy with 740 W, the energy cost of one hour of labor is

$$80 \times 75\text{W} \times 0.55 \times 60^2\text{s}/1000^2 = 11.9\ \text{MJ/h} \tag{3.22}$$

For a very poor economy with half the level of consumption, the cost per hour is 5.9 MJ/h. This value is almost three times the amount based on physiological requirements alone (2.1 MJ/h), underscoring the importance of taking into account population and social conditions when estimating the energy cost of labor.

3.1.2.3 Human Labor as Part of a Production System?

In Chap. 1, we mentioned that households in economics can be represented as a process with inputs such as food and outputs such as labor, respiration, and excreta. The approach to estimating the energy cost of human labor we have discussed can be seen as one variant of this process-based representation of labor supply, which explicitly considers the "standard of living," represented by E in watts, as a factor that determines the energy cost of labor. IE studies that focused on the energetic aspects of human labor include [22, 23]. Reference [22] compared the life cycle GHG emissions of US and Chinese manufacturing sectors considering the emissions associated with the supply of human labor in line with the above approach. Reference [23] provided a methodological framework to account for human labor in LCA and an application in several case studies.

3.1.2.4 Labor Input and Its Supply

Here, we consider the amount of labor required to produce rice and the total quantity of rice required to meet a given final demand for rice in society.

The Amount of Labor Hours Embodied in 1 kg of Rice

According to Eq. (3.5), producing 1 kg of rice requires 0.597 hours of human labor. However, this amount of labor is insufficient to meet the final demand for rice of 1 kg since a fraction of the output is reserved as seeds. Therefore, the required amount of labor (hours), $r_{\ell 1}$, is given by

$$\begin{aligned} r_{\ell 1} &= a_{\ell 1}(1 - a_{11})^{-1} \\ &= 0.597(1 - 0.021)^{-1} = 0.61 \end{aligned} \tag{3.23}$$

Here, $r_{\ell 1}$ represents the labor hours embodied in 1 kg of rice. Multiplying this by the amount of human power embodied in 1 hour of labor (3.22), we get the amount of human power embodied in 1 kg of rice

$$11.9\,\text{MJ/h} \times 0.61\,\text{h/kg-rice} = 7.26\,\text{MJ/kg-rice} \tag{3.24}$$

For the heat value of 15.19 MJ per kg of rice [1], the efficiency of this process (output/input) in terms of human energy was around two.

The Amount of Rice Needed to Produce Human Labor

Similar to the fuel needed to run machinery, human labor also requires nutrition in addition to that needed for basic metabolism. One hour of farm labor requires

$$75\,\text{W} \times 60^2\,\text{s} = 0.270\,\text{MJ} \tag{3.25}$$

Assuming that human bodies convert food energy into muscle energy with an efficiency of 25% [24], the amount of rice needed to provide this amount of energy is

$$0.270\,\text{MJ/h}/(15.19\,\text{MJ/kg-rice} \times 0.25) = 0.072\,\text{kg/h} \tag{3.26}$$

Thus, providing the human power required for one hour of farm work needs 72 g of additional rice.

The Total Amount of Rice Needed to Satisfy a Given Final Demand for Rice

We can now calculate the total amount of rice needed to satisfy a given amount of final demand, taking into account the fraction of the demand dedicated to producing labor. Let $a_{1\ell}$ denote the amount of rice required to provide one hour of labor (0.072 kg in our case). Since producing 1 kg of rice requires $a_{\ell 1}$ units of labor hour (0.597 in our example), the amount of rice required to produce the labor needed to produce 1 kg of rice is $a_{1\ell}a_{\ell 1}$, which is 0.0428 in our example. We can split the final demand for rice into two components: one required to support the basic metabolism of the population, denoted as y_0, and another needed to provide labor to produce rice by the amount x_1, which is $a_{\ell 1}a_{1\ell}x_1$. As a result, the balance of supply and demand (3.9) can now be expressed as

$$x_1 = a_{11}x_1 + a_{1\ell}a_{\ell 1}x_1 + y_0 \tag{3.27}$$

which can be solved as

$$\begin{aligned} x_1 &= (1 - a_{11} - a_{1\ell}a_{\ell 1})^{-1}y_0 \\ &= (1 - 0.021 - 0.0428)^{-1}y_0 = 1.0681\,y_0 \end{aligned} \tag{3.28}$$

Comparing this result with $l_{11} = (1 - 0.021)^{-1} = 1.0215$ shows that considering the labor needed to produce rice increases the amount of rice required to satisfy 1 kg of final demand by approximately 47 g.

In (3.27), the household's labor supply is endogenized by using the same representation as any production process for the household sector. Inputs (food) are converted into muscle energy, respiration, and excreta. Another way to interpret (3.27) is to view the term $a_{\ell 1}x_1$ as income generation and $a_{1\ell}a_{\ell 1}x_1$ as the final expenditure induced by this income. This concept was originally introduced by [25] in the context of an IO model based on monetary values, which we will consider in Chap. 4.

3.1.3 Environmental Extensions

The development of agriculture has lasting impacts on the environment as it replaces natural vegetation, increases species extinction rates, and alters biogeochemical cycles [26]. Even a simple staple food-based economy is not exempt from these impacts. In this section, we extend the one-sector IO model to include greenhouse gas (GHG) emissions, water consumption, land use, and waste. Section 2.6 provides a brief introduction to the basic science of GHG and climate change.

3.1.3.1 GHG Footprint of Rice Consumption

The economy we consider in this chapter does not use fossil fuel, but it still emits GHG due to the anaerobic decomposition of organic residue, stems, and roots in rice paddies, as mentioned in Sect. 2.2.1.2. The amount of CH_4 emissions from rice paddies varies widely depending on factors such as location, climate, and cultivation methods. Table 3.2 presents the CH4 emissions from rice paddies under different cultivation methods in Japan [27]. The emission per 1 kg rice grain ranges from around 19 g-CH_4 to 33 g-CH_4.

Denote by $f_{CH4,1}$ the emission of CH_4 (in kg) per kg of rice. Since puddling is the conventional way of rice cultivation, we take $f_{CH4,1} = 0.0330$ from Table 3.2 for our numerical example. Similar to the calculation of the labor embodied in 1 kg of rice, we can calculate the CH_4 embodied in 1 kg of rice, or the CH_4 footprint of 1 kg of rice, as

$$f_{CH_4,1} \times l_{11} = 0.0330 \times 1.0215 = 0.0337\,\text{kg-CH}_4\text{/kg-rice} \tag{3.29}$$

Since GHG gases differ in their impacts on climate change due to their differences in lifetimes and radiative forcing, we convert the CH_4 footprint to its equivalent CO_2 value by using the 100-year GWP of CH_4, which is 25. Therefore, the GHG footprint of 1 kg of rice is 0.8075 kg-CO_2e/kg rice, where CO_2e refers to the value made equivalent to CO_2 using the GWP. It is important to note that this simple economy, even without the use of fossil fuels, is not C-neutral. Consuming 1 kg of rice emits approximately 800 g of CO_2e, which cannot be neutralized by photosynthesis.

3.1.3.2 Water-and Land Footprints

Although water input is critically important in rice production, we have not yet considered it ([1] did not report any data on water use). However, [28] estimated the

Table 3.2 CH_4 emission in rice production

	Puddling	No tilling	No puddling	Unit
CH_4 average	179	102	182	kg-CH_4/ha
Straw yield	7123	7023	7207	kg/ha
Brown rice yield	5432	5488	5627	kg/ha
CH_4 per grain	0.0330	0.0186	0.0323	kg-CH_4/kg-rice

Source [27]

direct input of water for 1 kg of rice to be 55.63 m^3 based on their extensive study of water footprints for approximately 400 product items in Japan. For the purposes of this discussion, we will use this estimate.

Similar to the CH_4 footprint previously discussed, we can calculate the water footprint of rice production as

$$a_{\text{water1}}(1 - a_{11})^{-1} = 55.63 \times 1.0215 = 56.83\,\text{m}^3/\text{kg-rice} \tag{3.30}$$

In the literature on water footprint, it is usual to categorize water by its source into blue (surface water) and green (rainwater) [29]. Reference [28] differentiate water use into river water, rainwater, groundwater, and recovered water, with the respective shares in rice production being 0.66, 0.28, 0.06, and 0. Therefore, most of the above water footprint is attributed to rainwater. The water footprint is related to the actual water requirement by

$$\text{Water requirement} = a_{\text{water1}} x_1 = \underbrace{a_{\text{water1}}(1 - a_{11})^{-1}}_{\text{water footprint}} y_1 \tag{3.31}$$

Land use is widely recognized as a major driver of global biodiversity loss, and its environmental impact is extensively studied in LCA research [30]. For instance, land use change can disrupt co-evolved interactions between plants and their pollinators, thereby limiting plant reproduction through a reduction in pollen supply [31].

The IO-based land footprint is a frequently used method for estimating the land use embodied in a product [32, 33]. Similar to the water footprint, the land footprint can be defined as the area of land required directly and indirectly to produce 1 kg of rice for final demand and is calculated as

$$a_{\text{land1}}(1 - a_{11})^{-1} = 2.564 \times 1.0215 = 2.619\text{m}^2/\text{kg-rice} \tag{3.32}$$

As with the water requirement in (3.31), the land footprint is related to the actual land requirement for rice production as

$$\text{Land requirement} = a_{\text{land1}} x_1 = \underbrace{a_{\text{land1}}(1 - a_{11})^{-1}}_{\text{land footprint}} y_1. \tag{3.33}$$

3.1.3.3 *Residue, Human Excretion and Its Management: A First Course on Waste IO (WIO)

As discussed in Chap. 1, the operation of a production process results in the generation of by-products, waste, and emissions, most of which are undesirable and not part of the intended output of the process. The distinction between by-products and waste is not always clear and can vary depending on the context in which they are discussed. In fact, a universally accepted definition of waste that distinguishes it from by-products is currently lacking ([37], p. 90). In this book, we use the terms "by-products" and "waste" loosely to refer to materials and substances that are produced as a result of the production process and that are not part of the intended output.

Human Excreta as Fertilizer

In our single-product economy, where rice is the target product, by-products, and waste, such as noneligible fractions of rice (stem and roots) and human excreta, in addition to CH_4 emissions, are generated. In Table 3.1, rice straw occurs as an input used as fertilizer and soil conditioner: rice straw is a useful by-product. It can also be used as fuel, materials for cloth, roofing, etc.

Humans generate around 0.6 kg of excreta (urea and feces) for every 1 kg of food ingested [38]. Improper disposal of this waste can have serious consequences for hygiene. However, in the agricultural system of East Asia (in particular China and Japan) in the 17th century, human excreta, besides livestock manure, was an important source of fertilizer. The original data of Table 3.1 includes the input of manure originating from livestock (pigs) and humans, which was omitted for simplicity.

What is remarkable about the preindustrial practice (of using human excreta as fertilizer) in East Asia compared with Europe is that "exploitation of human sources of fertilizer was systematically carried out, whereas it was haphazard and informal in Europe [36]." An interesting quote on the subject reads ([35], p. 9)

> The most important difference between waste in Japan and in the West was that human excreta were as something that one paid to have removed, product with a positive economic value. The night soil of Japanese cities-and Chinese as well-was long used as fertilizer. With the growth of Japan's population, the limitation of arable land and the increasingly intensified use of land to feed the growing population, combined with the relative scarcity of animal wastes and other fertilizers, meant that human waste had a value as fertilizer that far exceeded its value in the West.

For people living in a "civilized" economy accustomed to flushing toilets and sewage systems from birth, using human waste for food production may seem horrifying. However, in preindustrial times without adequate sewage systems (not just dumping waste into nearby rivers), the systematic and extensive collection of human waste as a fertilizer source greatly helped maintain acceptable sanitary conditions in cities [35] (see [39] for a history of human waste treatment). Interestingly, a technologically advanced version of this process is now widely used in the form of the agricultural application of sewage sludge after treatment [40–44].

An IO Accounting System with By-products, Waste, and Emissions

The Balance of By-Products and Waste

The simple IO model discussed above is now augmented with the flow of by-products (rice straw) and human waste (excreta) to provide a more comprehensive picture. Let $w_{1,1}^{+}$ denote the amount of rice-straw generated in rice production, $w_{1,1}^{-}$ the fraction used in rice production (as compost), $w_{1,y}^{-}$ the fraction used in the final demand (as fuel, materials for cloth and roof, etc.), w_{1}^{+} the total amount of rice straw generated, and w_{1}^{-} the total amount of rice straw used. Since rice production is the only supply source of rice-straw, we have $w_1^+ = w_{1,1}^+$. On the other hand, since rice straw is used to satisfy both intermediate and final demand, we have $w_1^- = w_{1,1}^- + w_{1,y}^-$.

Similarly, let $w_{2,y}^+$ denote the amount of human waste generated in households, $w_{2,1}^-$ its fraction used in rice production, w_2^+ the total amount of human waste generated, and w_2^- the total amount of human waste used. In reality, human waste has to undergo treatments such as fermentation before being applied to agricultural land as fertilizer. For the sake of simplicity, we omitted this process and assumed that human waste was directly applicable to agricultural land. The same simplification applies to rice and rice straw, which, in reality, require some treatment, such as drying and refining, before consumption and use.

The WIO Representation of Rice, By-Products, and Waste

The upper panel of Table 3.3 shows a schematic representation of the inputs and outputs associated with rice production, including by-products and waste, using the Waste Input-Output (WIO) accounting framework [45]. In this representation, a

Table 3.3 The flow of the rice grain, residue, and waste in a one-sector WIO

The conceptual table

	Rice	Consumption	Total
Rice (seed)	z_{11}	y_1	x_1
The gross flows of by-products and waste			
Crop residue generation	$w_{1,1}^+$		w_1^+
Crop residue use	$w_{1,1}^-$	$w_{1,y}^-$	w_1^-
Human excreta generation		$w_{2,y}^+$	w_2^+
Human excreta use	$w_{2,1}^-$		w_2^-
The net flows of by-products and waste			
Crop residue	$w_{1,1}^+ - w_{1,1}^-$	$-w_{1,y}^-$	$w_1^+ - w_1^-$
Human excreta	$-w_{2,1}^-$	$w_{2,y}^+$	$w_2^+ - w_2^-$

A numerical example

	Rice	Consumption
Rice (seed)	0.015	1
The gross flows of by-products and waste		
Crop residue generation	1.4	
Crop residue use	1.0	0.2
Human excreta generation		0.6
Human excreta use[a]	0.24	
The net flows of by-products and waste		
Rice crop residue	0.4	−0.2
Human excreta	−0.24	0.6

[a] The water content of human excreta was assumed to be 74.6% [34]. Human excreta was assumed applicable as manure. In the East Asia of the 17th century, human excreta was traded as a valuable commodity [35, 36]

new column representing household consumption has been introduced. By reading the entries column-wise, one can see the flow of goods, by-products, and waste. Alternatively, reading the entries row-wise, one can observe the balance of supply and demand for each item. Since our main focus is on the integration of by-products and waste, we have omitted other items such as labor, land, water, and emissions.

We calculate the net flows of by-products and waste, denoted as w_{ij}, by subtracting the amounts used from those generated

$$w_{ij} = w_{i,j}^{+} - w_{i,j}^{-} \tag{3.34}$$

Since the sum of $w_{i,j}^{-}$ for all j cannot exceed the sum of w_{ij}^{+}, the net output is always nonnegative, as shown by the inequality

$$\sum_{j} \left(w_{i,j}^{+} - w_{i,j}^{-} \right) = w_{i}^{+} - w_{i}^{-} = w_i \geq 0 \tag{3.35}$$

The lower panel of Table 3.3 gives a numerical example of the net flow of by-products and waste.

The Net Waste (and By-Product) Generation Coefficients

Analogous to the input coefficients a_{ij}, we define the "net waste (and by-product) generation coefficients", g_{ij}, which refers to the net amount of waste i generated per unit of product j, as

$$g_{ij} = \frac{w_{ij}}{x_j} = \frac{w_{ij}^{+}}{x_j} - \frac{w_{ij}^{-}}{x_j} = g_{ij}^{+} - g_{ij}^{-} \tag{3.36}$$

where g_{ij}^{+} and g_{ij}^{-} refer to the amount of waste i generated and used per unit of product j.

The Amounts of Waste Induced by the Final Demand

Analogous to the supply and demand balance of rice (3.9), we can represent the supply and demand balance of waste as

$$\begin{aligned} g_{11}x_1 + w_{1,y} &= w_1, \text{ for rice straw} \\ g_{21}x_1 + w_{2,y} &= w_2, \text{ for human waste} \end{aligned} \tag{3.37}$$

Recalling that $x_1 = l_{11}y_1$, we can obtain the net amounts of waste induced by a given final demand y_1 as

$$\begin{aligned} w_1 &= g_{11}l_{11}y_1 + w_{1y} \\ w_2 &= g_{21}l_{11}y_1 + w_{2y} \end{aligned} \tag{3.38}$$

Table 3.3 (the lower panel) gives a numerical example of the entries in the Rice and Consumption columns of the upper panel. The Rice column shows the inputs and outputs required for producing a unit of rice. In this example, the amount of seeds used per unit of output is set to a smaller value compared to the previous case mentioned in (3.4). This reduction represents a hypothetical scenario where rice production productivity increases due to the application of additional fertilizer. Specifically, the value of a_{11} decreases from 0.021 to 0.015, resulting in a corresponding decrease in l_{11} from 1.0215 to 1.0152. Consequently, satisfying one kilogram of final demand for rice now requires producing approximately 60 g less, thanks to the improved productivity. It is important to note that this example is purely hypothetical and is used solely for illustrative purposes.

The Consumption column gives the inputs and outputs of household consumption normalized by the amount of rice consumption. It shows that consuming a unit of rice requires 0.2 units of rice straw (for example, as fuel for cooking the rice) and generates 0.6 units of post-consumption waste (excreta).

Application of (3.38) to the input and waste generation coefficients in Table 3.3 gives the net amount of waste generated per 1 kg of rice consumption

$$\begin{aligned} w_1 &= 0.4 \times 1.0152 - 0.2 = 0.206, \text{ for rice straw} \\ w_2 &= -0.24 \times 1.0152 + 0.6 = 0.356, \text{ for human waste} \end{aligned} \tag{3.39}$$

If waste recycling were not practiced, that is, $g_{ij}^- = 0$ and $w_{iy}^- = 0$, the amount of waste induced by 1 kg of the final demand for rice would be

$$\begin{aligned} w_1 &= g_{11}^+ l_{11} = 1.4 \times 1.0152 = 1.4213 \\ w_2 &= w_{2y} = 0.6 \end{aligned} \tag{3.40}$$

Comparison with (3.39) reveals that waste recycling reduces the net amounts of waste in this hypothetical economy by around 85% for rice straw and by 40% for human waste while increasing the yield, resulting in a win-win situation in terms of both environment and economy.

3.2 The Two-Sector IO Model: Exposition Without Matrices

In the previous section, we discussed a hypothetical one-sector economy where people could only afford to consume grain. Now, we will consider a two-sector economy where people can afford to consume animal protein as well. This economy has two sectors, one producing grain and the other producing animal meat.

3.2.1 *The Balance Between the Supply and Demand of Products*

Unlike the one-sector model, the two-sector IO model has intersectoral flows among the sectors. This means that in addition to using their own products as inputs, each sector can also use the product of the other sector as an input. We denote by z_{ij} the amount of input of product i used in sector j to produce product j.

3.2.1.1 The System of Balance Equations

Denoting by x_2 the amount of product 2 (produced by sector 2), by z_{22} the amount of product 2 used in sector 2, by z_{12} the amount of product 1 used in sector 2, by z_{21} the amount of product 2 used in sector 1, and by y_2 the amount of final demand for product 2, the balance between the supply and demand of products becomes

$$\begin{aligned} x_1 &= z_{11} + z_{12} + y_1 \\ x_2 &= z_{21} + z_{22} + y_2 \end{aligned} \tag{3.41}$$

3.2.1.2 The Input Coefficients

Generalizing the input coefficients involving self-products only in the one-product economy (3.2) to those involving inter-sectoral flows, we have

$$a_{ij} = z_{ij}/x_j, \; i, j \in 1, 2, \quad x_j > 0 \tag{3.42}$$

where a_{ij} refers to the input of i per unit output of product j. The 2×2 array of a_{ij}s (Table 3.4, the upper panel) represents the processes available in this economy in terms of the input-output relationships involving the two products only, with the first column referring to the process of sector 1 producing product 1 and the second column referring to the process of sector 2 producing product 2.

In terms of the input coefficients thus defined, the supply-demand balance Eq. (3.41) can be rewritten as

$$\begin{aligned} x_1 &= a_{11}x_1 + a_{12}x_2 + y_1 \\ x_2 &= a_{21}x_1 + a_{22}x_2 + y_2 \end{aligned} \tag{3.43}$$

Solving (3.43) for given final demand and input coefficients, we are able to obtain the amounts of products 1 and 2 that are required to meet the final demand, provided the system is solvable. Stated otherwise, it refers to finding the amounts of products needed to meet the final demand (support the population) for a given set of technology represented by the a_{ij}s. Before discussing how to solve (3.43) under general conditions, let us look at some numerical examples.

Table 3.4 Two production processes represented by a_{ij}s

	Sector (product) 1	Sector (product) 2
Sector (product) 1	a_{11}	a_{12}
Sector (product) 2	a_{21}	a_{22}
A numerical example of rice-pig economy		
	Rice[a]	Pig[b]
Rice	0.021	2.805
Pig	0	0.033
A numerical example of rice-fish economy		
	Rice	Fish
Rice	0.021	1.103[d]
Fish	0.071[c]	0.017[e]

Source [a] Table 3.1
[b] Table 2.4
[c] Table 2.2, with the sum of fertilizer assumed to be provided by fish
[d] Table 2.6, with the sum of feed assumed to be provided by rice
[e] Table 2.6

3.2.2 *The Two-Sector IO Model with Numerical Examples*

In our previous numerical example, we looked at a one-sector economy based on rice production in 17th century China. Now, we consider a two-sector IO model with a numerical example based on the production of rice and pigs. We must note that our numerical example is for illustrative purposes only and should not be used for a serious analysis of rice and livestock production.

3.2.2.1 A Rice-Pig Economy

We begin by assuming that waste, by-products, and emissions are not present and focus on the interaction between the two sectors: the rice sector and the pig sector. In ecological terms, pigs are consumers and need to be fed with feed. To simplify our example, we assume that pigs are fed with rice, which is becoming increasingly common worldwide [46–48]. In contrast, rice is a producer in ecological terms and can be grown without pigs.

Table 3.4, the middle panel, gives a numerical example of the a_{ij}s in the upper panel. To simplify the calculation, we assumed that the sum of the different feed items in Table 2.4 could be replaced with an equal amount of rice, resulting in the input of 2.8 kg of rice per kg of pig for slaughter. Remarkable in the process representation in Table 3.4, the middle panel, is the absence of input from the pig sector to the rice sector, $a_{21} = 0$: the pig has to be fed with rice, but rice does not need any input of pig. As a result, the array (matrix) of a_{ij}s has a triangular form.

With the values of a_{ij} given, the balance Eq. (3.41) become

$$\begin{aligned} x_1 &= 0.021x_1 + 2.805x_2 + y_1 \\ x_2 &= 0.033x_2 + y_2 \end{aligned} \tag{3.44}$$

We now consider solving this system of equations for x_1 and x_2, when a_{ij}s and y_is are given, and rearrange it as

$$\begin{aligned} (1 - 0.021)x_1 - 2.805x_2 &= y_1 \\ (1 - 0.033)x_2 &= y_2 \end{aligned} \tag{3.45}$$

Since this system has a triangular structure, it is easy to obtain the solution. Solving the second equation for x_2, we obtain

$$x_2 = y_2/(1 - 0.033) = 1.034y_2 \tag{3.46}$$

Substitution into the first equation of (3.45) gives

$$x_1 = (y_1 + y_2 \times 2.805/(1 - 0.033))/(1 - 0.021) = 1.021y_1 + 2.963y_2 \tag{3.47}$$

A notable difference from the one-sector IO model is that the amount of rice, x_1, now depends not only on the final demand for rice, y_1, but also on the final demand for pig, y_2, as rice is needed to produce a pig. For example, even if there is no final demand for rice, (3.47) shows that 2.963 units of rice have to be produced to satisfy one unit of the final demand for pig. It is noteworthy that producing one unit of pig requires almost three times as much rice (2.963 units) as producing one unit of rice (1.021 units).

The numerical values attached to y_1 and y_2 in (3.47) refer to the two-sector counterpart of the Leontief inverse coefficients we discussed previously, which is now given by a 2×2 array, denoted by $\boldsymbol{L}$

$$\boldsymbol{L} = \begin{pmatrix} l_{11} & l_{12} \\ l_{21} & l_{22} \end{pmatrix} = \begin{pmatrix} 1.021 & 2.963 \\ 0 & 1.034 \end{pmatrix} \tag{3.48}$$

Here, l_{12} refers to the amount of x_1 directly and indirectly needed to satisfy one unit of y_2, while l_{21} refers to the amount of x_2 directly and indirectly needed to satisfy a unit of y_1.

Before discussing the features of the Leontief inverse coefficients (the array $\boldsymbol{L}$) in a two-sector IO model and the productive conditions thereof, it is helpful to introduce another numerical example where all the input coefficients are positive. One such economy is based on rice and fish production, which gives a rough approximation of the preindustrialized Japanese economy before the middle of the 19th century. In this period, meat consumption was discouraged by religious authorities, and seafood was the main source of animal protein. In 16th–19th century Japan, fish (dried sardine) was also widely used to produce fertilizer [49]. Since fish can be fed with rice, using fish-based fertilizer provides a two-sector economy with full interdependence between the sectors.

3.2.2.2 A Rice-Fish Economy

The lower panel of Table 3.4 gives a numerical example of the a_{ij} values of an economy based on rice and fish production, demonstrating full interdependence between the sectors. The balance of supply and demand is given by

$$\begin{aligned} x_1 &= 0.021x_1 + 1.103x_2 + y_1 \\ x_2 &= 0.071x_1 + 0.017x_2 + y_2 \end{aligned} \tag{3.49}$$

Arranged analogous to (3.45), we obtain

$$\begin{aligned} (1 - 0.021)x_1 - 1.103x_2 &= y_1 \\ -0.071x_1 + (1 - 0.017)x_2 &= y_2 \end{aligned} \tag{3.50}$$

Solving (3.50) is not as simple as (3.45) because it is no longer triangular. However, elementary calculus of the junior-high school level allows us to find the solution as follows

$$\begin{aligned} x_1 &= \frac{1 - a_{22}}{d} y_1 + \frac{a_{12}}{d} y_2 = 1.112y_1 + 1.248y_2 \\ x_2 &= \frac{a_{21}}{d} y_1 + \frac{1 - a_{11}}{d} y_2 = 0.080y_1 + 1.107y_2 \end{aligned} \tag{3.51}$$

where

$$d = (1 - a_{11})(1 - a_{22}) - a_{12}a_{21} \tag{3.52}$$

Satisfying a unit of final demand for rice needs 1.112 units of rice and 0.080 units of fish, whereas satisfying one unit of final demand for fish needs 1.248 units of rice and 1.107 units of fish. Similar to the previous example of the rice-pig economy, it is noteworthy that more rice is needed to satisfy one unit of final demand for fish compared to the amount needed for rice.

The Leontief inverse coefficients, l_{ij}, are all positive in (3.51), indicating that for any positive y_i, $i = 1, 2$, the solution is $x_i > 0$, $i = 1, 2$. This important result, which we will discuss below in great detail, is a consequence of the specific properties of the a_{ij}s. However, these properties are not trivial. To illustrate this, let us consider the following case

$$\begin{pmatrix} a_{11} & a_{12} \\ a_{21} & a_{22} \end{pmatrix} = \begin{pmatrix} 0.1 & 1.7 \\ 1.6 & 0.2 \end{pmatrix} \tag{3.53}$$

which gives

$$L = \begin{pmatrix} -0.40 & -0.85 \\ -0.80 & -0.45 \end{pmatrix} \tag{3.54}$$

In this case, negative outputs are obtained for any positive final demand. Note that all the a_{ij}s are nonnegative in this example, and the productive condition relevant for the one-sector model, $1 - a_{ii} > 0$ as stated in (3.11), is satisfied. However, this example indicates that the productive condition for a one-sector IO model is not sufficient to ensure the productiveness of a two-sector model.

What are the features of a_{ij}s that result in all elements of $\boldsymbol{L}$ being positive in (3.51)? Note that all the numerators in the middle term of (3.51) are positive. This is due to the fact that $a_{12} > 0$, $a_{21} > 0$, and the productive condition of the one-sector economy, $1 - a_{ii} > 0$, $i = 1, 2$, is satisfied. In order for the elements of $\boldsymbol{L}$ to be positive, an additional condition is required, which is that the denominator of the expression, denoted as d, takes a positive value. In the case of Eq. (3.51), it is indeed the case that $d = 1.13$. Conversely, in the example given by (3.53), $d = -3.0$. Therefore, it can be concluded that $d > 0$ is the additional condition needed to ensure the productiveness of a two-sector economy.

Before we delve into the general properties of the solution of a two-sector IO model, we make a transition from an analysis based on elementary calculus to one based on matrix algebra. This transition is necessary to develop practical skills in IO for IE. For a quick course on the necessary elementary topics of matrix algebra, refer to Appendix A.

3.3 The IO Model Based on Matrices

So far, we have used the term "array" to refer to a collection of elements, such as a_{ij} or l_{ij} arranged in a two dimensional form consisting of rows and columns, with each element identified with its location at the intersection of the relevant row and column. We will now use the term "matrix" instead of "array." Let $\boldsymbol{A}$ be the 2×2 matrix with elements a_{ij} given in Table 3.4

$$\boldsymbol{A} = \begin{pmatrix} a_{11} & a_{12} \\ a_{21} & a_{22} \end{pmatrix} = \begin{pmatrix} 0.021 & 1.103 \\ 0.071 & 0.017 \end{pmatrix} \tag{3.55}$$

If you are familiar with spreadsheet software, such as ® Microsoft Excel or ® OpenOffice Calc, you will recognize that naming an array is equivalent to naming a range of cells in a spreadsheet. This simplifies calculation since you do not need to select the range of cells manually by scrolling through a large number of cells.

We use boldface letters to denote matrices, including those with only one column or row, which we call a vector. To make the elements of a matrix explicit, we use the notation $\boldsymbol{A} = [a_{ij}]$. Lowercase boldface letters represent column vectors unless otherwise stated. For a matrix with two columns, we denote the elements of the first column by $\boldsymbol{a}_{.1}$ and those of the second column by $\boldsymbol{a}_{.2}$. Similarly, we denote the elements of the first row by $\boldsymbol{a}_{1.}$ and those of the second row by $\boldsymbol{a}_{2.}$. Thus, we can write

$$\boldsymbol{A} = (\boldsymbol{a}_{.1}, \boldsymbol{a}_{.2}) = \begin{pmatrix} \boldsymbol{a}_{1.} \\ \boldsymbol{a}_{2.} \end{pmatrix} \tag{3.56}$$

We denote the 2×1 arrays of output and final demand by $\boldsymbol{x}$ and $\boldsymbol{y}$, respectively

$$\boldsymbol{x} = \begin{pmatrix} x_1 \\ x_2 \end{pmatrix}, \quad \boldsymbol{y} = \begin{pmatrix} y_1 \\ y_2 \end{pmatrix} \tag{3.57}$$

Using matrix addition and multiplication, we can represent (3.49) as

$$\boldsymbol{x} = \boldsymbol{A}\boldsymbol{x} + \boldsymbol{y} \tag{3.58}$$

Before we can proceed with the matrix-based representation of (3.50), we need to introduce another matrix, denoted by $\boldsymbol{I}$. Matrix $\boldsymbol{I}$ is a 2×2 diagonal matrix with all off-diagonal elements equal to zero and diagonal elements equal to one, defined as

$$\boldsymbol{I} = \begin{pmatrix} 1 & 0 \\ 0 & 1 \end{pmatrix} \tag{3.59}$$

This matrix is known as the identity matrix and is analogous to 1 in scalar arithmetic. Using the rule of matrix multiplication, we have

$$\boldsymbol{x} = \boldsymbol{I}\boldsymbol{x} \tag{3.60}$$

Substitution of this expression into (3.58) gives

$$\boldsymbol{I}\boldsymbol{x} = \boldsymbol{A}\boldsymbol{x} + \boldsymbol{y} \tag{3.61}$$

which can be rearranged as

$$(\boldsymbol{I} - \boldsymbol{A})\boldsymbol{x} = \boldsymbol{y} \tag{3.62}$$

3.3.1 The Analogy Between the Scalar-Based One-Sector Model and the Matrix-Based Two-Sector Model

Before discussing how to solve (3.62), it is worthwhile to compare the matrix-based expressions (3.58) and (3.62) with their one-sector counterparts, (3.9) and (3.10), reproduced below for the sake of quick reference

$$x_1 = a_{11}x_1 + y_1 \tag{3.9}$$

$$(1 - a_{11})x_1 = y_1 \tag{3.10}$$

We notice that except for the use of $\boldsymbol{A}$ instead of a_{11}, $\boldsymbol{x}$ instead of x_1, $\boldsymbol{y}$ instead of y_1, and $\boldsymbol{I}$ instead of 1, the expressions are formally identical: (3.9) and (3.10) refer to the special case of (3.62) and (3.58) corresponding to an economy with only one-sector. It then follows that by analogy to that (3.10) was solved for x_1 by inverting $(1 - a_{11})$

$$x_1 = (1 - a_{11})^{-1}y_1 \tag{3.13}$$

Equation (3.62) can be solved by multiplying the "inverse" of $\boldsymbol{I} - \boldsymbol{A}$ from the left-hand side

$$\boldsymbol{x} = (\boldsymbol{I} - \boldsymbol{A})^{-1}\boldsymbol{y} \tag{3.63}$$

This approach of following the analogy to the one-sector case is perfectly correct. The only caveat is that one needs to ensure that the inverse of $\boldsymbol{I} - \boldsymbol{A}$ does exist and provides a nonnegative solution for any nonnegative final demand. The latter point relates to the productive condition discussed in Sect. 3.1.1.2 and touched upon above.

3.3.2 *The Leontief Quantity Model*

We will now focus on discussing some important details regarding $(\boldsymbol{I} - \boldsymbol{A})^{-1}$, including its definition, the conditions under which it exists, and if it does exist, how to calculate it. It is worth noting that although our numerical example represents a two-sector economy ($n = 2$), once we represent it in matrix form, the size of an IO model that can be represented is not limited to $n = 2$. In fact, it is applicable to any $n > 0$.

3.3.2.1 The Inverse Matrix

A square matrix (a matrix with an equal number of rows and columns) $\boldsymbol{C}$ is called an inverse matrix of another square matrix of the same order, $\boldsymbol{B}$, when their multiplication gives an identity matrix

$$\boldsymbol{C}\boldsymbol{B} = \boldsymbol{I} \tag{3.64}$$

which is nothing but the matrix counterpart of $c = 1/b$. It is customary to write $\boldsymbol{C} = \boldsymbol{B}^{-1}$.

For a square matrix $\boldsymbol{B}$ of order $n = 2$, $\boldsymbol{B}^{-1}$ exists when the determinant of the matrix, denoted by $|\boldsymbol{B}|$ or $\det \boldsymbol{B}$

$$|\boldsymbol{B}| = b_{11}b_{22} - b_{12}b_{21} \tag{3.65}$$

is nonzero. The inverse matrix can then be computed as

$$\boldsymbol{B}^{-1} = \frac{1}{b_{11}b_{22} - b_{12}b_{21}} \begin{pmatrix} b_{22} & -b_{12} \\ -b_{21} & b_{11} \end{pmatrix} \tag{3.66}$$

For matrices of order greater than 2, the determination of the inverse matrix is generally more complex and is beyond the scope of this discussion.

3.3.2.2 The Leontief Inverse Matrix

The matrix $(\boldsymbol{I} - \boldsymbol{A})^{-1}$ is the inverse of $(\boldsymbol{I} - \boldsymbol{A})$, and it is commonly known as the Leontief inverse matrix. For the case of $n = 2$, the Leontief inverse matrix is given by

$$(\boldsymbol{I} - \boldsymbol{A})^{-1} = \frac{1}{(1 - a_{11})(1 - a_{22}) - a_{12}a_{21}} \begin{pmatrix} 1 - a_{22} & a_{12} \\ a_{21} & 1 - a_{11} \end{pmatrix} \tag{3.67}$$

which reproduces the Leontief inverse coefficients in (3.51), with $d = |\boldsymbol{I} - \boldsymbol{A}|$. For $(\boldsymbol{I} - \boldsymbol{A})^{-1}$ to exist, the matrix $\boldsymbol{I} - \boldsymbol{A}$ has to be nonsingular, that is, its determinant, $|\boldsymbol{I} - \boldsymbol{A}|$, must not be zero. In other words, the columns of $\boldsymbol{I} - \boldsymbol{A}$ have to be linearly independent. For instance, the following matrix is singular

$$\begin{pmatrix} b_{11} & \alpha b_{11} \\ b_{21} & \alpha b_{21} \end{pmatrix} \tag{3.68}$$

where α is a scalar.

It is important to note that the mere existence of the solution $\boldsymbol{x}$ in (3.62) is not enough to ensure its economic relevance. The solution must also be nonnegative, $\boldsymbol{x} \geq 0$, for any $\boldsymbol{y} \geq 0$ to make economic sense. A necessary and sufficient condition for this is the nonnegativity of all the elements of $\boldsymbol{L}$. From (3.67), we observe that the elements in the numerator are all nonnegative due to $a_{ij} \geq 0$ and $1 - a_{ii} > 0$, which is the productive condition of a one-sector economy. Thus, for (3.67) to yield a nonnegative solution, the determinant of $\boldsymbol{I} - \boldsymbol{A}$, $|\boldsymbol{I} - \boldsymbol{A}|$, must be positive. The conditions under which (3.62) has a nonnegative solution $\boldsymbol{x} \geq 0$ for any $\boldsymbol{y} \geq 0$ are referred to as the productive conditions.

3.3.2.3 Productive Conditions and *A*

We begin with the definition of a productive $\boldsymbol{A}$ matrix, as given by [15].

Definition 3.1 The matrix $\boldsymbol{A}$ is productive if there exists a nonnegative $\boldsymbol{x}$ such that $(\boldsymbol{I} - \boldsymbol{A})\boldsymbol{x} > \boldsymbol{0}$.

The following theorem, also from [15], is important:

Theorem 3.1 ([15], p. 296) *If $\boldsymbol{A}$ is productive then for any $\boldsymbol{y} \geq 0$ the equation*

$$(\boldsymbol{I} - \boldsymbol{A})\boldsymbol{x} = \boldsymbol{y} \tag{3.69}$$

has a unique nonnegative solution.

We call an economy with a productive $\boldsymbol{A}$ a productive economy.

To determine if a given $\boldsymbol{A}$, with $a_{ij} \geq 0$ for all i and j, is productive, the following theorem from [14] provides an answer:

Theorem 3.2 ([14], p. 90 and p. 95) *The following four conditions are mutually equivalent*

1. *Weak solvability (WS)* *Equation* (3.69) *has a set of nonnegative solutions $\boldsymbol{x} \geq 0$ for some set of positive $\boldsymbol{y} > 0$.*
2. *Strong solvability (SS)* *Equation* (3.69) *is solvable in the nonnegative unknowns $\boldsymbol{x} \geq 0$ for any set of nonnegative $\boldsymbol{y} \geq 0$.*
3. *HS* *The square matrix $\boldsymbol{B} = [b_{ij}] = \boldsymbol{I} - \boldsymbol{A}$ has n positive upper left-hand corner (or "leading") principal minors*

$$M_{kk} = \begin{vmatrix} b_{11} & \cdots & b_{1k} \\ \vdots & & \vdots \\ b_{k1} & \cdots & b_{kk} \end{vmatrix} > 0 \quad (k = 1, \ldots, n) \tag{3.70}$$

4. *HS'* *All the principal minors of $\boldsymbol{B}$, M_{ij}, $i, j = 1, \ldots n$, are positive.*
5. *The positiveness of the inverse* *All the elements of $\boldsymbol{B}^{-1}$ are nonnegative.*

Conditions 3 or 4 are known as the Hawkins–Simon condition [13] and are denoted HS. Since HS is easier to compute than HS', the latter is not used in the rest of this book.

For a given matrix $\boldsymbol{A}$ to be considered productive, it must satisfy the HS condition. This condition can be determined by computing all the leading principal minors of the matrix (a minor of $\boldsymbol{A}$ of order k is principal if it is obtained by deleting $n - k$ rows and $n - k$ columns). An important implication of the above theorem is that if matrix $\boldsymbol{A}$ was obtained from data on a productive economy, the satisfying of the HS condition can be verified without computing any minors. A productive economy is characterized by a positive final demand $\boldsymbol{y} > 0$ and nonnegative outputs $\boldsymbol{x} \geq 0$, which implies the fulfillment of condition 1 (WS). Therefore, according to the theorem, the matrix $\boldsymbol{A}$ obtained from a productive economy always satisfies the HS condition.

In the case of a productive $\boldsymbol{A}$ matrix, we can ensure that

$$\boldsymbol{x} = (\boldsymbol{I} - \boldsymbol{A})^{-1}\boldsymbol{y} \tag{3.71}$$

gives $\boldsymbol{x} \geq 0$ for any $\boldsymbol{y} \geq 0$. However, it is crucial to note that this result is not valid when there are negative elements in $\boldsymbol{A}$, which can occur due to the presence of by-products. This issue will be discussed further in Sect. 3.4.3.

Additionally, there is a sufficient condition, proposed by [50], that allows for a fast check of the productiveness of a given matrix $\boldsymbol{A}$. This condition is particularly useful when the matrix was not obtained from actual IO data. Let $\boldsymbol{\iota}$ be a vector of order n with all elements equal to one.

Corollary 3.1 ([50])

(i) If $\boldsymbol{A}\boldsymbol{\iota} < \boldsymbol{\iota}$*, then (WS), (SS), and (HS) hold.*
(ii) If $\boldsymbol{\iota}^\top \boldsymbol{A} < \boldsymbol{\iota}^\top$*, then (WS), (SS), and (HS) hold.*

A proof can be found in [37]. We will see in Chap. 4 that the second condition is particularly useful when dealing with IO tables in monetary units, which is typically the case in practice.

3.3.3 *Human Labor as a Power Source and Its Energy Requirements*

Similar to the one-sector economy discussed in Sect. 3.1.2, in our two-sector (pig-rice) economy, human labor (muscle power) remains the only source of energy for operating the production processes. We analyze the energy aspects of human labor, extending our analysis from the single-product economy in Sect. 3.1.2. For numerical illustration, we use Table 3.5.

3.3.3.1 The Labor Hours Embodied in Products

Following our analysis in Sect. 3.1.2.4, we can calculate the required amount of labor hours per kg of rice and pig as

$$\boldsymbol{a}_{\iota}.(\boldsymbol{I}-\boldsymbol{A})^{-1} = (0.597\ 3.981)\begin{pmatrix}1.021 & 2.963\\0.000 & 1.034\end{pmatrix} = (0.610\ 5.963) \tag{3.72}$$

Table 3.5 The rice, pig, and fish process with the input of human power

	Rice	Pig	Fish
	a_{ij}: intermediate input coefficients		
Rice	0.021	2.805	1.103
Pig	0.000	0.033	0.017
	$\boldsymbol{a}_{\iota}.$; labor input coefficients (hours per kg)		
Hours of work[a]	0.597	3.981	0.533

Source [a] Reference [1]. See Table 3.4 for the $\boldsymbol{A}$ matrix where L is taken from (3.48). The results show that 1 kg of rice embodies 0.61 hours of human labor, while 1 kg of pig embodies 5.96 hours of labor: producing 1 kg of pig requires 9.8 times the amount of labor required to produce the same amount of rice.

The required amounts of labor hours per kg of products in the rice-fish economy are obtained as follows

$$\boldsymbol{a}_{\ell}.(\boldsymbol{I}-\boldsymbol{A})^{-1} = \begin{pmatrix}0.597 & 0.533\end{pmatrix}\begin{pmatrix}1.112 & 1.248\\ 0.080 & 1.107\end{pmatrix} = \begin{pmatrix}0.707 & 1.335\end{pmatrix} \tag{3.73}$$

where L is taken from (3.51). The results show that 1 kg of rice embodies 0.707 hours of human labor, while 1 kg of fish embodies 1.335 hours of labor. The slight increase in the embodied amount of labor hours in rice (from 0.610 to 0.707 hours of human labor) compared with the rice-pig economy is attributed to the use of fish as fertilizer in rice production. In this numerical example, the fish-based economy requires fewer labor hours than one based on pigs.

3.3.3.2 *The Energy Embodied in Products

Recalling that the energy cost of one hour of labor is 11.9 MJ (3.22), the amount of human power embodied in 1 kg of pig is given by

$$11.9\ \text{MJ/h} \times 5.963\ \text{h/kg-pig} = 70.57\ \text{MJ/kg-pig} \tag{3.74}$$

Since there is no input from the pig sector into the rice sector, the amount of human power embodied in 1 kg of rice remains unchanged (7.26 MJ) from the result in Sect. 3.1.2.4.

For the rice-fish economy, the corresponding results are

$$11.9\ \text{MJ/h} \times 0.707\ \text{h/kg-rice} = 8.413\ \text{MJ/kg-rice} \tag{3.75}$$

$$11.9\ \text{MJ/h} \times 1.335\ \text{h/kg-fish} = 15.89\ \text{MJ/kg-fish} \tag{3.76}$$

The slight increase in the embodied amount of human power in rice (from 7.26 to 8.41 MJ) is attributed to the use of fish as fertilizer in rice production. As far as this numerical example is concerned, the fish-based economy requires less human power than one based on pigs.

3.3.3.3 *The Amount of Food Needed to Provide Human Muscle Power

In Sect. 3.1.2.4, we calculated that one hour of farm labor requires 270 kJ of energy input. Now, we want to know how much rice and pig meat are required to provide this energy. According to [1, 2], the food consumption in the region consists of 73% rice, 15% wheat, 4% pig, and 7% fish. We assume that 80% of the energy required for human labor is provided by rice and the remaining 20% by pig meat [51]. The absence of data on the heat value of fish in [1, 2] prevents us from considering fish meat as a source of human power. Assuming a conversion efficiency of 25% for food to human muscle energy [24], the amount of rice and pig meat required to provide the required human power per hour, $a_{i\ell}$, is

$$\begin{aligned} a_{1\ell} &= 0.270 \text{ MJ/h} \times 0.8/(15.19 \text{ MJ/kg} \times 0.25) = 0.057 \text{ kg/h} \\ a_{2\ell} &= 0.270 \text{ MJ/h} \times 0.2/(18.83 \text{ MJ/kg} \times 0.25) = 0.011 \text{ kg/h} \end{aligned} \tag{3.77}$$

Thus, providing the human power needed for one hour of farm work requires an additional 57 g of rice and only 11 g of pig meat.

3.3.3.4 *The Miyazawa Model: Endogenizing the Input for Human Power Production

We are now ready to calculate the amount of food needed to satisfy a given final demand for food, considering the amount needed to produce human labor required for food production. Let $\boldsymbol{y}_\ell$ denote the amount of food needed to produce the human labor required for food production, which depends on the amount of food production $\boldsymbol{x}$. While $\boldsymbol{y}_\ell$ is a component of $\boldsymbol{y}$, it is no longer exogenously given but endogenized. Therefore, the balance of supply and demand, involving both goods and labor hours, can be rewritten as

$$\begin{pmatrix} \boldsymbol{x} \\ \ell \end{pmatrix} = \begin{pmatrix} \boldsymbol{A} & \boldsymbol{a}_{.\ell} \\ \boldsymbol{a}_{\ell.} & 0 \end{pmatrix} \begin{pmatrix} \boldsymbol{x} \\ \ell \end{pmatrix} + \begin{pmatrix} \boldsymbol{y}_0 \\ y_{\ell *} \end{pmatrix} = \begin{pmatrix} \boldsymbol{x} \\ \ell \end{pmatrix} \tag{3.78}$$

where ℓ refers to labor hours, $\boldsymbol{y}_0$ refers to the final demand net of $\boldsymbol{y}_\ell$, that is, $\boldsymbol{y}_0 = \boldsymbol{y} - \boldsymbol{y}_\ell$, and $\boldsymbol{y}_{\ell *}$ refers to the final demand for human power, say servants, which we assume is equal to null. Rearranging gives

$$\begin{pmatrix} \boldsymbol{I} - \boldsymbol{A} & -\boldsymbol{a}_{.\ell} \\ -\boldsymbol{a}_{\ell.} & 1 \end{pmatrix} \begin{pmatrix} \boldsymbol{x} \\ \ell \end{pmatrix} = \begin{pmatrix} \boldsymbol{y}_0 \\ y_{\ell *} \end{pmatrix} \tag{3.79}$$

Applying the rule of inversion of a partitioned matrix, we obtain (with $y_{\ell *} = 0$)

$$\begin{aligned} \begin{pmatrix} \boldsymbol{x} \\ \ell \end{pmatrix} &= \begin{pmatrix} (\boldsymbol{I} - \boldsymbol{A} - \boldsymbol{a}_{.\ell}\boldsymbol{a}_{\ell.})^{-1} & (\boldsymbol{I} - \boldsymbol{A})^{-1}\boldsymbol{a}_{.\ell}(1 - \boldsymbol{a}_{\ell.}(\boldsymbol{I} - \boldsymbol{A})^{-1}\boldsymbol{a}_{.\ell})^{-1} \\ \boldsymbol{a}_{\ell.}(\boldsymbol{I} - \boldsymbol{A} - \boldsymbol{a}_{.\ell}\boldsymbol{a}_{\ell.})^{-1} & (1 - \boldsymbol{a}_{\ell.}(\boldsymbol{I} - \boldsymbol{A})^{-1}\boldsymbol{a}_{.\ell})^{-1} \end{pmatrix} \begin{pmatrix} \boldsymbol{y}_0 \\ 0 \end{pmatrix} \\ &= \begin{pmatrix} (\boldsymbol{I} - \boldsymbol{A} - \boldsymbol{a}_{.\ell}\boldsymbol{a}_{\ell.})^{-1} \\ \boldsymbol{a}_{\ell.}(\boldsymbol{I} - \boldsymbol{A} - \boldsymbol{a}_{.\ell}\boldsymbol{a}_{\ell.})^{-1} \end{pmatrix} \boldsymbol{y}_0 \end{aligned} \tag{3.80}$$

The upper submatrix reduces to (3.28) when $n = 1$.

The Eq. (3.80) is formally referred to as the Miyazawa model of endogenized labor supply [25]. However, our derivation of this model is different from that of [25] as we have considered the physiological, demographic, and living standards of the population for an economy based on muscle power. In contrast, the original Miyazawa model defined $\boldsymbol{a}_{\ell.}$ as the income (value added) per output and $\boldsymbol{a}_{.\ell}$ as the consumption basket of households.

For our numerical example, the matrix $\boldsymbol{a}_{.\ell}\boldsymbol{a}_{\ell.}$ is given by

$$\boldsymbol{a}_{.\ell}\boldsymbol{a}_{\ell.} = \begin{pmatrix} 0.057 \\ 0.011 \end{pmatrix} \begin{pmatrix} 0.597 & 3.981 \end{pmatrix} = \begin{pmatrix} 0.034 & 0.227 \\ 0.007 & 0.044 \end{pmatrix} \tag{3.81}$$

The human power needed to produce 1 kg of rice requires 34 g of rice and 7 g of pig meat to provide the power, while 227 g of rice and 44 g of pig meat are needed to provide the human power to produce 1 kg of pig meat.

The amount of food needed to satisfy a given final demand, considering the amount needed to produce the human labor required for food production, is then given by

$$\boldsymbol{x}_0 = (\boldsymbol{I} - \boldsymbol{A} - \boldsymbol{a}_{.\ell}\boldsymbol{a}_{\ell.})^{-1}\boldsymbol{y}_0 = \begin{pmatrix} 1.083 & 3.557 \\ 0.008 & 1.108 \end{pmatrix} \boldsymbol{y}_0 \tag{3.82}$$

Comparison of the Leontief inverse in (3.82) with that in (3.48), without the endogenized final demand component, shows endogenization of (a part of) the final demand increases the magnitudes of sectoral inter-dependence. For instance, the amount of rice needed to produce one unit of pig increases by 20%. What is remarkable is the emergence of the dependence of rice on pig as a result of the endogenization, whereas there is no such dependence when the final demand is exogenous.

The amount of labor required to produce the output given by (3.82) is

$$\begin{aligned} \ell &= \boldsymbol{a}_{\ell.}(\boldsymbol{I} - \boldsymbol{A} - \boldsymbol{a}_{.\ell}\boldsymbol{a}_{\ell.})^{-1}\boldsymbol{y}_0 \\ &= \begin{pmatrix} 0.597 & 3.981 \end{pmatrix} \begin{pmatrix} 1.083 & 3.557 \\ 0.008 & 1.108 \end{pmatrix} \boldsymbol{y}_0 = \begin{pmatrix} 0.677 & 6.536 \end{pmatrix} \boldsymbol{y}_0 \end{aligned} \tag{3.83}$$

Comparison with (3.72) indicates that the total amount of work hours required to meet a given final demand increases due to the inclusion of the hours of work needed to produce food for the required human muscle power.

3.3.4 Cost and Price

In Sect. 3.1.2, we calculated the hours of work embodied in products. If labor is homogeneous and the only primary factor of production (an input that is not produced within the system and is exogenously given to the system), this calculation can be used to value the products using the labor theory of value. This method of evaluation dates back to classical economists like Adam Smith and David Ricardo and is commonly associated with Marxian economics [52]. In this section, we discuss the basics of cost and price in the standard IO to prepare for dealing with IO tables measured in monetary units in subsequent chapters.

3.3.4.1 Valuing the Products

To value products based on the amount of labor embodied, we consider a value vector $\boldsymbol{t} = [t_i]$ where t_i is the value of product i based on the amount of labor embodied. We use the following cost balance equation, derived by [52]

$$\boldsymbol{t} = \boldsymbol{t}\boldsymbol{A} + \boldsymbol{a}_{\ell}. \tag{3.84}$$

Assuming $\boldsymbol{A}$ satisfies the HS conditions, we can solve (3.84) for $\boldsymbol{t}$ as

$$\boldsymbol{t} = \boldsymbol{a}_{\ell}.(\boldsymbol{I} - \boldsymbol{A})^{-1} \tag{3.85}$$

In the rice-pig economy discussed above, we have from (3.83)

$$\boldsymbol{t} = \begin{pmatrix} 0.677 & 6.536 \end{pmatrix} \tag{3.86}$$

It follows that the value of pigs is around 6.5 hours per kg, while rice is valued at around 0.7 hours per kg.

We next consider valuing the products based on the monetary wage per hour of work. Denote by q_{ℓ} the monetary wage per hour of work, which is assumed equal across all the sectors of the economy because of the homogeneity of labor hours, and by p_i the monetary value of product i. If labor is the only primary factor of production, we have from (3.84) the following cost balance equation in terms of a $1 \times n$ vector $\boldsymbol{p} = [p_i]$

$$\boldsymbol{p} = \boldsymbol{p}\boldsymbol{A} + q_{\ell}\boldsymbol{a}_{\ell}. \tag{3.87}$$

which can be solved for $\boldsymbol{p}$ as

$$\boldsymbol{p} = q_{\ell}\boldsymbol{a}_{\ell}.(\boldsymbol{I} - \boldsymbol{A})^{-1} \tag{3.88}$$

3.3.4.2 Monetary Cost and Price of Products

In (3.87), labor hours were the only primary factor of production considered. However, as discussed in Sect. 3.4.2, the land is another primary factor of production, even in a simple two-sector economy. To account for the cost of land use, we introduce

q_{land}, the price of land use (rent or the price of land services) per square meter per year, assumed to be equal across all land use for simplicity. With this added factor cost, the monetary value of products can be represented as follows

$$\boldsymbol{p} = \boldsymbol{p}\boldsymbol{A} + \underbrace{(q_{\ell}\boldsymbol{a}_{\ell.} + q_{land}\boldsymbol{a}_{land})}_{\boldsymbol{v}} \tag{3.89}$$

This equation shows that the value of products is the sum of the cost of intermediate inputs and the cost of primary factors of production. The second term on the right-hand side represents the compensation for labor and land services per unit output, also known as the value-added ratio. Denoting by $\boldsymbol{v} = [v_i]$ the vector of value-added ratios, we can express the monetary price as

$$\boldsymbol{p} = \boldsymbol{v}(\boldsymbol{I} - \boldsymbol{A})^{-1} \tag{3.90}$$

This equation provides a general expression for the monetary price, taking into account the value of intermediate inputs and the cost of primary factors of production.

3.4 The Two-Sector IO Model: Environmental Extensions

With the basics of a two-sector IO behind us, we now turn to the environmental extension of the two-sector model. We start our discussion with GHG emissions and proceed to other items such as water, land, and waste.

3.4.1 GHG Emissions

In Sect. 3.1.3.1, we discussed CH_4 emissions from rice agriculture in the context of a one-sector IO model. Now we extend our analysis to a two-sector IO model, starting with the example of a rice and fish economy.

3.4.1.1 The Rice-Fish System

In our two-sector economy example, which consists of rice and fish sectors, there is no use of fossil fuel, and the production of rice, fish, and fertilizer relies on human power. Furthermore, cooking rice and fish at home is done by burning biomass. Consequently, the emission of CH_4 due to the production of rice can be regarded as the only direct source of GHG emissions (CH_4 emissions from fish farm are not considered). From (3.29), the amount of CH_4 emission is proportional to rice production

$$e_{CH4} = f_1 x_1 = 0.033 x_1 \tag{3.91}$$

where f_1 represents the CH_4 emission per unit of rice.

However, in contrast to the one-sector case presented in Sect. 3.1.3.1, the amount of rice production in this two-sector economy depends not only on the final demand for rice but also on the final demand for fish. Therefore, the CH_4 emission in this economy is not only dependent on the final demand for rice but also on the final demand for fish. Thus, the emission can be expressed as

$$e_{CH4} = f_1 x_1 = f_1 (l_{11} y_1 + l_{12} y_2) = r_1 y_1 + r_2 y_2 = 0.037 y_1 + 0.041 y_2 \tag{3.92}$$

where $r_i = f_1 l_{1,i}$ refers to the total emission coefficients, and f_i refers to direct emission coefficients. This implies that per unit of final demand, fish generates a slightly larger amount of CH_4 (0.041 units) than rice, even though fish production generates no CH_4 directly.

3.4.1.2 Direct and Indirect Emissions and Scopes 1, 2, and 3

The difference between total and direct emissions is called indirect emissions, which arise from sector interdependence (or value-chain emissions in the language of the GHG protocol). Since there is no direct emission in fish production, the value of $r_2 = 0.041$ is solely due to indirect effects. The GHG protocol [53] distinguishes emissions associated with the activity of an organization (e.g., corporations or cities) into three scopes: scope 1 (direct emissions), scope 2 (emissions from power use), and scope 3 (upstream and downstream value chain emissions). We will discuss the concept of scopes 1, 2, and 3 in detail in Sect. 5.1.4.

Our definition of direct emissions is similar to scope 1 but differs slightly from it because we count the emission associated with producing seeds as indirect emissions. Since there is no scope 1 emission from fish production, the indirect emission of fish production corresponds to scope 3 emissions. In the present case, scope 2 is irrelevant because of the absence of power sectors.

Denoting the row vector of direct emission coefficients as $\boldsymbol{f} = (f_1, f_2)$ and that of total emission coefficients as $\boldsymbol{r} = (r_1, r_2)$, a general matrix-based expression for (3.92) is given by (we omit the subscript $_{CH4}$ for simplicity)

$$e = \boldsymbol{r}\boldsymbol{y} = \boldsymbol{f}\boldsymbol{L}\boldsymbol{y} = \boldsymbol{f}(\boldsymbol{I} + \boldsymbol{A} + \boldsymbol{A}^2 + \cdots)\boldsymbol{y} \tag{3.93}$$

Based on this representation, [16, 54] consider $\boldsymbol{f}(\boldsymbol{I} + \boldsymbol{A})\boldsymbol{y}$ as the direct effects. However, we choose to consider $\boldsymbol{f}\boldsymbol{y}$ as the direct emission because it is closer to scope 1. As a result, we define the difference $\boldsymbol{e} - \boldsymbol{f}\boldsymbol{y}$ as the indirect emissions. In Chap. 5, we will discuss seemingly different definitions of direct emissions based on more detailed modeling of the technology.

3.4.1.3 The Rice-Pig System: Direct Emissions from Both Sectors

We next consider the case where GHG emissions occur in both sectors of a two-sector economy. Our numerical example is the rice-pig economy we considered above, augmented with the CH_4 emissions associated with pig production, mostly from manure treatment. Table 2.4 shows that per unit of pig production (in sector 2), around 0.02 units of CH_4 are generated, that is, $f_2 = 0.02$. From (3.93) the total emission coefficients $\boldsymbol{r}$ are given by

$$\boldsymbol{r} = \boldsymbol{f}\boldsymbol{L} = \begin{pmatrix} 0.033 & 0.02 \end{pmatrix} \begin{pmatrix} 1.021 & 2.963 \\ 0.000 & 1.034 \end{pmatrix} = \begin{pmatrix} 0.034 & 0.118 \end{pmatrix} \tag{3.94}$$

We notice that the indirect emissions from pig production are around six times larger than the direct emissions, most of which are attributable to the use of rice to feed pigs. On the other hand, the total emissions from the rice sector are almost equal to the direct emissions because the sector uses none of the output of the pig sector.

3.4.2 *Land Footprint*

Similar to (3.94), the amount of an exogenous input required to satisfy the final demand can be obtained. For instance, producing one kg of rice and pig requires 2.564 m^2 (3.5) and 4.400 m^2 of land (Table 2.4), respectively. Since the relevant data on land requirements are not available for fish production in [1, 2], the land requirement for fish was not considered. Accordingly, the total amount of land required per unit of final demand for rice and pig is given by

$$\boldsymbol{a}_{\boldsymbol{land}}\boldsymbol{L} = \begin{pmatrix} 2.564 & 4.400 \end{pmatrix} \begin{pmatrix} 1.021 & 2.963 \\ 0.000 & 1.034 \end{pmatrix} = \begin{pmatrix} 2.619 & 12.147 \end{pmatrix} \tag{3.95}$$

where $\boldsymbol{a}_{\boldsymbol{land}} = [a_{land,j}]$ is a 1×2 matrix of land input coefficients. Similar to the GHG footprint obtained above, the total land footprint of a pig is around three times larger than its direct requirement, most of which is attributable to the use of rice to feed pigs.

3.4.3 *Waste Generation and Recycling: A Two-Sector WIO*

Properly managed pig manure is a valuable source of nutrients for crop production. Here, we extend our discussion from Sect. 3.1.3.3 to consider the integration of waste flows into the IO framework, specifically, the integration of pig manure into a two-sector IO model. Moreover, building on our previous discussion about using human waste as fertilizer in Sect. 3.1.3.3, we assume that human waste can also be used as

Table 3.6 The rice-pig system with the flow of waste

	Rice	Pig	Final demand
	$a_{.1}$	$a_{.2}$	y
Rice	0.021	2.805	y_1
Pig	0.000	0.033	y_2
Waste net generation			
	$g_{.1}$	$g_{.1}$	$g_{.y}$
Manure (dry)	−0.16[a]	0.53[b]	$0.152^c \times \sum y_i$

Source [a] The pig manure [2]
[b] Reference [55]
[c] The value taken from [38] adjusted for the water content (74.6% [34])

manure. To exemplify this integration, we present a numerical example in Table 3.6. This table integrates the flow of pig manure into Table 3.4 while including a column that refers to the final demand, which accounts for the generation of human waste. It is worth noting that producing 1kg of pig generates 0.53 kg of manure directly, and rice production requires 0.16 kg of manure per kg of rice grain.

3.4.3.1 The Balance of Products and Waste

Denoting by $\boldsymbol{g} = [g_{1j}]$ the 1×2 vector of net waste (manure) input coefficients, the balance of products and waste in this economy is represented by

$$\boldsymbol{Ax} + \boldsymbol{y} = \boldsymbol{x} \tag{3.96}$$

$$\boldsymbol{gx} + w_y = w \tag{3.97}$$

Here, w_y represents the generation of human waste and is given by $w_y = 0.152 \times \sum y_i$, while w refers to the net output of manure (in dry weight). The net amount of manure generated in this economy can be related to the final demand $\boldsymbol{y}$ as

$$\begin{aligned} w &= \boldsymbol{gx} + w_y = \boldsymbol{g}(\boldsymbol{I} - \boldsymbol{A})^{-1}\boldsymbol{y} + w_y \\ &= (-0.163 + 0.152)y_1 + (0.074 + 0.152)y_2 = -0.011y_1 + 0.226y_2 \end{aligned} \tag{3.98}$$

In a preindustrial economy, open dumping was a common way of disposing of animal and human waste, which applies to our two-sector economy. If animal and human waste were not recycled as fertilizer, this economy would end up with $0.152y_1 + 0.226y_2$ units of waste openly dumped.

3.4.3.2 A WIO with Open Dumping as a Treatment Process

The net amount of waste for disposal was obtained in two steps as follows: first, we solved (3.96) to obtain the output, which was then inserted into (3.97) in the next step to calculate $\boldsymbol{w}$. This process can be simplified by introducing a new treatment process for open dumping and associating its output (activity) with the net amount of waste (manure in this case), as shown in Table 3.7.

By redefining the net amount of waste as the amount of waste for treatment, we can integrate the flow of waste into an extended IO framework. You may wonder if it is appropriate to call open dumping a treatment process. However, since waste has mass and does not disappear on its own, it must be treated in some way if we want it to be removed, even if it is through open dumping. From a modeling perspective, even illegal dumping must be considered a process if it accounts for a significant share of the waste flow.

It is important to note that the Open Dumping (OD) column in Table 3.7 contains null elements, as neither rice nor pig is required for open dumping. The coefficient matrix in Table 3.7 is symmetric, meaning that the same list of sectors occurs in both its columns and rows. This is not the case in Table 3.6. Denoting the 3×3 matrix of input coefficients in Table 3.7 as $\boldsymbol{A}^*$, we have

$$\boldsymbol{A}^* = \begin{pmatrix} 0.021 & 2.805 & 0 \\ 0 & 0.033 & 0 \\ -0.104 & 0.53 & 0 \end{pmatrix} \tag{3.99}$$

Let x_3 be the output of the OD process (the amount of manure for open dumping). The overall balance of products and waste can then be expressed as

$$\boldsymbol{A}^*\boldsymbol{x}^* + \boldsymbol{y}^* = \boldsymbol{x}^* \tag{3.100}$$

where

$$\boldsymbol{x}^* = \begin{pmatrix} x_1 \\ x_2 \\ x_3 \end{pmatrix}, \quad \boldsymbol{y}^* = \begin{pmatrix} y_1 \\ y_2 \\ w_y \end{pmatrix} \tag{3.101}$$

Table 3.7 A WIO representation of the rice-pig economy with a waste disposal process

	Rice	Pig	Open dumping	Final demand
Rice	0.021	2.805	0	y_1
Pig	0	0.033	0	y_2
Open Dumping	−0.16	0.53	0	w_y

Source Table 3.6

This equation can be solved as

$$\begin{aligned} \boldsymbol{x}^* &= (\boldsymbol{I} - \boldsymbol{A}^*)^{-1}\boldsymbol{y}^* \\ &= \begin{pmatrix} 1.021 & 2.963 & 0 \\ 0.000 & 1.034 & 0 \\ -0.163 & 0.074 & 1 \end{pmatrix} \begin{pmatrix} y_1 \\ y_2 \\ w_y \end{pmatrix} \end{aligned} \tag{3.102}$$

This implies that the net generation of manure is

$$w_1 = -0.163y_1 + 0.074y_2 + w_y \tag{3.103}$$

It is worth noting that the $\boldsymbol{A}^*$ matrix in (3.99) contains a negative element, which violates the preconditions of Theorem 3.2. As a result, the theorem is no longer applicable, and a nonnegative solution is not guaranteed. The resulting Leontief inverse matrix also contains a negative element, indicating that for y_2, which is large relative to y_1 and w_y, w_1 can take a negative value. However, since negative w_1 is not feasible, it suggests that the relevant configuration of final demand is not feasible in this economy.

References

1. Dazhong, Wen, and David Pimentel. 1986. Seventeenth century organic agriculture in China: I. Cropping systems in Jiaxing region. *Human Ecology* 14 (1): 1–14.
2. Dazhong, Wen, and David Pimentel. 1986. Seventeenth century organic agriculture in China: II. Energy flows through an agroecosystem in Jiaxing region. *Human Ecology* 14 (1): 15–28.
3. Koopmans, Tialing C. 1951. An analysis of production as an efficient combination of activities (Chap. III). In *Activity analysis of production and allocation*, ed. Tialing C. Koopmans, 33–97. New York: Wiley.
4. Heijungs, Reinout, and Sangwon Suh. 2002. *The computational structure of life cycle assessment*, vol. 11. Springer Science & Business Media.
5. Hauschild, Michael Z., Ralph K. Rosenbaum, and Stig Irvin Olsen, eds. 2018. *Life cycle assessment theory and practice*. Springer.
6. Shephard, Ronald W. 1970. Proof of the law of diminishing returns. *Zeitschrift für Nationalökonomie Journal of Economics* 30 (1–2): 7–34.
7. Shephard, Rolf, and Ronald W. Faere. 1973. The law of diminishing returns. *Zeitschrift für Nationalökonomie Journal of Economics* 34: 69–90.
8. Calvo, Guiomar, Gavin Mudd, Alicia Valero, and Antonio Valero. 2016. Decreasing ore grades in global metallic mining: A theoretical issue or a global reality? *Resources* 5 (4): 36.
9. Tribe, M.A., and R.L.W. Alpine. 1986. Scale economies and the "0.6 rule". *Engineering Costs and Production Economics* 10 (4): 271–278.
10. Salter, W.E.G. 1960. *Productivity and technical change*. Cambridge University Press.
11. Johansen, Leif. 1972. *Production functions: An integration of micro and macro, short run and long run aspects*, vol. 75. North-Holland.
12. Benestad, Rasmus E. 2017. A mental picture of the greenhouse effect: A pedagogic explanation. *Theoretical and Applied Climatology* 128 (3–4): 679–688.
13. Hawkins, David, and Herbert Simon. 1949. Note: Some conditions of macroeconomic stability. *Econometrica* 13 (2): 556.

14. Nikaido, Fukukane. 1968. *Convex structurers and economic theory*. New York: Academic.
15. Gale, David. 1989. *The theory of linear economic models*. University of Chicago Press.
16. Lave, Lester B., Elisa Cobas-Flores, Chris T. Hendrickson, and Francis McMichael. 1995. Using Input-Output Analysis to estimate economy-wide discharges. *Environmental Science & Technology* 29 (9): 4220–426.
17. Oppenheimer, Clive. 2003. Climatic, environmental and human con sequences of the largest known historic eruption: Tambora volcano (Indonesia) 1815. *Progress in Physical Geography* 27 (2): 230–259.
18. Diamond, Jared. 2011. *Collapse: How societies choose to fail or succeed*. Revised edition. Penguin.
19. Morris, Julian. 2005. Confuse: How jared diamond fails to convince. *Energy and Environment* 16 (3–4): 395–421.
20. Gause, Emma. 2014. A critique: Jared diamond's collapse put in perspective. *Papers from the Institute of Archaeology* 24 (1): 1–7.
21. Giampietro, Mario, and David Pimentel. 1990. Assessment of the energetics of human labor. *Agriculture, Ecosystems and Environment* 32 (3–4): 257–272.
22. Ming, Xu, Eric Williams, and Braden Allenby. 2010. Assessing environmental impacts embodied in manufacturing and labor input for the China-U.S. trade. *Environmental Science and Technology* 44 (2): 567–573.
23. Rugani, Benedetto, Daryna Panasiuk, and Enrico Benetto. 2012. An input-output based framework to evaluate human labour in life cycle assessment. *The International Journal of Life Cycle Assessment* 17 (6): 795–812.
24. Murphy, Tom. 2011. MPG of a Human: https://dothemath.ucsd.edu/2011/11/mpg-of-a-human/
25. Miyazawa, Kenichi. 1968. Input-output analysis and interrelational income multiplier as a matrix. *Hitotsubashi Journal of Economics* 8 (2): 39–58.
26. Lewis, Simon L., and Mark A. Maslin. 2015. Defining the anthropocene. *Nature* 519 (7542): 171.
27. Harada, Hisatomi, Hitomi Kobayashi, and Hayato Shindo. 2007. Reduction in greenhouse gas emissions by no-tilling rice cultivation in Hachirogata polder, northern Japan: Life-cycle inventory analysis. *Soil Science and Plant Nutrition* 53 (5): 668–677.
28. Ono, Yuya, Masaharu Motoshita, and Norihiro Itsubo. 2015. Development of water footprint inventory database on Japanese goods and services distinguishing the types of water resources and the forms of water uses based on input-output analysis. *International Journal of Life Cycle Assessment* 20 (10): 1456–1467.
29. Chapagain, Ashok, and Arjen Hoekstra. 2010. The blue, green and grey water footprint of rice from both a production and consumption perspective. Technical Report 40, UNESCO-IHE.
30. De Baan, Laura, Rob Alkemade, and Thomas Koellner. 2013. Land use impacts on biodiversity in LCA: A global approach. *International Journal of Life Cycle Assessment* 18 (6): 1216–1230.
31. Bennett, Joanne M., Janette A. Steets, Jean H. Burns, Laura A. Burkle, Jana C. Vamosi, Marina Wolowski, Gerardo Arceo-Gómez, Martin Burd, Walter Durka, Allan G. Ellis, Leandro Freitas, Junmin Li, James G. Rodger, Valentin Stefan, Jing Xia, Tiffany M. Knight, and Tia Lynn Ashman. 2020. Land use and pollinator dependency drives global patterns of pollen limitation in the Anthropocene. *Nature Communications* 11 (1): 1–6.
32. Steen-Olsen, Kjartan, Jan Weinzettel, Gemma Cranston, A. Ertug Ercin, and Edgar G. Hertwich. 2012. Carbon, land, and water footprint accounts for the european union: Consumption, production, and displacements through international trade. *Environmental Science and Technology* 46 (20): 10883–10891.
33. Bruckner, Martin, Günther. Fischer, Sylvia Tramberend, and Stefan Giljum. 2015. Measuring telecouplings in the global land system: A review and comparative evaluation of land footprint accounting methods. *Ecological Economics* 114: 11–21.
34. Rose, C., Alison Parker, Bruce Jefferson, and Elise Cartmell. 2015. The characterization of feces and urine: A review of the literature to inform advanced treatment technology. *Critical reviews in environmental science and technology* 45 (17): 1827–1879.

35. Hanley, Susan B. 1987. Urban sanitation in preindustrial Japan. *The Journal of Interdisciplinary History* 18 (1): 1–26.
36. Ferguson, Dean T. 2014. Nightsoil and the 'Great Divergence': Human waste, the urban economy, and economic productivity, 1500–1900. *Journal of Global History* 9 (3): 379–402.
37. Nakamura, Shinichiro, and Yasushi Kondo. 2009. *Waste input-output analysis*. Dordrecht: Springer Science & Business Media.
38. Muñoz, Ivan, Llorenc i Canals, and Clift Roland. 2008. Consider a spherical man a simple model to include human excretion in life cycle assessment of food products. *Journal of Industrial Ecology* 12: 521–538.
39. Harold Farnswprth Gray. 1940. Sewerage in ancient and mediaeval times. *Sewage Works Journal* 12 (5): 939–946.
40. Tervahauta, Taina, Sonia Rani, Lucía Hernández Leal, Cees J. N. Buisman, and Grietje Zeeman. 2014. Black water sludge reuse in agriculture: Are heavy metals a problem? *Journal of Hazardous Materials* 274: 229–236.
41. Singh, R.P., and Manindra Agrawal. 2008. Potential benefits and risks of land application of sewage sludge. *Waste Management* 28 (2): 347–358.
42. Johansson, Kristin, Maria Perzon, Morgan Fröling, Agnes Mossakowska, and Magdalena Svanström. 2008. Sewage sludge handling with phosphorus utilization-life cycle assessment of four alternatives. *Journal of Cleaner Production* 16 (1): 135–151.
43. Donatello, Shane, and Christopher R. Cheeseman. 2013. Recycling and recovery routes for incinerated sewage sludge ash (ISSA): A review. *Waste Management* 33 (11): 2328–2340.
44. Aleisa, Esra, Abdalrahman Alsulaili, and Yasmeen Almuzaini. 2021. Recirculating treated sewage sludge for agricultural use: Life cycle assessment for a circular economy. *Waste Management* 135: 79–89.
45. Nakamura, Shinichiro, and Yasushi Kondo. 2002. Input-output analysis of waste management. *Journal of Industrial Ecology* 6 (1): 39–63.
46. Yoo, Daekyum, Muhammad Mahboob Ali Hamid, Hanbeen Kim, Joonbeom Moon, Jaeyong Song, Seyoung Lee, and Jakyeom Seo. 2020. Substitution effects of rice for corn grain in total mixed ration on rumen fermentation characteristics and microbial community in vitro. *Journal of Animal Science and Technology* 62 (5): 638–647.
47. Scheibler, R.B., J. Schafhäuser, F.A. Rizzo, J.L. Nörnberg, D.P. Vargas, J.L.S. Silva, A.C. Fluck, and V.I. Fioreze. 2015. Replacement of corn grain by brown rice grain in dairy cow rations: Nutritional and productive effects. *Animal Feed Science and Technology* 208: 214–219.
48. Roll, Aline Arassiana Piccini, Edenilse Gopinger, Martha Lopes Schuch De Castro, Jorge Schafhäuser Junior, Victor Fernando Büttow Roll, and Fernando Rutz. 2017. Brown rice, selenium yeast and alpha-tocopherol acetate in chicken's diet: Effects on meat quality. *Semina: Ciencias Agrarias* 38 (2): 957–968.
49. Ito, Y. 1941. The dried sardine and dregs market of Edo. *Mita Journal of Economics* 35 (11): 1362–1380.
50. Solow, Robert. 1952. On the structure of linear models. *Econometrica* 20 (1): 29–46.
51. Astudillo, Miguel F., Gunnar Thalwitz, and Fritz Vollrath. 2015. Modern analysis of an ancient integrated farming arrangement: Life cycle assessment of a mulberry dyke and pond system. *International Journal of Life Cycle Assessment* 20 (10): 1387–1398.
52. Okishio, Nobuo. 1963. A mathematical note on marxian theorems. *Weltwirtschaftliches Archiv* 91: 287–299.
53. WBCSD and WRI. 2012. The Greenhouse Gas Protocol A Corporate Accounting and Reporting Standard. Technical report, WBCSD WRI.
54. MacLean, Heather L., and Lester B. Lave. 2003. Life cycle assessment of automobile/fuel options. *Environmental Science & Technology* 37: 5445–5452.
55. Nguyen, Thu Lan T., John E. Hermansen, and Lisbeth Mogensen. 2011. Environmental Assessment of Danish Pork.

Chapter 4
Standard Input-Output: Single and Multi-regional Models

Abstract This chapter shifts focus from hypothetical IO tables/models to actual IO tables used by statistical offices worldwide. These tables involve up to 600 sectors and are measured in monetary units. Defining sectors and products in these tables is complex due to the presence of by-products, and secondary products. The commodity-by-industry approach is commonly employed to handle this complexity. Measurement in monetary units requires selecting appropriate prices and considering transport and trade services. Regional extensions are necessary to account for the interconnectedness of countries and regions, including the use of global multi-regional IO (MRIO) models and tables. The chapter also explores the embodiment of processes in capital stock and capital goods, discussing the construction phase and associated flows. Finally, the chapter concludes with a brief discussion of the cost and price model and the impacts of changes in import prices. Overall, this chapter provides a comprehensive understanding of actual IO tables, their complexities, and their practical applications.

4.1 From Processes in Physical Units to IO Tables in Monetary Units

The original concept of the Input-Output (IO) model by Leontief was initially based on physical units [1, 2]. However, the currently available IO tables are predominantly compiled in monetary units. This shift can be attributed to two key factors. Firstly, IO tables have become an integral part of national accounts, which are expressed in monetary terms. These accounts aim to capture all economic activities within a specific territorial boundary, and thus, it is natural for IO tables to align with the monetary framework. Secondly, the limited resolution of sector classification poses a challenge. There are numerous similar but distinct products within a sector that need to be aggregated for practical purposes. In such cases, monetary value serves as the only feasible unit of measurement to facilitate meaningful aggregation. The adoption of monetary units in IO tables allows for consistent integration with

S. Nakamura, *A Practical Guide to Industrial Ecology by Input-Output Analysis*,
https://doi.org/10.1007/978-3-031-43684-0_4

national accounts and enables the consolidation of similar products within sectors. This transition enhances the practicality and applicability of IO analysis in real-world economic contexts.

However, in environmental applications, IO tables in physical units are generally more preferable, as environmental phenomena are inherently physical rather than monetary in nature. It is important to acknowledge that statistical offices developing IO tables typically prioritize economic and monetary aspects rather than environmental considerations. While there are exceptions to this general trend, such as the Dutch NAMEA (National Accounting Matrix including Environmental Accounts) [3] and the Japanese EEIO (Environmentally Extended Input-Output) tables [4], these cases are not widespread. Environmental applications of IO analysis require additional considerations and specialized frameworks beyond the scope of traditional IO tables, which we discuss in detail in Chap. 5.

IO tables in monetary units can be considered equivalent to physical data if the price of a product is the same across all its users [5]. We discuss this equivalence in Sect. 4.2.3 below. The likelihood of maintaining price equivalence increases with higher sector resolution in an IO table. The highest level of resolution typically found in IO tables is around 400 production sectors. For example, the US has IO tables with 405 industry sectors [6], Japan has tables with 391 column sectors and 509 row sectors [7], and Korea has tables with 381 sectors [8]. However, the majority of IO tables have less than 150 production sectors, with China having 149 sectors, Australia having 114 sectors, the UK having 105 sectors, and the EU-MRIO tables having 64 sectors. Reference [9] gives an extensive list of national IO tables as of 2010. Even at the highest resolution level, it is crucial to understand that a "sector" in an IO table still represents an aggregation of multiple underlying processes. Similarly, the "output" of a sector is an aggregation of various products resulting from those processes. Addressing these issues of aggregation constitutes the first topic that will be discussed.

4.1.1 Industry and Its Product

This section mostly refers to [10]. In an IO table, a sector is defined based on an industry and its product. We use the term "industry" to refer to a production sector consisting of processes producing the same or similar products. The term "production sector" is used to distinguish it from the "final demand sector". The production activity is divided into three activities, and the resulting products into four categories.

4.1.1.1 Principal, Secondary, and Ancillary Activities and Their Products

An entity's (a production unit's) activity and product can be classified as follows:

1. *Principal activity and product* This refers to the activity that contributes the most to the entity's value-added, resulting in principal products and by-products. A principal product is one that contributes the most to the value-added of the entity.
2. *By-products* These are products that are necessarily produced together with principal products for technological reasons. Some common examples of by-products include feathers from poultry processing, sawdust and bark from the processing of logs into lumber, gypsum from flue gas desulfurization, fly ash from the combustion of coal, sludge from wastewater treatments, electric power from waste incineration plants, as well as gold and silver from copper- and lead electrolysis.
3. *Secondary activity and products* A secondary activity is the production activity of the entity that results in products for third parties and is not its principal product. The principal product of a secondary activity is a secondary product. Different from by-products, a secondary activity does not have to occur with the principal activity. Some common examples of secondary products include motorcycles made by auto manufacturers, a cake made by a bakery, hotels run by airlines, and extension schools run by universities.
4. *Ancillary activity and products* Ancillary activity refers to miscellaneous supporting and managing activities within an entity, such as bookkeeping, storage, purchasing, and sales promotion, indispensable for its principal and secondary production activities but not for third parties.

4.1.1.2 Industry and Its Classification

Industries are categorized based on their principal products. The degree to which the production processes and resulting principal products within a particular industry are considered "similar" depends on the level of resolution of the relevant sector classification.

4.1.1.3 Secondary Activities and Products in IO Tables

The data used to construct IO tables is typically collected through surveys or censuses of engaged in production activities within a region of interest. However, for establishments that undertake multiple production activities, financial data may only be available for the establishment as a whole and not for individual activities, even if they are technologically independent and separable [10]. As a result, actual IO tables are likely to include secondary activities within industries, even if they are technically separable.

Therefore, applying the rule of mechanically classifying establishments into an industry may result in the total output of an industry being recorded as the sum of all establishments assigned to that industry, including both their principal and secondary products ([11], p. 159). However, the presence of significant secondary products in an economy of interest can lead to misleading results unless certain measures are taken to address this issue.

4.1.2 The Commodity-by-Industry Approach

Since establishments are the basic unit of industry, the number of principal products (m) is typically greater than the number of industries (n). This results in a rectangular IO flow matrix with columns referring to industries, which are smaller than its rows referring to principal products. Even when the IO matrix is square with $m = n$, which is commonly assumed in the literature [11, 12], the presence of secondary products necessitates a reformulation of the original IO model.

A widely used approach to this reformulation is the commodity-by-industry approach, a brief outline of which is given below.

4.1.2.1 The Use and Make Matrices: $U(B)$ and $V(D)$

Henceforth, the term commodities is used synonymously for products. Let there be m products and n industries. Denote by q_i the output of product i, by x_j the output of industry j, and by y_i the final demand for product i. The products are measured in monetary units. Note that the final demand $\boldsymbol{y} = [y_i]$ is defined for $\boldsymbol{q} = [q_i]$ but not for $\boldsymbol{x} = [x_j]$.

The Use Matrix $\boldsymbol{U}$

Denote the use matrix by an $m \times n$ matrix $\boldsymbol{U} = [u_{ij}]$, with u_{ij} referring to the input of product i into industry j. The supply and demand balance of products is given by

$$\boldsymbol{q} = \boldsymbol{U}\boldsymbol{\iota}_n + \boldsymbol{y} \tag{4.1}$$

Defining $\boldsymbol{B} = [b_{ij}]$ as

$$\boldsymbol{B} = \boldsymbol{U}\hat{\boldsymbol{x}}^{-1}. \tag{4.2}$$

Equation (4.1) becomes

$$\boldsymbol{q} = \boldsymbol{U}\boldsymbol{\iota}_n + \boldsymbol{y} = \boldsymbol{B}\boldsymbol{x} + \boldsymbol{y} \tag{4.3}$$

The Make Matrix V

Denote the make matrix by an $n \times m$ matrix, $\boldsymbol{V} = [v_{ij}]$, with v_{ij} referring to the amount of product j produced by industry i. By definition

$$\begin{aligned} \boldsymbol{x} &= \boldsymbol{V}\boldsymbol{\iota}_m \\ \boldsymbol{q} &= \boldsymbol{V}^\top \boldsymbol{\iota}_n \end{aligned} \tag{4.4}$$

If there are no secondary products, $\boldsymbol{V}$ reduces to a diagonal matrix when $m = n$. Define an $n \times m$ matrix $\boldsymbol{D} = [d_{ij}]$ as

$$\boldsymbol{D} = \boldsymbol{V}\hat{\boldsymbol{q}}^{-1} \tag{4.5}$$

with $d_{ij} = v_{ij}/q_j$ giving the fraction of product j produced by industry i. By definition

$$\boldsymbol{\iota}_n^\top \boldsymbol{D} = \boldsymbol{\iota}_n^\top. \tag{4.6}$$

From (4.4), we have

$$\boldsymbol{D}\boldsymbol{q} = \boldsymbol{V}\boldsymbol{\iota}_m = \boldsymbol{x} \tag{4.7}$$

The Product Mix Matrix C

Finally, define the product mix matrix by an $m \times n$ matrix $\boldsymbol{C} = [c_{ij}]$ as

$$\boldsymbol{C} = \boldsymbol{V}^\top \mathbf{diag}(\boldsymbol{x})^{-1} = \boldsymbol{V}^\top \mathbf{diag}(\boldsymbol{V}\boldsymbol{\iota}_m)^{-1} \tag{4.8}$$

with $c_{ij} = v_{ij}/x_i$ giving the fraction of commodity i in the output of industry j. Each column of $\boldsymbol{C}$ gives the product mix of the output of the relevant industry.

4.1.2.2 Industry-Based Technology

We now consider how the original IO model can be modified to account for secondary products. Note that if there were no secondary products, the $\boldsymbol{B}$ matrix reduces to the $\boldsymbol{A}$ matrix of the original IO model. There are two major approaches distinguished by how the $\boldsymbol{B}$ matrix is weighted to accommodate secondary products. We start with one called industry-based technology, which uses $\boldsymbol{D}$ as the weights. Starting from (4.3) and (4.7), we have

$$\boldsymbol{q} = \boldsymbol{B}\boldsymbol{D}\boldsymbol{q} + \boldsymbol{y} \tag{4.9}$$

If matrix $(\boldsymbol{I} - \boldsymbol{B}\boldsymbol{D})$ is nonsingular, this equation can be solved as

$$\boldsymbol{q} = (\boldsymbol{I} - \boldsymbol{B}\boldsymbol{D})^{-1}\boldsymbol{y} \tag{4.10}$$

The matrix $(\boldsymbol{I} - \boldsymbol{BD})^{-1}$ is called a commodity-by-commodity total requirements matrix and connects the final demand for a product to product output. This matrix plays the same role as $(\boldsymbol{I} - \boldsymbol{A})^{-1}$ in the original IO model, with $\boldsymbol{BD}$ playing the role of $\boldsymbol{A}$. The assumption behind this approach is that all commodities produced in an industry have the same input structure, given by that industry's column in the $\boldsymbol{B}$ matrix ([11], p. 192). Therefore, this approach is termed industry-based technology. Please refer to Sect. 4.1.2.4 for an illustration.

4.1.2.3 Commodity-Based Technology

An alternative approach to the above is based on $\boldsymbol{C}$ and is derived as follows. From (4.8) and (4.4), we have

$$\boldsymbol{C}\,\mathrm{diag}(\boldsymbol{x})\boldsymbol{\iota}_n = \boldsymbol{C}\boldsymbol{x} = \boldsymbol{V}^\top \boldsymbol{\iota}_n = \boldsymbol{q} \tag{4.11}$$

If $\boldsymbol{C}$ is square and nonsingular, we have

$$\boldsymbol{x} = \boldsymbol{C}^{-1}\boldsymbol{q} \tag{4.12}$$

Substitution into (4.3) gives

$$\boldsymbol{q} = \boldsymbol{B}\boldsymbol{C}^{-1}\boldsymbol{q} + \boldsymbol{y} \tag{4.13}$$

with solution (provided it exists)

$$\boldsymbol{q} = (\boldsymbol{I} - \boldsymbol{B}\boldsymbol{C}^{-1})^{-1}\boldsymbol{y} \tag{4.14}$$

This approach assumes that a commodity has the same input structure (coefficients) in all industries that produce it, which is the opposite of the industry-based technology approach. Please see Sect. 4.1.2.4 for an illustration. For this feature, this approach is called commodity-based technology. For a product whose input structure varies little with the location/surroundings, this approach is more suitable than the industry-based approach.

However, a caveat to this approach is that the inverse matrix $\boldsymbol{C}^{-1}$ can contain negative elements, which can result in $\boldsymbol{B}\boldsymbol{C}^{-1}$ having negative elements. This is undesirable when $\boldsymbol{B}\boldsymbol{C}^{-1}$ is considered the counterpart to the $\boldsymbol{A}$ matrix in the original IO model. Furthermore, an input-coefficients matrix ($\boldsymbol{B}\boldsymbol{C}^{-1}$ in the current case) with negative elements can make Theorem 3.2 inapplicable, and the positiveness of the solution cannot be guaranteed. Note that even if the HS is satisfied, the positiveness of $\boldsymbol{C}^{-1}$ is not guaranteed because the condition under which the theorem applies is not satisfied. Another weakness of this approach is that it requires the stringent condition $m = n$. On the other hand, the industry-based approach does not require the matrix $\boldsymbol{D}$ to be square and nonsingular, even when it happens to be square.

Table 4.1 A numerical example of products-industry data

		Products				Industry				y	Output
		Car	Bike	Steel	Al	Car	Bike	Steel	Al		
Products	Car					0	0	0	0	100	100
	Bike					0	0	0	0	20	20
	Steel					60	2	20	0	8	90
	Al					10	8	0	2	2	22
Industry	Car	100	10	0	0						110
	Bike	0	10	0	0						10
	Steel	0	0	90	0						90
	Al	0	0	0	22						22

Note Bike refers to a motorbike. Al refers to aluminum. The numbers are purely hypothetical, expressed in a hypothetical monetary unit

4.1.2.4 A Numerical Example of the Two Approaches

We illustrate the two approaches to accommodating secondary products by using a numerical example. In this example, we consider the car industry, which produces both cars and motorbikes, such as BMW, Honda, and Suzuki, using steel and aluminum as inputs (see Table 4.1).

For this data, the matrices $\boldsymbol{B}, \boldsymbol{BD}, \boldsymbol{C}^{-1}$, and $\boldsymbol{BC}^{-1}$ are respectively given by

$$\boldsymbol{B} = \boldsymbol{U}\,\mathbf{diag}(\boldsymbol{x})^{-1} = \begin{pmatrix} 0 & 0 & 0 & 0 \\ 0 & 0 & 0 & 0 \\ 60 & 2 & 20 & 0 \\ 10 & 8 & 0 & 2 \end{pmatrix} \begin{pmatrix} 110 & 0 & 0 & 0 \\ 0 & 10 & 0 & 0 \\ 0 & 0 & 90 & 0 \\ 0 & 0 & 0 & 22 \end{pmatrix}^{-1}$$

$$= \begin{pmatrix} 0.000 & 0.000 & 0.000 & 0.000 \\ 0.000 & 0.000 & 0.000 & 0.000 \\ 0.545 & 0.200 & 0.222 & 0.000 \\ 0.091 & 0.800 & 0.000 & 0.091 \end{pmatrix} \tag{4.15}$$

$$\boldsymbol{D} = \boldsymbol{V}\,\mathbf{diag}(\boldsymbol{q})^{-1} = \begin{pmatrix} 100 & 10 & 0 & 0 \\ 0 & 10 & 0 & 0 \\ 0 & 0 & 90 & 0 \\ 0 & 0 & 0 & 22 \end{pmatrix} \begin{pmatrix} 100 & 0 & 0 & 0 \\ 0 & 20 & 0 & 0 \\ 0 & 0 & 90 & 0 \\ 0 & 0 & 0 & 22 \end{pmatrix}^{-1}$$

$$= \begin{pmatrix} 1.00 & 0.50 & 0.00 & 0.00 \\ 0.00 & 0.50 & 0.00 & 0.00 \\ 0.00 & 0.00 & 1.00 & 0.00 \\ 0.00 & 0.00 & 0.00 & 1.00 \end{pmatrix} \tag{4.16}$$

$$\boldsymbol{BD} = \begin{pmatrix} 0.000 & 0.000 & 0.000 & 0.000 \\ 0.000 & 0.000 & 0.000 & 0.000 \\ 0.545 & 0.373 & 0.222 & 0.000 \\ 0.091 & 0.445 & 0.000 & 0.091 \end{pmatrix} \tag{4.17}$$

$$\boldsymbol{C}^{-1} = \begin{pmatrix} 0.909 & 0.000 & 0.000 & 0.000 \\ 0.091 & 1.000 & 0.000 & 0.000 \\ 0.000 & 0.000 & 1.000 & 0.000 \\ 0.000 & 0.000 & 0.000 & 1.000 \end{pmatrix}^{-1} = \begin{pmatrix} 1.10 & 0.00 & 0.00 & 0.00 \\ -0.10 & 1.00 & 0.00 & 0.00 \\ 0.00 & 0.00 & 1.00 & 0.00 \\ 0.00 & 0.00 & 0.00 & 1.00 \end{pmatrix} \tag{4.18}$$

$$\boldsymbol{BC}^{-1} = \begin{pmatrix} 0.000 & 0.000 & 0.000 & 0.000 \\ 0.000 & 0.000 & 0.000 & 0.000 \\ 0.580 & 0.200 & 0.222 & 0.000 \\ 0.020 & 0.800 & 0.000 & 0.091 \end{pmatrix} \tag{4.19}$$

Note that $\boldsymbol{C}^{-1}$ contains negative elements.

We now compare the $\boldsymbol{A}$ equivalents of car and motorbike sectors represented by $\boldsymbol{B}$, $\boldsymbol{BD}$, and $\boldsymbol{BC}^{-1}$.

Industry-Based Approach

In the industry-based approach, $\boldsymbol{BD}$, the input structure of the car industry (the first column) remains the same as in $\boldsymbol{B}$, while the input structure of the bike industry (the second column) becomes the weighted average of the first column and the second column of $\boldsymbol{B}$ with weights based on the share of motorbike production.

$$\begin{pmatrix} 0.000 \\ 0.000 \\ 0.373 \\ 0.445 \end{pmatrix} = 0.5 \times \begin{pmatrix} 0.000 \\ 0.000 \\ 0.545 \\ 0.091 \end{pmatrix} + 0.5 \times \begin{pmatrix} 0.000 \\ 0.000 \\ 0.200 \\ 0.800 \end{pmatrix} \tag{4.20}$$

As the car industry produces motorbikes differently from the motorbike industry (using more steel and less aluminum), the input structure of the motorbike industry needs to be adjusted to reflect the fact that 50% of motorbikes are produced using the method employed by the car industry, while the remaining 50% is produced using the method specific to the original motorbike industry.

Commodity-Based Approach

In the industry-based approach, $\boldsymbol{BC}^{-1}$, the input structure of the motorbike industry (the second column) remains the same as in $\boldsymbol{B}$. However, for the car industry (the first column in $\boldsymbol{B}$), the input structure needs to be adjusted since it is a weighted average of the input structures specific to cars and motorbikes. The weights are based on the product mix, $\boldsymbol{C}$. The first column in $\boldsymbol{BC}^{-1}$ gives the input structure of cars isolated from the mixture with the input structure of motorbikes. This is demonstrated by observing that the weighted average of the first and second columns of $\boldsymbol{BC}^{-1}$ with the weights based on the product mix of the car industry (90.9% cars and 9.1% motorbikes) reproduces the "mixed" input structure in the first column of $\boldsymbol{B}$

$$0.909 \times \begin{pmatrix} 0.000 \\ 0.000 \\ 0.580 \\ 0.020 \end{pmatrix} + 0.091 \times \begin{pmatrix} 0.000 \\ 0.000 \\ 0.200 \\ 0.800 \end{pmatrix} = \begin{pmatrix} 0.000 \\ 0.000 \\ 0.545 \\ 0.091 \end{pmatrix} \tag{4.21}$$

4.1.2.5 Which Approach to Use?

No Consensus on the Preferred Method

A vast body of literature explores the advantages and drawbacks of various models, including many variations of the basic ones mentioned above, in a commodity-industry framework . Despite this extensive research, there is no consensus on the preferred method for choosing between an industry-based and a commodity-based approach [11–13].

The significance of secondary activity depends on the level of sector resolution. At a low level of sector resolution, a sector represents an aggregate of processes with distinct principal products that contribute varying shares to total output value. In this case, the principal products of processes with small shares in the total output are classified as secondary products of the aggregated sector. However, when the level of sector resolution is increased, the importance of secondary production diminishes because the original flows of distinct principal products are restored. For example, in the WIOD database of multi-regional IO (MRIO, Sect. 4.3.4) with 56 industry sectors [14], electronic, electrical and optical products are combined into a single sector called Electrical and Optical Equipment, whereas in the more granular EXIOBASE3 with 163 industry sectors [15], they are classified into three separate sectors.

Activity-Based Approach: Japanese IO Tables

The Japanese IO tables are widely recognized for having one of the world's most finely detailed sector classifications. The first IO table for Japan was created in 1951 and contained 200 production sectors, followed by a 1955 table with over 300 producing sectors. The distinguishing feature of the Japanese IO tables was that their sector classification was based on activities rather than establishments, which was possible due to the abundant availability of regularly compiled primary data [16–18].

Although the Japanese IO tables have been developed using the commodity × activity approach, the growing prominence of service sectors, globalization, and reduced government funding for data development may undermine this "unique advantage" [18].

One potential solution for developing activity-based IO tables is to utilize LCA inventory data. By integrating detailed process-based data into the IO framework, it is possible to obtain high-resolution IO data and address the issue of secondary products simultaneously.

How Do We Proceed?

To facilitate a practical approach, we will assume that the number of producing sectors is equal to the number of principal products, with each sector exclusively generating a single principal product. We will also assume that the sector resolution is sufficiently high. In essence, we will begin with a given symmetric IO table, without delving too deeply into the specifics of how the table was compiled.

4.1.3 Transport and Trade

In any economy that goes beyond barter exchange and extends beyond a small geographic area where transport is not an issue, the flow of inputs and outputs depends on trade and transport services. Consumers purchase products in retail markets, while producers obtain inputs from wholesale markets. The actual flow of inputs and outputs involves the sectors providing trade and transport services. However, including these transactions directly in an IO table results in a table where all products are first delivered to trade and transport sectors before being distributed to individual users ([19], p. 44). Such a representation undermines the very feature of an IO table that tries to capture technological, not commercial, interdependence among producing sectors.

To address this issue, most IO tables adopt a convention where the flows of products are recorded as if they were directly delivered from and distributed to the sectors where they are produced and used. The transport services that deliver inputs to a sector are summed and recorded as the input entry for "transportation" in that column. Similarly, all trade services on inputs to a sector are summed and recorded as the input entry for "trade" in that column. The trade and transportation sectors are not treated as producing and consuming sectors in the economy but only as "pass-through sectors" (see [11], p. 146 and [19], p. 44, for further details).

The quantity of transport services can be represented physically, for instance, in kgkm, enabling their sum to be recorded as the input entry for transport. However, measuring the physical amount of trade services poses a challenge. Therefore, a measure based on the value term is used to record the input entry for trade. The convention is to use the value of trade margins as a measure of trade services.

4.1.4 By-Products and Waste

In Sect. 4.1.1.1, the definition of by-products was provided, but for analytical purposes, it is necessary to further classify by-products into two categories: competitive by-products and noncompetitive by-products.

4.1.4.1 Competitive By-Products and Noncompetitive By-Products

A by-product is classified as a competitive by-product when it is produced as the primary product of a sector. On the other hand, a by-product is classified as a non-competitive by-product when no sector specifically produces it as its primary product.[1] Examples of competitive by-products include silver produced in the smelting process of copper or lead, hydrogen produced in the production process of sodium hydroxide, and electricity generated in a waste incineration plant with heat recovery. On the other hand, examples of noncompetitive by-products include slag and sludge from metal smelting processes, ash and flyash from a waste incineration plant, and animal residues from slaughtering and meat processing facilities.

However, it is important to note that the distinction between competitive and noncompetitive by-products is not absolute and may depend on technological, economic, and social conditions. For instance, until the invention of the automobile, crude oil was distilled to produce kerosene for lighting, with gasoline, a by-product of kerosene, discarded because there was no use for it [21].

Moreover, according to the Basel Convention on the Control of Transboundary Movements of Hazardous Wastes and Their Disposal, wastes are substances or objects that are disposed of or intended to be disposed of by the provisions of national law. This definition of waste does not exclude the possibility of its reuse and recycling, and thus the distinction between noncompetitive by-products and wastes may fluctuate depending on market conditions. This distinction is discussed in more detail in Chap. 5, while this section focuses on competitive by-products.

4.1.4.2 The IO Model with Competitive By-Products

In Sect. 3.3, we obtained the matrix $\boldsymbol{I} - \boldsymbol{A}$ as an intermediate result while solving the balancing Eq. (3.58). Despite its importance, the matrix $\boldsymbol{I} - \boldsymbol{A}$ has received less attention in the IO literature compared with its inverse $\boldsymbol{L}$ and does not have a specific name. However, in LCA, the matrix $\boldsymbol{I} - \boldsymbol{A}$ is referred to as the "technology matrix" and plays a crucial role in representing technology. Its positive elements represent outputs, while negative elements represent inputs [22, 23]. While $\boldsymbol{A}$ provides information about inputs only, $\boldsymbol{I} - \boldsymbol{A}$ provides richer information about both inputs and outputs. To avoid confusion, we will refer to $\boldsymbol{I} - \boldsymbol{A}$ and its variant as the "technology matrix" from now on, while we refer to $\boldsymbol{A}$ as the "input-coefficients matrix".

In $\boldsymbol{I} - \boldsymbol{A}$, the identity matrix $\boldsymbol{I}$ signifies that each sector produces one unit of its principal product (all its diagonal elements are unity) and no competitive by-products (all its off-diagonal elements are zero). Competitive by-products can be represented by substituting $\boldsymbol{I}$ with a general output matrix $\boldsymbol{D}$, where $d_{ij} > 0, i \neq j$, denotes the generation of product i as a by-product per unit of producing j. $\boldsymbol{D}$ is a square matrix because, for each competitive by-product, there is a column in $\boldsymbol{D}$ referring to the

[1] We find this labeling is more informative than "type I" and "type II" used in [19]. In [20], the former is called "exclusive by-products" and the latter "ordinary by-products".

sector where the by-product occurs as the principal product. Replacing $\boldsymbol{I}$ with $\boldsymbol{D}$, we obtain the technology matrix with competing by-products as $\boldsymbol{D} - \boldsymbol{A}$. This method of representing technology was used by von Neumann's seminal article [24] (originally published in German in 1938) before the advent of LCA.

With the matrix $\boldsymbol{I} - \boldsymbol{A}$ replaced by this generalized one, the Leontief quantity model that allows for the presence of by-product can be given by

$$\boldsymbol{x} = (\boldsymbol{D} - \boldsymbol{A})^{-1}\boldsymbol{y}. \tag{4.22}$$

Alternatively, we can use the negative input method of Stone to introduce by-products into the input-coefficients matrix. This involves defining the matrix $\breve{\boldsymbol{A}}$ as

$$\breve{\boldsymbol{A}} = \boldsymbol{A} - \boldsymbol{D} + \boldsymbol{I}, \tag{4.23}$$

with the Leontief quantity model with by-products represented as

$$\boldsymbol{x} = (\boldsymbol{I} - \breve{\boldsymbol{A}})^{-1}\boldsymbol{y}. \tag{4.24}$$

In this method, an increase in the final demand for the principal product of a sector with a competitive by-product increases the supply of the by-product at the expense of the sector that produces it as the principal product. This method was introduced by Stone [25] and is named after him.

4.1.4.3 A Numerical Example

For illustration, we consider an example where $\boldsymbol{A}$ is given by

$$\boldsymbol{A} = \begin{pmatrix} 0.0 & 0.0 & 0.15 \\ 0.4 & 0.0 & 0.3 \\ 0.1 & 0.1 & 0.0 \end{pmatrix}, \tag{4.25}$$

with

$$\boldsymbol{I} - \boldsymbol{A} = \begin{pmatrix} 1 & 0.0 & -0.15 \\ -0.4 & 1 & -0.3 \\ -0.1 & -0.1 & 1 \end{pmatrix}, \tag{4.26}$$

and the Leontief inverse matrix given by

$$(\boldsymbol{I} - \boldsymbol{A})^{-1} = \begin{pmatrix} 1.02 & 0.01 & 0.15 \\ 0.45 & 1.03 & 0.37 \\ 0.14 & 0.10 & 1.05 \end{pmatrix}. \tag{4.27}$$

Suppose now that 0.3 units of product 1 are obtained as a by-product per unit of production of product 2:

$$D = \begin{pmatrix} 1 & 0.3 & 0 \\ 0 & 1 & 0 \\ 0 & 0 & 1 \end{pmatrix} \tag{4.28}$$

From (4.22) and (4.24) the input-output coefficients matrix then becomes

$$D - A = I - \breve{A} = \begin{pmatrix} 1 & 0.3 & -0.15 \\ -0.4 & 1 & -0.3 \\ -0.1 & -0.1 & 1 \end{pmatrix}, \tag{4.29}$$

with the associated Leontief inverse matrix given by

$$(I - \breve{A})^{-1} = \begin{pmatrix} 0.89 & -0.26 & 0.05 \\ 0.39 & 0.91 & 0.33 \\ 0.12 & 0.06 & 1.03 \end{pmatrix} \tag{4.30}$$

While the off-diagonal elements of $I - A$ are nonpositive, this is no longer the case for its counterpart with by-products $I - \breve{A}$. This leads to a significant difference in the Leontief inverse matrix, where all elements of $(I - A)^{-1}$ are nonnegative, but $(I - \breve{A})^{-1}$ contains a negative element. Specifically, the final delivery of a unit of product 2 generates 0.26 units of product 1 as a by-product and reduces its production in Sector 1 by the same amount. This situation may occur, for example, if Sector 1 is a coal-fired power sector and Sector 2 is a waste incineration sector, and a portion of power from coal is replaced by energy recovery from waste incineration. Such a change could reduce CO_2 emissions into the atmosphere compared with the case where waste is incinerated without energy recovery.

4.1.4.4 Implications for Productive Conditions

While the above procedure represents a simple way to introduce competitive by-products into the IO model, this generalization can lead to problems. Specifically, the Leontief inverse matrix $(I - \breve{A})^{-1}$ now contains negative elements, rendering Theorem 3.2 in Sect. 3.3 inapplicable. Consequently, the nonnegativity of the solution (4.22) (the productiveness of $\breve{A}$) can no longer be guaranteed. For instance, in the example above, if the final demand for product 2 is five times larger than that for product 1, the output of product 1 can become negative:

$$\begin{pmatrix} x_1 \\ x_2 \\ x_3 \end{pmatrix} = \begin{pmatrix} 0.89 & -0.26 & 0.05 \\ 0.39 & 0.91 & 0.33 \\ 0.12 & 0.06 & 1.03 \end{pmatrix} \begin{pmatrix} 1 \\ 5 \\ 4 \end{pmatrix} = \begin{pmatrix} -0.20 \\ 6.30 \\ 4.61 \end{pmatrix}. \tag{4.31}$$

However, negative output is not possible in reality and thus represents a weakness of the model. In the model (4.22), negative output occurs when there is an excess supply of by-product over the demand for it. This is because in the model, the by-product supply is driven not by its demand but by the demand for the primary product alone. In reality, excess supply of competing by-products is typically subject to waste management through an adjustment mechanism, which is not considered in the present model. Therefore, the weakness of the model is not the occurrence of a nonzero off-diagonal element in $\boldsymbol{D}$, i.e., the Stone method, but the lack of consideration for the adjustment process of waste management. This point will be discussed in greater detail in Sect. 5.4.

4.1.4.5 Pollution Abatement and the Closure of Sulfur Mines in Japan

As we saw above, the output of a sector whose principal product competes with a competitive by-product can become negative when the supply of the competitive by-product exceeds the demand for the product. While, in reality, the output cannot become negative, the excess supply of a by-product can cause the sector whose principal product faces competition with the by-product to be destroyed.

A good example is the sulfur mining industry in Japan. Japan is rich in sulfur resources because of its many active volcanoes. Japanese sulfur mines were the largest in Asia from the beginning of the twentieth century to the end of the 1960s. With the tightening of regulation on the emission of SO_2 from petroleum refining, metal smelting, and power generation industries in the 1960s, these industries started installing desulfurization facilities, with sulfur generated as a by-product in forms such as sulfuric acids or gypsum (see Sect. 2.2.4.3). Since the sulfur mine could not compete with the by-product-based sulfur, which was available at prices significantly lower than those of the mine, all the once flourishing sulfur mines were closed by the beginning of the 1970s.

A remarkable example of this "industrial drama" is the Matsuo mine located in Northern Japan, which once accounted for 30% of Japan's sulfur production and employed around 14,000 workers. The mine facilities were closed in 1972, leaving behind the once-flourishing mining town as a ghost town. The mining facilities were replaced with a water treatment facility to neutralize a large amount of strongly acidic water flowing out of the mine into the local river (Fig. 4.1).

4.2 Physical Versus Monetary Units

Once an IO table is assembled based on monetary units, the next step is to determine which price to use for evaluating the input-output flow. There are three options for valuing the input-output flow: producer prices, purchaser prices, and basic prices ([26], p. 92).

Fig. 4.1 The present state of the former Matsuo sulfur mine, Iwate, Japan. The foreground is the mine wastewater treatment facility, with the ruins of apartment blocks for the mine workers in the background. Courtesy of JOGMEC

4.2.1 Basic, Producer and Consumer Prices

The producer price refers to the amount receivable by the producer from the purchaser for a unit of goods or services produced as output, minus any value-added tax (VAT) or similar deductible tax invoiced to the purchaser, and it excludes any transport charges invoiced separately by the producer.

The purchaser price is the amount paid by the purchaser, excluding any deductible VAT or similar deductible tax, to take delivery of a unit of a good or service at the time and place required by the purchaser. The purchaser's price of a good includes any transport charges paid separately by the purchaser to take delivery at the required time and place.

The basic price is the amount receivable by the producer from the purchaser for a unit of goods or services produced as output, minus any tax payable and plus any subsidy receivable, on that unit as a consequence of its production or sale, and it excludes any transport charges invoiced separately by the producer.

Although the basic price is considered the best option from a theoretical point of view, as it reflects more accurately than other price concepts the costs of all elements inherent in the product [26], the valuation at producer's prices is often used, including in the Japanese IO table ([11], p. 144). A simpler definition of producer's prices, sometimes referred to as free-on-board prices, is the prices at which the seller completes the transaction, while purchaser's prices include trade and transportation margins and often excise taxes ([11], p. 144).

For simplicity, we assume that an IO table in monetary units is based on either producer's prices or basic prices, without going into the details of the adopted valuation option. Therefore, we use the term "producer's prices" synonymously with "basic prices" in this context.

4.2.2 *Transport- and Trade Services in Value Terms*

In Sect. 4.1.3, we discussed how the input entry for transport and trade is recorded in an IO table. We can further elaborate on this convention using the concept of producer prices and purchaser prices introduced above.

Let p_{ij} denote the producer price of product i that applies to user j. We assume that the producer price of a product is the same across all its users

$$p_{ij} = p_i, \; \forall j \tag{4.32}$$

Denoting by T_{ij} the sum of transport- and trade margins associated with the input of product i in sector j, that is, z_{ij}, the actual (purchase) cost for sector j of inputting z_{ij} is given by

$$p_i z_{ij} + T_{ij} \tag{4.33}$$

Let t_{ij} be the rate of sum of transport- and trade margins per unit of z_{ij}:

$$t_{ij} = T_{ij}/z_{ij} \tag{4.34}$$

Note that t_{ij} is user specific because it refers to transport charges paid by the purchaser to "take delivery at the required time and place [26]". Substituting (4.34) into (4.33) gives

$$\text{the purchase cost of } z_{ij} = (p_i + t_{ij})z_{ij} \tag{4.35}$$

Dividing both sides by z_{ij} then gives the purchase cost per unit of z_{ij}, the purchasers' price of product i for sector j:

$$p_{ij}^{\text{purchaser}} = p_i + t_{ij} \tag{4.36}$$

Note that the purchasers' prices are generally user-dependent because the amounts of transport- and trade services needed for the actual transaction are likely to differ depending on the location and size of the user. The conventional method, as described in Sect. 4.1.3, is to record $p_i z_{ij}$ as the input of i into sector j, and $\sum_i T_{ij}$ (or $\sum_i t_{ij} z_{ij}$) as the input entry for transport- and trade services.

4.2.3 *Physical Versus Monetary IO*

Let $\tilde{Z}$ denote the $n \times n$ matrix of the inter-sector flow of products in monetary units, given by

$$\tilde{Z} = \hat{p}Z \tag{4.37}$$

where $p = [p_i]$ is the $1 \times n$ vector of (producers') prices. Denoting the matrix of input coefficients obtained from the IO table in value units by $\tilde{A}$, we obtain

$$\tilde{A} = \hat{p}Z(\hat{p}\hat{x})^{-1} = \hat{p}Z\hat{x}^{-1}\hat{p}^{-1} = \hat{p}A\hat{p}^{-1} \tag{4.38}$$

The associated Leontief inverse matrix is

$$\tilde{L} = (I - \hat{p}A\hat{p})^{-1} = (\hat{p}\hat{p}^{-1} - \hat{p}A\hat{p})^{-1} = (\hat{p}(I - A)\hat{p}^{-1})^{-1} = \hat{p}(I - A)^{-1}\hat{p}^{-1} \tag{4.39}$$

with the value version of the quantity model (3.71) given by

$$\tilde{L}\hat{p}y = \hat{p}(I - A)^{-1}\hat{p}^{-1}\hat{p}y = \hat{p}Ly \tag{4.40}$$

It follows that when the same price p_i applies across all its users, as in (4.32), the results obtained from a monetary IO table are the same up to the difference in the unit [5]. Henceforth in this book, we assume that the condition (4.32) is satisfied and we treat IO tables in monetary units as equivalent to those in physical units up to the units of measurement, except when otherwise stated. We will discuss IO models without making a distinction between those in monetary units and those in physical units.

4.3 Regionally Extended IO Models

The global economy is composed of production processes and consumers that span various spatial units such as nations, states, cities, and villages. Due to the interdependence of these processes, the operation of a production or consumption activity in one part of the world can have ripple effects across many different regions. This interconnection has been growing with the liberalization of international trade, economic specialization, and the increasing importance of emerging economies, aided by advancements in communications and transport technologies.

This section focuses on extending previous models to account for the interdependence among different spatial units, referred to as regions, which can encompass a range of spaces, such as a nation like India, a group of nations like the EU, a city like Tokyo, or a state like Connecticut (US).

4.3.1 A Three Region Model

We consider a multi-regional IO (MRIO) model where the world economy consists of three regions denoted as a, b, and c, although this model can easily be extended to include any number of regions. Table 4.2 shows a MRIO table for the three regions, where $\boldsymbol{Z}^{ab}$ represents the flow of products produced in region a and used by region b for intermediate use, $\boldsymbol{y}^{ab}$ represents the final demand in region b for the products produced in region a, and $\boldsymbol{x}^{a}$ represents the output of region a. Similarly, $\boldsymbol{Z}^{bc}$ and $\boldsymbol{y}^{bc}$ represent the import of products from region b by region c for intermediate use and final use, respectively.

Defining by $\boldsymbol{A}^{ab}$ the regionally extended matrix of input coefficients

$$\boldsymbol{A}^{ab} = \boldsymbol{Z}^{ab}\hat{\boldsymbol{x}}_b^{-1} \tag{4.41}$$

we obtain the following supply-demand balance equations

$$\begin{pmatrix} \boldsymbol{x}^a \\ \boldsymbol{x}^b \\ \boldsymbol{x}^c \end{pmatrix} = \begin{pmatrix} \boldsymbol{A}^{aa} & \boldsymbol{A}^{ab} & \boldsymbol{A}^{ac} \\ \boldsymbol{A}^{ba} & \boldsymbol{A}^{bb} & \boldsymbol{A}^{bc} \\ \boldsymbol{A}^{ca} & \boldsymbol{A}^{cb} & \boldsymbol{A}^{cc} \end{pmatrix} \begin{pmatrix} \boldsymbol{x}^a \\ \boldsymbol{x}^b \\ \boldsymbol{x}^c \end{pmatrix} + \begin{pmatrix} \sum_{r\in a,b,c} \boldsymbol{y}^{ar} \\ \sum_{r\in a,b,c} \boldsymbol{y}^{br} \\ \sum_{r\in a,b,c} \boldsymbol{y}^{cr} \end{pmatrix}, \tag{4.42}$$

which can be solved to obtain the output for each region as

$$\begin{aligned} \begin{pmatrix} \boldsymbol{x}^a \\ \boldsymbol{x}^b \\ \boldsymbol{x}^c \end{pmatrix} &= \begin{pmatrix} \boldsymbol{I}-\boldsymbol{A}^{aa} & -\boldsymbol{A}^{ab} & -\boldsymbol{A}^{ac} \\ -\boldsymbol{A}^{ba} & \boldsymbol{I}-\boldsymbol{A}^{bb} & -\boldsymbol{A}^{bc} \\ -\boldsymbol{A}^{ca} & -\boldsymbol{A}^{cb} & \boldsymbol{I}-\boldsymbol{A}^{cc} \end{pmatrix}^{-1} \begin{pmatrix} \sum_{r\in a,b,c} \boldsymbol{y}^{ar} \\ \sum_{r\in a,b,c} \boldsymbol{y}^{br} \\ \sum_{r\in a,b,c} \boldsymbol{y}^{cr} \end{pmatrix}, \\ &= \begin{pmatrix} \boldsymbol{L}^{aa} & \boldsymbol{L}^{ab} & \boldsymbol{L}^{ac} \\ \boldsymbol{L}^{ba} & \boldsymbol{L}^{bb} & \boldsymbol{L}^{bc} \\ \boldsymbol{L}^{ca} & \boldsymbol{L}^{cb} & \boldsymbol{L}^{cc} \end{pmatrix} \begin{pmatrix} \sum_{r\in a,b,c} \boldsymbol{y}^{ar} \\ \sum_{r\in a,b,c} \boldsymbol{y}^{br} \\ \sum_{r\in a,b,c} \boldsymbol{y}^{cr} \end{pmatrix}, \end{aligned} \tag{4.43}$$

Table 4.2 A three region IO table

	Intermediate demand			Final demand			Output
	Region a	Region b	Region c	Region a	Region b	Region c	
Region a	$\boldsymbol{Z}^{aa}$	$\boldsymbol{Z}^{ab}$	$\boldsymbol{Z}^{ac}$	$\boldsymbol{y}^{aa}$	$\boldsymbol{y}^{ab}$	$\boldsymbol{y}^{ac}$	$\boldsymbol{x}^{a}$
Region b	$\boldsymbol{Z}^{ba}$	$\boldsymbol{Z}^{bb}$	$\boldsymbol{Z}^{bc}$	$\boldsymbol{y}^{ba}$	$\boldsymbol{y}^{bb}$	$\boldsymbol{y}^{bc}$	$\boldsymbol{x}^{b}$
Region c	$\boldsymbol{Z}^{ca}$	$\boldsymbol{Z}^{cb}$	$\boldsymbol{Z}^{cc}$	$\boldsymbol{y}^{ca}$	$\boldsymbol{y}^{cb}$	$\boldsymbol{y}^{cc}$	$\boldsymbol{x}^{c}$

We can derive a region's imports from other regions. For instance, region a's imports from region b, $\mathbf{x}^{ba}$, can be obtained as follows

$$\mathbf{x}^{ba} = \mathbf{A}^{ba} \sum_{s \in a,b,c} \mathbf{L}^{as} \left(\sum_{r \in a,b,c} \mathbf{y}^{sr} \right) + \mathbf{y}^{ba} \tag{4.44}$$

This equation implies that $\mathbf{x}^{ba}$ can be influenced not only by the final demand in region a but also by the final demand in regions b and c. Therefore, the effects of inter-regional repercussions resulting from the import of intermediate inputs can be fully accounted for.

4.3.2 The National IO Model with Exogenous Foreign Trade

A regional model is termed "regionally closed" when all the regions within the model are endogenous, meaning that the flows of products between regions are explained similarly to the flows of products within each region. On the other hand, if the model does not consider the interregional flows of products in the same manner as the flows within regions, it is termed " regionally open".

The regional IO model presented above is a closed model because it endogenizes foreign trade with respect to intermediate inputs, and accounts for the fact that the exports from region a to region b are the imports of region b from a. Since the model involves three regions, it is termed a three-region closed model.

However, national IO tables compiled by national statistical offices do not provide regional details of imports and exports and treat all exports as exogenous. As a result, the inter-regional repercussions via the import of intermediate inputs are not considered in national IO tables. These tables are regionally open.

Table 4.3 shows a prototype of such a national table, which considers the import demand of the rest of the world (RoW) for the products of region a as exogenously given. The entries in the table are related to those in Table 4.2 as

$$\mathbf{Z}^{*a} = \sum_{r=b,c} \mathbf{Z}^{ra}, \ \mathbf{y}^{*a} = \sum_{r=b,c} \mathbf{y}^{ra}, \ \mathbf{y}^{a*} = \sum_{r=b,c} \mathbf{Z}^{ar}\boldsymbol{\iota} + \mathbf{y}^{ar}. \tag{4.45}$$

Table 4.3 A national IO table with foreign trade

	Region a		Other regions (RoW)	Output
Region a	$\mathbf{Z}^{aa}$	$\mathbf{y}^{aa}$	$\mathbf{y}^{a*}$	$\mathbf{x}^{a}$
RoW	$\mathbf{Z}^{*a}$	$\mathbf{y}^{*a}$		

This national IO table is a two-region model, comprising region a and the RoW. Since it treats exports as exogenous, it is termed a two-region open model.

The total import of region a, $\boldsymbol{m}^a$, is given by

$$\boldsymbol{m}^a = \boldsymbol{Z}^{*a}\boldsymbol{\iota} + \boldsymbol{y}^{*a} \tag{4.46}$$

The national counterpart of (4.42) is given by

$$\begin{aligned} \boldsymbol{x}^a &= \boldsymbol{A}^{aa}\boldsymbol{x}^a + \boldsymbol{y}^{a*} \\ \boldsymbol{m}^a &= \boldsymbol{A}^{*a}\boldsymbol{x}^a + \boldsymbol{y}^{*a} \end{aligned} \tag{4.47}$$

with the national counterpart to (4.43) given by the noncompetitive import model (or Isard-type [27]) model:

$$\begin{aligned} \boldsymbol{x}^a &= \boldsymbol{L}^{aa}\boldsymbol{y}^{a*} \\ \boldsymbol{m}^a &= \boldsymbol{A}^{*a}\boldsymbol{L}^{aa}\boldsymbol{y}^{a*} + \boldsymbol{y}^{*a} \end{aligned} \tag{4.48}$$

Here, the term noncompetitive import refers to treating domestic products and imports as distinct products. It does not necessarily imply that the model deals with heterogeneous products; rather, it means that the model considers the domestic and imported versions of a product as separate entities.

The Chenery-Moses-type model [28, 29], also known as the competitive import model, is an alternative model that can be expressed as

$$\begin{aligned} \boldsymbol{x}^a &= \left(\boldsymbol{I} - (\boldsymbol{I} - \hat{\boldsymbol{\mu}}^a)\boldsymbol{A}^a\right)^{-1}(\boldsymbol{I} - \hat{\boldsymbol{\mu}}^a)\boldsymbol{y}^a, \\ \boldsymbol{m}^a &= \hat{\boldsymbol{\mu}}^a\left((\boldsymbol{I} - (\boldsymbol{I} - \hat{\boldsymbol{\mu}}^a)\boldsymbol{A}^a)^{-1}(\boldsymbol{I} - \hat{\boldsymbol{\mu}}^a) + \boldsymbol{I}\right)\boldsymbol{y}^a, \end{aligned} \tag{4.49}$$

where

$$\begin{aligned} \boldsymbol{A}^a &= (\boldsymbol{Z}^{aa} + \boldsymbol{Z}^{*a})\hat{\boldsymbol{x}}^{a^{-1}}, \\ \boldsymbol{y}^a &= \boldsymbol{y}^{aa} + \boldsymbol{y}^{a*}, \\ \mu_i^a &= m_i^a/(x_i^a + m_i^a) \end{aligned} \tag{4.50}$$

In this model, products are not differentiated based on their origin of production. The term 'Competitive' indicates that imports and domestic products are considered as substitutes. The parameter μ_i represents the share of imports in the total supply of product i. It is calculated by taking the ratio of imports to the sum of imports and domestic production (total supply in region a) of product i.

4.3.3 International IO Tables

There was a hybrid of MRIO and national IO tables, called international IO tables (IIO tables), pioneered by a study on US–Canada bilateral trade [30]. Before the emergence of global MRIO tables, which is the subject of the next section, many IIO tables had been developed, particularly in Asia-Pacific regions, including bilateral IIO tables for US–Japan, Thailand–Japan, China–Japan, Singapore–Japan, Malaysia–Japan, Indonesia–Japan, Philippines–Japan, and Taiwan–Japan, as well as multilateral IIO tables involving all these countries [31]. A bilateral IIO table is a type of three-region table (Table 4.2), where region c (RoW) is treated exogenously, with the columns $\mathbf{Z}^{\bullet c}$ moved to final demand and added with $\boldsymbol{y}^{\bullet c}$, indicating that the import demand of RoW is given exogenously.

However, the growing demand from users for large country coverage and diverse satellite accounts, such as emissions, coupled with the challenges faced by IIO tables in keeping up with the accelerating globalization of the world economy, has led to the end of IIO tables ([32], Sect. 4.3).

4.3.4 Global MRIO Tables

Table 4.4 provides an overview of the key characteristics of global MRIO databases that include environmental satellite accounts. WIOD stands out as being the only database that is solely based on official and publicly available data, mostly from

Table 4.4 Comparison of major MRIO database with environmental satellite accounts

Database	Coverage		Satellite accounts[c]	Source
	Regions[a]	Sectors[b]		
EXIOBASE3	44 + 5	163	GHG, PM, Water, Land	[15]
Gloria	160 + 4	97	62 materials	[33]
Eora	187 + 0	26–500	GHG, PM, Water	[9]
GTAP	121 + 20	65	GHG, Land	[34]
WIOD	43 + 1	56	CO_2	[14]

Source Based on Table 1 of [35]. The coverage of years is not shown because the end year changes constantly

[a] The number of countries + the number of RoW regions (geographically aggregated countries)

[b] EXIOBASE is also available in a supply-use form, with 200 products and 163 industries, providing more detail on co-produced products, such as refinery fuels. The sector resolution of Eora ranges between 26 and 500, depending on countries. When harmonized over all 187 countries, the resolution becomes 26

[c] Excluding databases with incomplete coverage. EXIOBASE3 and GTAP include the data on labor. PM refers to particulate matter. The list of environmental factors is not exclusive. See the relevant literature mentioned as Source for details. The 62 materials considered in Gloria consist of 23 types of biomass, 15 types of metal ores, 14 types of nonmetallic minerals, and 10 types of fossil fuels

statistical institutes. This approach ensures a high level of data quality but also limits regional coverage. Countries with no official data are approximated by the RoW region, and exports to this region are residually defined to maintain consistency of global trade flows.

In contrast, all the other MRIO databases rely on estimation methods to obtain data for countries with less developed statistical systems. The estimation methods vary across the databases. EXIOBASE, Eora, and Gloria centrally and transparently estimate the missing data, and the data sources and algorithms are published in peer-reviewed articles. In contrast, the GTAP database is constructed in a decentralized manner, with IO tables contributed by GTAP Network members. For countries without official IO tables, the GTAP database uses various imputation methods, including proxying by the IO table of a similar country or estimating from scratch based on a top-down approach from national accounts (see the GTAP web page for further details).

We excluded OECD-ICIO [36] and GRAM [37] due to their low sector resolution (less than 50 sectors). Currently, EXIOBASE3 and Gloria (the successor of Eora) are the only publicly available databases with harmonized country and sector resolution, covering a broad range of environmental pressures. While the GTAP database has limited environmental coverage, it is subject to license fees.

Due to the lack of data on the destination of international trade flows, most global MRIO databases, including EXIOBASE3 and Eora/Gloria, rely on the import proportionality assumption. This means that imported products are proportionally distributed over the target sectors of an imported region, based on the share of the region in the total import of the product [38]. For instance, the input-coefficient matrix of the imports of region s from region r is obtained as [38]

$$\mathbf{A}^{rs} = \hat{\boldsymbol{\mu}}^{rs}\mathbf{A}^{*s}, \quad \mathbf{y}^{rs} = \hat{\boldsymbol{\mu}}^{rs}\mathbf{y}^{*s} \tag{4.51}$$

where $\mathbf{A}^{*s}$ refers to the import matrix in region s, and $\mu_i{}^{rs}$ to the share of region r in the total import of i in region s.[2]

WIOD adopted a slightly generalized version of the standard proportionality assumption by allowing for different import sharers among two use categories [39]: intermediate use (Z), final consumption (C). However, it is important to note that WIOD has a lower sector resolution of 56 sectors compared with EXIOBASE3, meaning that a sector's output in WIOD likely includes more heterogeneous products than in EXIOBASE3. Therefore, there is a need to address this issue by applying different import shares to different categories.

For countries where the national statistical agencies regularly publish IO tables, compiling the MRIO table breaks down to a problem of (dis)aggregation of products and industrial sectors between countries and satellite accounts and balancing the trade between countries [40]. The bilateral trade is balanced by reconciling trade data, such as the BACI (based on the UN Comtrade and balanced) and IEA trade data. In theory,

[2] To be exact, EXIOBASE uses the proportionality of import only to the underlying import use table but not to the symmetric IO table.

the balancing is straightforward because the sum of global imports equals the sum of global exports. In reality, however, the situation is far from straightforward because of issues such as bilateral asymmetries, missing values, and outliers [41]. Another complication for trade data is the use of different pricing for imports and exports (e.g., CIF versus FOB). Once the bilateral trade has been balanced, the amount of product available on the market is obtained, and national tables are balanced. Since MRIO databases use data from many diverse sources at the global level, sophisticated optimization algorithms are used to minimize data conflicts [32].

MRIO databases differ in how they deal with countries lacking national IO tables. EXIOBASE3 and Eora/Gloria have their own approaches to this issue, which are discussed further below.

4.3.4.1 Eora/Gloria

Eora

Eora stands out for its remarkable coverage of 189 countries, ranging from Afghanistan to Zimbabwe, with a regional resolution that practically covers all individual countries in the world [9]. However, 118 of these countries, referred to as "common-classed countries," did not have official IO tables available. To address this, Eora estimated IO tables with 26 sectors for these countries, using United Nations' SNA data and large-scale optimization algorithms to balance global supply and demand while accounting for trade among the 189 countries.

For the "separately-classed countries" that have official IO tables, Eora used the data in its original form, which had varying sector resolutions, ranging from 41 to 500. The level of sector resolution was harmonized based on the countries selected for analysis. When the "common-classed countries" were included, the sector resolution was at its lowest level, 26.

To convert national currencies used in individual national IO tables into current US dollars, Eora used exchange rates based on IMF Official Exchange Rates [9]. Potential regional differences in relative prices were not considered.

GLORIA

Addressing the issue of Eora's low sectoral resolution, [42] developed a new MRIO database, GLORIA (Global Resource Input-Output Assessment), which is characterized by the resolution of 97 sectors: GLORIA has superseded Eora, which will not be updated in the future. This increase in resolution was achieved by resorting to a method similar to that used for Eora to estimate IO tables for countries with no national IO tables. The imposition of "engineering exclusions" can be mentioned as a distinguishing feature of the method, which was used to avoid illogical or incorrect intermediate input-output transactions. Using GLORIA, [33] examined the global material footprints of final demand for 62 material types and found the unlikeliness of absolute decoupling to occur over the next few decades.

4.3.4.2 EXIOBASE3

EXIOBASE was developed to answer sustainability questions for the EU, its trading partners, and major global economies, covering about 90% of the global GDP [15]. EXIOBASE3 is the latest version and has a high sectoral resolution of 163 sectors, providing detailed information on sectors such as agriculture, energy, mining, and transport, which can have varying impact intensities.

For some countries, the resolution of 163 sectors is higher than their official IO tables, which were then converted using "generic coefficients" that represent the expected technology adopted in each industry. EXIOBASE3's underlying structure is based on these generic coefficients, which are obtained by combining physical input per unit output data from life cycle inventory databases, the most disaggregated IO tables available (Japan (>500), US (429), Canada (322), Australia (123)), and information from industrial and agricultural sources [43]. Large-scale optimization algorithms are used to balance inter-regional trade and other flows, similar to as Eora did [15].

Countries without official IO tables are aggregated into five regions based on their geographical location, and IO tables for these regions are obtained by choosing the coefficients of a country that is considered a suitable representation of the region for a particular sector. The values in local currencies are converted to euros using exchange rates from Eurostat, and the conversion is complemented with detailed information on the price of mining and mineral products [15].[3]

4.3.4.3 Improving the Spatial/Sectoral Resolution of EXIOBASE3 and Eora

While EXIOBASE3 stands out with its high sectoral resolution, its spatial resolution is limited compared with Eora. We close this section on MRIOs by briefly mentioning some of the latest developments (at the time of writing) to improve their spatial/sectoral resolution.

EXIOBASE 3rx: Improving the Spatial Resolution of EXIOBASE3

In their experimental attempt to improve the spatial resolution of EXIOBASE3, [45] increased the regional resolution from 49 to 214 by disaggregating the five RoW regions to 170 individual countries, applying the coefficients from the respective RoW region to each of them. While the coverage of associated satellite accounts was limited to land use only, their results indicate significant differences in the land footprint of sectors with high biomass demand due to spatial disaggregation.

Another Attempt to Improve the Spatial Resolution of EXIOBASE 3

Observing that current MRIO databases are limited in their spatial (e.g., EXIOBASE3) or sectoral resolution (e.g., Eora26 and GTAP), as well as their

[3] Using the BACI trade database based on the UN Comtrade Database, [44] found significant geographical variation in export prices.

indicator coverage, [46] developed an automated, transparent, and comparatively time-efficient approach to improve the resolution, quality, and indicator coverage of an existing MRIO database. Applied to EXIOBASE3, they disaggregated and improved its limited spatial resolution by weighting each element with country and sector-specific shares derived from Eora26, FAOSTAT, and previous studies. The resolved database covers 189 countries, 163 sectors, and a cutting-edge set of environmental and socio-economic indicators from 1995 to 2015. Their improvements revealed a significant increase in the EU's water stress and biodiversity loss footprint due to spatial disaggregation and regionalized assessment.

4.4 Capital Goods, Capital Stock, and Capital Services

In an IO table, final demand, or final use transactions, consist of the transactions that make up the final-expenditure components of GDP: personal consumption expenditures; fixed investment; change in inventories; exports of goods and services; and government consumption expenditures and fixed investment (including investment by government enterprises) [47]. On the other hand, intermediate consumption includes goods and services, such as energy, materials and purchased services, that are used up by producers during production to produce output of goods and services within the accounting period. These inputs are also referred to as current-account expenditures and do not include capital-account purchases, which are durable goods used over a long period of time [47, 48]. This definition coincides with that used in LCA, where operation is distinguished from capital goods, and it is the operation that occurs as flows in a process [49].

So far, our discussion on IO has been focused on intermediate or current account flows. This section is devoted to the basics and some recent developments in IE of representing and modeling the flow and stock of capital goods in IO.

4.4.1 Capital Goods in IO Tables

Capital goods refer to the goods destined for the final demand category, fixed capital formation, such as structures, equipment, software, and livestock that are held in inventory (Appendix B.1 gives a detailed definition of capital goods). While the purchase of capital goods in production activities by any sector is recorded in the domestic gross fixed capital formation, there are exceptions as follows where the purchase is recorded as intermediate inputs [50]:

1. *Machine built-in*. Being incorporated into other machines, a capital good becomes a part of a new and different machine.
2. *Construction detour*. Capital goods, such as elevators and boilers, become part of the building: construction sectors use these capital goods as intermediate inputs.

3. *Civil engineering bypass*. Civil engineering work is necessary to install capital goods, such as bridges and flood gates: civil engineering uses these goods as intermediate inputs.
4. *Shipbuilding detour*. In shipbuilding, capital goods such as boilers and communications equipment become a part of the ship.

Reference [49] gives examples taken from the ecoinvent database where the distinction between capital goods and intermediate goods is not straightforward.

4.4.1.1 Capital Goods in the Construction and Operation of Renewable Power Technologies

Table 4.5 gives an actual example of *Machine built-in* and *Construction detour* for renewable power technologies, where capital goods are counted as intermediate inputs. The table is based on the Renewable Energy-Focused Input-Output (REFIO) Table, which was developed by [51] using detailed information from construction design documents and the balance sheets of individual plants. It shows the ten largest inputs per unit of output for three power technologies (solar, wind, and geothermal), divided into the construction and operation phases.

The inputs required and their amounts are remarkably different between the two phases for all three power technologies, with services such as insurance, wholesale, and business services being major inputs in the operation phase, and facility components being major inputs in the construction phase. The full list of power sectors covered by REFIO can be found in ([51], Table A1).

The coefficients in the first three columns, which refer to inputs in the construction phase, are known as "capital coefficients" and represent the inputs required per unit of productive capacity. We will discuss these coefficients in more detail in Sect. 4.4.3.

4.4.2 *Capital Flow Table*

The inter-industry flow in an IO table represents the flow of inputs associated with the operation phase of the processes. A production process is embodied in a production facility consisting of capital goods such as machinery, equipment, software, R&D, structures, and other durable goods. These capital goods, in turn, are composed of numerous parts and components. The flow of capital goods is associated with the construction phase of production processes. In an IO table, the quantities of capital goods used in the construction phase are included in the final demand item called fixed capital formation, aggregated for all users (with the exceptions mentioned above).

Let $I_{ij}(t)$ denote the amount of capital good i used in fixed capital formation to construct a certain productive capacity of process j in year t. Summing $I_{ij}(t)$ over all j gives the $n \times 1$ vector of the amount of i used in fixed capital formation, $\boldsymbol{I} = [I_i]$:

Table 4.5 The top 10 inputs required for construction and operation of renewable power generation technologies in Japan

Large-scale industrial photovoltaic installation[a]		Wind power generation facility construction[a]		Construction of large-scale geothermal power generation		Large-scale industrial photovoltaic power generation		Wind power		Large-scale geothermal power generation	
Solar cell module	0.358	Nacelle	0.283	Production well drilling	0.133	Insurance	0.018	Insurance	0.026	Production well drilling	0.022
Other industrial electrical equipment	0.061	Blade	0.132	Injection well drilling	0.069	Other industrial electrical equipment	0.006	Other general-purpose machines	0.013	Injection well drilling	0.013
Wholesale	0.040	Tower	0.109	Steam pipeline construction	0.051	Switching control device	0.004	Goods leasing business	0.012	Insurance	0.007
Electric wire cable	0.039	Goods leasing business	0.032	Pump compressor	0.046	Transformer	0.004	Other industrial electrical equipment	0.012	Steel pipe	0.002
Aluminum rolled products	0.032	Civil engineering services	0.031	Wholesale	0.037	Business services	0.003	Business services	0.011	Wholesale	0.001
Plated steel	0.025	Wholesale	0.019	Laying of hot water transport pipes	0.036	Wholesale	0.003	Wholesale	0.010	Civil engineering services[b]	0.001
Transformer	0.024	Transformer	0.019	Turbine	0.032	Fixed telecommu-nications	0.001	Power transmission and distribution	0.008	Other business services	0.001
Wiring device	0.022	Switching control device	0.017	Switching control device	0.029	Office supplies	0.001	Office supplies	0.006	Business services	0.001
Hot rolled steel	0.017	Electric wire/cable	0.011	Rotary electric machine	0.023	Private transportation	0.001	Light bulbs	0.005	Pump compressor	0.001
Road freight transport	0.016	Ready-mixed concrete	0.010	Road freight transport	0.022	Publication	0.001	Switching control device	0.005	Waterworks	0.001

Source [51]. *Units* Japanese Yen/Japanese Yen at 2013 prices

[a] Ground installation

[b] Includes construction services

$$I_i(t) = \sum_j I_{ij}(t) \tag{4.52}$$

The capital flow table, $[I_{ij}]$, expands the data in $\boldsymbol{I}$ and shows the inter-industry flow of capital goods. Below, we will discuss some real examples of the capital flow table.

The Capital Flow Table Accompanying the Japanese IO Table 2015

In Japanese IO tables, the estimate of the capital flow table is routinely reported as subsidiary information. The 2015 capital flow table consists of 122 rows (capital goods at the most detailed sector resolution) and 152 columns (capital formation sectors at the intermediate sector resolution), expanding the data contained in the public- and private fixed capital formation columns of the Japanese IO table (see Table B.1 for the full list of 122 capital goods; the original table is available in Japanese only). The first column of Table 4.6 gives the share of the capital goods that make 80% of capital formation. It is noteworthy that R&D occurs as the largest item occupying 10% of total capital formation, followed by Construction and civil engineering. Of machinery and equipment, cars occupy the largest share, followed by electric devices and equipment (the counting of R&D as a capital good started with the 2015 flow table). The columns in the middle and right give the capital goods that make up 90% of capital formation in Electronic devices and Passenger cars. Semiconductor manufacturing equipment occupies 50% of capital formation in Electric devices, followed by In-house R&D, Wholesale, and Software, with these four items occupying 80%. Finally, In-house R&D makes 51% of capital formation in Passenger cars, followed by Software, Metal machine tools, Wholesale, Mold, and Robots, with these six items occupying three quarters of capital formation.

The Capital Flow Table Accompanying the US IO Table 1997

The BEA once published a capital flow table of private fixed investment for the US in 1997, which showed the investment in equipment, software, and structures by industries. However, the outdated 1997 table is still the most recent table published by the BEA. Despite this, the US capital flow table is valuable because it includes sectors not available in the Japanese economy, such as Aerospace and Oil and Gas Extraction. Table 4.7 presents the US counterpart of Table 4.6, with Aerospace and Oil and Gas Extraction replacing Electronic Devices and Passenger Cars. Compared with the Japanese capital flow table, the US flow has broader items of capital goods that make up 80% of capital formation, which can be attributed to the concentration of the Japanese economy into a limited number of industries: Japan has no significant resource extraction-, aerospace-, and military industries. It is interesting to see that in the Aircraft industry, Custom computer programming services is the largest component of capital formation, three times the amount of Special tools, dies, jigs, and fixtures. In Oil and gas extraction, Support activities for oil and gas operations and Drilling of gas wells occupy 70% of capital formation. It is noteworthy that while the BEA's official capital flow table for the US is no longer updated since its

Table 4.6 Capital flow Japan 2015 (excerpts)

All sectors	C.S.[a]	Electronic devices	C.S.	Passenger cars	C.S.
In-house R&D	0.10	Semiconductor manufacturing equipment	0.58	In-house R&D	0.51
Nonresidential construction[b]	0.19	In-house R&D	0.70	Software industry	0.58
Software industry	0.25	Wholesale	0.79	Metal machine tools	0.64
Residential building[c]	0.32	Software industry	0.85	Wholesale	0.68
Construction repair	0.37	Vacuum equipment/Vacuum equipment	0.87	Mold	0.71
Residential construction[b]	0.43	Natural science research institute	0.88	Robot	0.74
Road-related public works	0.47	Electrical measuring instrument	0.89	Metalworking machinery	0.77
Other civil eng. construction	0.51	Other civil eng. construction	0.91	Material handling machine	0.80
Wholesale	0.55	Construction repair	0.92	Other business services	0.81
Rivers, sewers, and other public works	0.59			Other industrial electrical equipment	0.83
Real estate brokerage	0.61			Nonresidential construction[1]	0.84
Passenger car	0.63			Packaging and Packing machinery	0.86
Natural science research institution[d]	0.64			Construction repair	0.87
Trucks, buses and other automobiles	0.66			Plastic processing machinery	0.88
Other business services	0.68			Switching control device	0.89
Retail	0.69			Measuring equipment	0.90
Railway track construction	0.70				
Switching control device	0.71				
Electronic computer accessory	0.72				
Entertainment equipment	0.73				
Mold	0.74				
Aircraft	0.75				
Semiconductor manufacturing equipment	0.76				
Construction and mining equipment	0.77				
Personal computer	0.78				
Medical equipment	0.79				
Wireless telecommunications equipment[e]	0.79				
Refrigerator/Temperature and humidity adjustment device	0.80				

[a] Cumulative share
[b] Nonwooden construction
[c] Wooden construction
[d] National and public
[e] Excluding mobile phones

Table 4.7 Capital flow US 1997 (excerpts)

Total		Aircraft		Oil and gas extraction	
New residential 1-unit nonfarm structures construction	0.13	Custom computer programming services	0.21	Support activities for oil and gas operations	0.38
Automobiles and light trucks	0.23	New industrial plants construction	0.38	Drilling oil and gas wells	0.71
Wholesale trade	0.28	Special tools, dies, jigs, and fixtures	0.45	New industrial plants construction	0.79
Custom computer programming services	0.33	Wholesale trade	0.50	Wholesale trade	0.81
New nonfarm residential additions and alterations	0.37	New office building construction	0.55	Oil and gas field machinery and equipment	0.83
New office building construction	0.40	Electronic computers	0.58	Ship building and repairing	0.85
New commercial structures construction	0.43	Software publishers	0.62		
Electronic computers	0.46	Aircraft	0.65	Custom computer programming services	0.87
Software publishers	0.48	Analytical laboratory instruments	0.67	Software publishers	0.88
Retail trade	0.50	Metal cutting machine tools	0.70	Pumps and pumping equipment	0.89
Telephone apparatus	0.53	Engineering services	0.72	Aircraft	0.90
New industrial plants construction	0.54	Other computer peripheral equipment	0.74		
Residential maintenance and repair construction	0.56	Broadcast and wireless communications equipment	0.76		
Broadcast and wireless communications equipment	0.58	Rolling mill and other metalworking machinery	0.77		
Engineering services	0.59	Metal forming machine tools	0.78		
New electric utility construction	0.61	Retail trade	0.80		
Other computer peripheral equipment	0.62	Computer storage devices	0.81		
New hotels and motels construction	0.63	New warehouse construction	0.82		

(continued)

Table 4.7 (continued)

Total		Aircraft		Oil and gas extraction	
New residential garden apartments construction	0.64	Conveyors and conveying equipment	0.82		
Heavy duty trucks	0.65	Packaging machinery	0.83		
Construction machinery	0.66	Welding and soldering equipment	0.84		
Aircraft	0.67	Laboratory apparatus and furniture	0.85		
Farm machinery and equipment	0.68	Office furniture, except wood	0.86		
Support activities for oil and gas operations	0.69	Industrial process variable instruments	0.86		
Search, detection, and navigation instruments	0.70	Air and gas compressors	0.87		
New warehouse construction	0.71	Scales, balances, and miscellaneous general purpose machinery	0.88		
New hospital construction	0.72	Photographic and photocopying equipment	0.88		
New telephone and telegraph construction	0.72	Line number and Industry code	0.89		
Manufactured homes, mobile homes	0.73	Industrial molds	0.90		
Computer storage devices	0.74				
Drilling oil and gas wells	0.75				
New academic facilities construction	0.75				
Surgical and medical instruments	0.76				
Special tools, dies, jigs, and fixtures	0.77				
Electricity and signal testing instruments	0.77				
Photographic and photocopying equipment	0.78				
Office furniture, except wood	0.78				
New gas utility facilities construction	0.79				
Semiconductor machinery	0.79				
Electromedical apparatus	0.79				
Metal cutting machine tools	0.80				

1997 table, [52] constructed a new capital flow matrix, which was used by [53] to provide a capital-inclusive database of carbon, energy, and material footprints and multipliers.

4.4.3 Capital Stock and Capital Matrix

Building upon the inter-industry flow of capital goods in an IO table broken down by products, we now shift our focus to endogenizing the flow in the IO model, starting with capital coefficients.

4.4.3.1 Capital Coefficients Matrix $\boldsymbol{B}$

Assuming that investment in year t becomes new or additional productive capacity in year $t + 1$, the amount of stock of capital good i in sector j at the end of year t, $K_{ij}(t)$, can be obtained as

$$K_{ij}(t) = I_{ij}(t) + (1 - \delta_{ij})K_{ij}(t-1), \tag{4.53}$$

where δ_{ij} is a fixed rate of replacement (for detailed information on how this equation is derived, please refer to Sect. B.2). Let $\bar{x}_j$ denote the productive capacity of sector j at the end of year t, referring to the amount of product j that can be obtained by operating the existing capital stock installed in the past. The capital coefficients matrix $\boldsymbol{B} = [b_{ij}]$ is given by

$$b_{ij} = K_{ij}(t)/\bar{x}_j(t) \tag{4.54}$$

where b_{ij} represents the amount of capital good i required per unit capacity of j and is assumed to be time-invariant.

Substituting b_{ij} into (4.53) and rearranging yields

$$I_{ij}(t) = b_{ij}\bar{x}_j(t) + \left(\delta_{ij} - 1\right) b_{ij}\bar{x}_j(t-1) \tag{4.55}$$

which shows that the investment in i required to reach the productive capacity of j is equal to b_{ij} times $\bar{x}_j$, plus a term involving the replacement of existing capital stock. In matrix form, this becomes

$$\boldsymbol{I}(t) = \boldsymbol{B}\bar{\boldsymbol{x}}(t) + (\boldsymbol{\delta} - \boldsymbol{I})\boldsymbol{B}\bar{\boldsymbol{x}}(t-1), \tag{4.56}$$

which relates the inter-industry flow of capital goods to the productive capacity of each sector.

4.4.3.2 Capital Flow Table Versus Capital Coefficients

It is important to differentiate between capital flows I_{ij} and capital coefficients b_{ij}. Capital flows represent the distribution of sectoral fixed investment across different products, while capital coefficients provide information about the quantities of products needed to create a unit of productive capacity for a specific sector. For instance, Table 4.6 displays the amount of Semiconductor manufacturing equipment purchased by the electric device sector in 2015. However, it does not indicate the corresponding increase in productive capacity in 2016 resulting from that purchase. On the other hand, Table 4.5 provides the quantity of solar modules required to install a unit capacity of photovoltaic power, which represents capital coefficients.

4.4.3.3 The IO Model with Fixed Investment Endogenized

Denoting by $\boldsymbol{y}^*$ the vector of the sum of final demand categories except fixed capital formation and neglecting foreign trade, the basic balance equation of IO is (we take the liberty of representing $\boldsymbol{y}^I$ by $\boldsymbol{I}$ to keep the consistency of notations)

$$\boldsymbol{x}(t) = \boldsymbol{A}\boldsymbol{x}(t) + \boldsymbol{I}(t) + \boldsymbol{y}^*(t) \tag{4.57}$$

where ι is an identity vector used for summation. Before we can endogenize $\boldsymbol{I}(t)$ by use of $\boldsymbol{B}$, the relation has to be established between the level of output, $\boldsymbol{x}$, and the level of productive capacity, $\bar{\boldsymbol{x}}$. Assuming a 100% utilization of the productive capacity, we have

$$\boldsymbol{x}(t+1) = \bar{\boldsymbol{x}}(t) \tag{4.58}$$

where it is to be noted that $\bar{\boldsymbol{x}}(t)$ refers to the productive capacity at the end of t, while $\boldsymbol{x}(t)$ refers to the level of production during t; production in t is carried out using the capacity available at the beginning of t. With this assumption, (4.54) becomes

$$b_{ij} = K_{ij}(t)/x_j(t+1) \tag{4.59}$$

and (4.57) becomes

$$\boldsymbol{x}(t) = \boldsymbol{A}\boldsymbol{x}(t) + \boldsymbol{B}\boldsymbol{x}(t+1) - (\boldsymbol{I} - \boldsymbol{\delta})\boldsymbol{B}\boldsymbol{x}(t) + \boldsymbol{y}^*(t) \tag{4.60}$$

which can be rearranged as

$$\boldsymbol{B}\boldsymbol{x}(t+1) = (\boldsymbol{I} - \boldsymbol{A} + (\boldsymbol{I} - \boldsymbol{\delta})\boldsymbol{B})\,\boldsymbol{x}(t) - \boldsymbol{y}^*(t) \tag{4.61}$$

If $\boldsymbol{B}$ is nonsingular, (4.61) can be solved for $\boldsymbol{x}(t)$ in a “forward-looking” (or forward recursive) way as

$$\boldsymbol{x}(t+1) = \boldsymbol{B}^{-1}\left((\boldsymbol{I} - \boldsymbol{A} + (\boldsymbol{I} - \boldsymbol{\delta})\boldsymbol{B})\,\boldsymbol{x}(t) - \boldsymbol{y}^*(t)\right) \tag{4.62}$$

This equation is the Leontief dynamic model [54], which was once the subject of active research in economics within the context of multi-sectoral economic growth theory, which flourished in the 1960s (see [55–57]).[4]

However, the practical applicability of the Leontief dynamic model in general and in IE, in particular, is limited by the fact that $\boldsymbol{B}$ is singular, and its inverse does not exist because of the occurrence of rows with zero elements: all the elements of the rows referring to noncapital goods are zeros. The problem of singularity could be "solved" by consolidating (aggregating) $\boldsymbol{B}$ until all the rows have nonzero elements [58]. Since IO studies in IE require a high sectoral resolution, the deliberate reduction of the resolution of the original IO table for the sake of "theoretical convenience" is not a recommended procedure.

4.4.3.4 The Lenzen Model of Fixed Capital Formation

Although the Leontief dynamic model has limited practical applicability, it remains an ambitious attempt to describe the future path of economic growth within a simple framework. Nevertheless, the original model's main theoretical problem is guaranteeing the positiveness of the solution $\boldsymbol{x}(t)$. Reference [59] addressed this problem by devising an alternative dynamic model that does not require the inversion of $\boldsymbol{B}$ and ensures the positivity of $\boldsymbol{x}(t)$. However, for the purpose of explaining investment demand endogenously, a less ambitious approach with high relevance to IE is to use a similar method without deriving a model of economic growth. Reference [60] developed a static IO model, the Lenzen model, that endogenizes fixed capital formation to assign the impacts embodied in the capital goods to the goods and services destined for final consumption.

To "internalize capital investment into intermediate input", [60] assumed a proportional relationship between the level of fixed capital formation and the level of production in t as an alternative to (4.54):

$$I_{ij} = c_{ij} x_j \tag{4.63}$$

where c_{ij} is a proportional factor. Reference [60] referred to the resulting matrix $\boldsymbol{C} = [c_{ij}]$ as "a matrix of sectoral flows of gross fixed capital expenditures per unit output", while others have called it the "capital flow matrix [61]" or "capital requirement matrix [62]". Henceforth, we will use the term "capital flow matrix".

Substitution of (4.63) into (4.57) gives

$$\boldsymbol{x} = (\boldsymbol{A} + \boldsymbol{C})\boldsymbol{x} + \boldsymbol{y}^* \tag{4.64}$$

---

[4] In the theoretical literature in economics, the model is usually simplified by assuming $\boldsymbol{\delta} = \boldsymbol{0}$.

which yields

$$x = \left(I - (A + C)\right)^{-1} y^{*}. \tag{4.65}$$

Unlike the original Leontief dynamic model (4.60), the Lenzen model (4.65) does not require inverting the matrix C. Let $F = [f_{kj}]$ denote the unit environmental burden matrix, where f_{kj} represents the amount of environmental burden k per unit output of j. The environmental footprint of y^{*}, which incorporates the construction phase of production facilities, can then be obtained as

$$E = F\left(I - (A + C)\right)^{-1} y^{*}. \tag{4.66}$$

The Lenzen model, which is based on equation (4.66), has been extensively applied in the field of IE. For example, it has been used by [60] to analyze energy and GHG flows in the Australian economy, by [61] to calculate labor and energy multipliers for Australian industries, and by [63] to assess the environmental impacts of fixed capital formation and household consumption. However, despite its widespread use, there has been relatively little discussion about the theoretical underpinnings of the Lenzen model, with only a few exceptions such as [64]. Notably, the Lenzen model is not mentioned in the comprehensive book on IO [12], which runs to over 800 pages.

We want to explore the relationship between the Lenzen model and the original Leontief dynamic model. Substituting from (4.63), which refers to the Lenzen model's fundamental assumption, (4.53) becomes

$$I_{ij}(t) = \Delta K_{ij}(t) + \delta_{ij} K_{ij}(t-1) = c_{ij} x_j(t) \tag{4.67}$$

where $\Delta K_{ij}(t) = K_{ij}(t) - K_{ij}(t-1)$. Recall that $K_{ij}(t-1)$ represents the stock of capital goods i available for use in t. In this context, it is reasonable to assume a proportionality between $\delta_{ij} K_{ij}(t-1)$ and $x_j(t)$. However, the relationship between the net expansion of capital stock, $\Delta K_{ij}(t)$ and $x_j(t)$ is more complex.

To simplify the exposition and focus on the relationship between $\Delta K_{ij}(t)$ and $x_j(t)$, we neglect the term $\delta_{ij} K_{ij}(t-1)$ in the following (alternatively, $\delta_{ij} K_{ij}(t-1)$ can be removed from $I_{ij}(t)$ and modeled proportional to $x_j(t)$). Assuming a capacity utilization rate of 100%, we can use (4.59) to derive:

$$I_{ij}(t) = \Delta K_{ij}(t) = b_{ij}\left(x_j(t+1) - x_j(t)\right) \tag{4.68}$$

New facilities and equipment are constructed with the purpose of increasing productive capacity. As a result, the amount of fixed capital formation in a given year, denoted as $I_{ij}(t)$, is strongly linked to the planned expansion of productive capacity, represented by $\Delta x_j(t+1)$, while its correlation with the production level $x_j(t+1)$ is generally weaker. Assuming a fixed proportionality between annual fixed capital formation and production may not be easily justifiable under general conditions. This problem has been recognized in the IE literature. For example, [63] points out that "the fixed capital formation taking place in a given year will generally be working for production of commodities after that year" and it is difficult to determine how long specific instances of fixed capital formation in a given year will continue to work for commodity production, both directly and indirectly. Reference [62] acknowledges that the IO model endogenizing gross fixed capital formation (GFGC) "disregards the fact that capital goods are (per definition) used for a period of more than a year" and makes the allocation of emissions more sensitive to extreme events in the economy. Similarly, [52] notes that if GFCF is unsteady, as is often the case in reality, the Lenzen model will result in erratically fluctuating impacts.

Additionally, the assumption of a fixed proportionality between fixed capital formation and output level is not supported by standard economic models of fixed capital formation [65]. For example, in a production function of the KLEMS type, where the output x is determined by inputs of capital, labor, energy, materials, and services through a production function g as $x = g(K, L, E, M, S)$, the optimal level of capital is determined based on the relative prices of inputs and the desired output. The determination of investment demand involves considering adjustment or delay processes that prevent the immediate achievement of the optimal level of capital.

Although the Lenzen model is widely used in IE, it is difficult to justify it theoretically when $\boldsymbol{C}$ is based on (4.63). However, there is one exception where the Lenzen model with (4.63) can be used theoretically soundly. This is when (4.69) holds, which corresponds to zero productive capacity in the previous period ([19], p. 102):

$$I_{ij}(t) = b_{ij} x_j(t+1) \tag{4.69}$$

In this case, $b_{ij} = c_{ij}$ holds, too, because

$$b_{ij} = \frac{K_{ij}(t)}{x_j(t+1)} = \frac{I_{ij}(t) + K_{ij}(t-1)}{\Delta x_j(t+1) + x_j(t)} = \frac{I_{ij}(t)}{\Delta x_j(t+1)} = \frac{I_{ij}(t)}{x_j(t+1)} = c_{ij} \tag{4.70}$$

The Lenzen model (4.65) then gives the level of $\boldsymbol{x}$ that is required to achieve the realization of $\boldsymbol{y}^*$ from scratch, that is, from the situation where no production activity was available. The Lenzen model provides the output level necessary to satisfy all the requirements for the construction of plants before production starts, including the inputs required for their construction. Therefore, when used with these features in mind, the Lenzen model provides an extremely useful and theoretically sound framework for studying the life cycle and economy-wide impacts of a given $\boldsymbol{y}^*$, including both the production and construction phases of products. For instance, the data on inputs required for constructing new renewable power facilities developed by [51] (Table 4.5) would be ideal for using the Lenzen model.

4.4.3.5 The Augmentation Approach

Before moving on to the next topic, it is important to briefly mention another approach to endogenizing capital formation, known as the augmentation method [66]. In an IO table, gross fixed capital expenditure is represented as a column vector $\boldsymbol{y}^I$ as part of final demand $\boldsymbol{y}$, while Consumption Fixed capital Capital (depreciation), CFC, is represented as a row vector $\boldsymbol{v}^{\text{CFC}}$ as part of primary inputs. To apply the augmentation method, an $n \times 1$ vector $\boldsymbol{a}^I = [a_i^I]$ and a $1 \times n$ vector $\boldsymbol{a}^{\text{CFC}} = [a_i^{\text{CFC}}]$ are defined as follows:

$$a_i^I = y_i^I / \sum_j y_j^I, \quad a_j^{\text{CFC}} = v_j^{\text{CFC}} / x_j, \quad i, j = 1, \ldots, n \tag{4.71}$$

The augmentation method then obtains a simplified version of the matrix $\boldsymbol{A} + \boldsymbol{C}$ in (4.65) by using the $(n+1) \times (n+1)$ matrix:

$$\begin{pmatrix} \boldsymbol{A} & \boldsymbol{a}^I \\ \boldsymbol{a}^{\text{CFC}} & \boldsymbol{0} \end{pmatrix} \tag{4.72}$$

One advantage of the augmentation method is its minimal data requirements. However, it completely neglects the differences in the composition of capital goods required across sectors, as shown in Tables 4.5, 4.6, and 4.7. For example, aircraft manufacturing plants do not require drilling oil and gas wells, and a wind farm does not need a solar module. The composition of fixed capital formation greatly differs among sectors, and thus, the augmented method is not recommended for use in IE, where empirical relevance is crucial. It will not be discussed further in this book.

4.4.3.6 Endogenizing the Consumption of Fixed Capital

The IO models of capital formation we discussed above are concerned with endogenizing gross fixed capital formation (GFCF) that occurs as a distinctive final demand category, $\boldsymbol{y}^I$. However, assuming a fixed proportionality between annual GFCF and production may not be justifiable, under general conditions, as mentioned above. To address this shortcoming, [62] proposed an alternative approach by constructing the capital flow matrix using CFC instead of unsteady GFCF. They argue that endogenizing CFC is more sensible for footprint-type calculations since it incorporates the emissions associated with the production of the capital currently used by industries.

Conceptually, CFC is defined as "the decline, during the course of the accounting period, in the current value of the stock of fixed assets owned and used by a producer as a result of physical deterioration, normal obsolescence or normal accidental damage ([67] Chapter 6H)". This means that CFC can be used as a proxy for the portion of the capital stock that needs to be replaced in the current period, which is denoted by $\boldsymbol{\delta K}(t-1)$ in equation (4.53). Therefore, assuming that CFC is proportional to $\boldsymbol{x}(t)$ is reasonable because $\boldsymbol{K}(t-1)$ refers to the stock of capital goods being used in period t. In contrast, $\mathbf{\Delta K}(t)$ refers to the installation of new capital goods in period t, which increases the available productive capacity.

To obtain the capital flow matrix $\boldsymbol{C}$ using the CFC data, [52, 62] disaggregated it into individual capital goods using information about the CFC's composition in capital goods, denoted as $\boldsymbol{K}^{comp}$. Using the KLEMS data [68] to estimate $\boldsymbol{K}^{comp}$, they obtained $\boldsymbol{C}$ as follows

$$\underbrace{\boldsymbol{C}}_{n \times n} = \text{diag}(\underbrace{\boldsymbol{v}^{CFC}}_{1 \times n}) \times \underbrace{\boldsymbol{K}^{comp}}_{n \times n} \times \underbrace{\textbf{diag}(\boldsymbol{x})^{-1}}_{n \times n} \tag{4.73}$$

This method allowed them to endogenize capital and estimate the carbon footprint of final consumption, which was found to substantially increase (by up to 57% for some countries). They also found that the gap between production-based and consumption-based emissions increases for most countries (Chap. 5 provides more details about the concept of a "carbon footprint").

In subsequent research, [69] introduced a new indicator of material use called the capital-augmented material footprint (CAMF), which includes all the materials embedded in capital goods. They found that this indicator showed substantial increases in the material footprints of final consumption when capital is endogenized, particularly for mineral use. Other studies have also applied similar methods to the US data. Instead of using the KLEMS data, [52, 53] used capital asset investment and depreciation data, as well as other data sources from the US-BEA. Reference [52] found that by allocating capital based on CFC rather than GFCF, capital consumption is more reasonably allocated and trends more smoothly over time, representing a significant improvement over most existing flow matrix methods. Reference [53] estimated the carbon, energy, and material footprints and found that endogenizing capital leads to the biggest proportional increase in footprints of sectors with otherwise low environmental multipliers.

The Lenzen model and its variants discussed earlier are limited in that they are static and do not capture the growth trajectory of capital stock over time. Additionally, they assume that all capital goods of different vintages are homogeneous, with the only difference being the decline in efficiency over time. To overcome these limitations, recent studies have attempted to extend the original Lenzen model by considering the heterogeneity of capital goods based on their time and space of production. Here, we briefly mention [62, 70].

Capital compensation encompasses two main elements: CFC and business surplus, which is also referred to as gross operating surplus (Sect. B.2.2). Typically, the sum of CFC is smaller than the sum of GFCF. To bridge this gap, [62] assumed that the entire business surplus is reinvested in new capital and postulated the following relationship between CFC and GFCF

$$\text{CFC} + \text{gross operating surplus} = \text{GFCF} \tag{4.74}$$

As we saw earlier, CFC pertains to capital stocks in use that consist of capital goods produced in different periods reflecting the technology available at that time, whereas GFCF refers to the acquisition of new capital goods reflecting current technology. Based on the SUT framework, [62] extended identity (4.74) and developed a new IO model that can evaluate the impacts embodied in current and past GFCF simultaneously.

The KLEMS database is commonly used to estimate the capital stock in dynamic models that use the CFC approach, such as those discussed previously. This database uses the geometric efficiency distribution model to estimate the capital stock. Reference [70] used the geometric mortality distribution, which is detailed in Sect. B.2.1, to link past investments $I_i(t-r), r > 0$ to current capital consumption of those investments $\mathrm{CFC}(t, r)$ as

$$\mathrm{CFC}_i(t, r) = \delta_i (1 - \delta_i)^{t-r} I_i(t-r) \tag{4.75}$$

By doing so, they were able to obtain a CFC that distinguished the age of investment. This method allowed for the assessment of the global greenhouse gas impacts of China's GFCF, fully considering the time and regions where the capital goods were produced and the differences in the technology used. Using a similar method as [62] and embedded in the MRIO framework, they were able to extend the analysis to consider the impacts embodied in both current and past GFCF simultaneously.

4.4.3.7 Modeling CFC Versus GFCF

Although the approach to estimating the capital flow matrix based on CFC is conceptually sound and likely to yield stable results, it suffers from the complexity of obtaining reliable data on CFC, particularly when trying to distinguish between different types of product or industry output, such as buildings versus machinery. The CFC is most often estimated using depreciation from business accounts, which may not accurately reflect physical capital usage or production levels [62]. While the 1993 SNA treats CFC as a synonym for depreciation ([71], p. 44), the information provided in business accounts may not be suitable for calculating the consumption of fixed capital ([67], Chap. 6H). In contrast, estimating a capital flow matrix from data on GFCF is straightforward because it refers to the actual purchase of capital goods, albeit aggregated over different users. Reference [52] identifies future research directions to address the data issues associated with the CFC approach.

4.4.3.8 A Dynamic Model with Different Age Cohorts of Capital Stock

Capital goods produced in different years may embody different technologies, reflecting technological changes over time. For instance, today's cars use more electronics and plastics than those built in the 1970s. This means that the stock of capital goods at any given time is likely to be a mixture of technologies, each with different input coefficients.

We saw above that [70] considered the age composition of capital stock in their assessment of the environmental impacts of capital formation. Before closing this section on capital goods, we briefly mention [72], which is noteworthy for explicitly considering the age distribution of capital stock embodying various technologies. This model is similar to one developed by [73].

Here, we consider the evolution of production capacity and output in sector j over time. Let r, s, t be integers representing years with $r \leq s \leq t$. We assume that investments made in t become production capacity at the start of $t+1$, with a one year gestation period of investment. We define $\eta_{1,j}(t,r)$ as the age efficiency of production capacity installed in year r in year t. This measures how effectively capacity installed in year r can be used in year t, accounting for the effects of aging. We also define $\lambda_j(t,r)$ as the mortality rate of production capacity installed in year r in year t. This measures the proportion of capacity installed in year r that is decommissioned in year t due to aging or other factors.

Let $\bar{x}_{in,j}(t)$ denote newly added production capacity in year t. Summing up the amount of capacity installed in years $r \leq t$ that is decommissioned in year t, we obtain production capacity decommissioned in year t, $\bar{x}_{out,j}(t)$:

$$\bar{x}_{out,j}(t) = \sum_{r \leq t} \bar{x}_{in,j}(r)\lambda_j(t,r) \tag{4.76}$$

Subtracting from the amount of capacity installed in year r the proportion that is decommissioned in years $r \leq s \leq t$, we obtain the amount of capacity installed in year r that is still operational in year t, $\bar{x}_{o,j}(t,r)$

$$\bar{x}_{o,j}(t,r) = \bar{x}_{in,j}(r)\left(1 - \sum_{r \leq s \leq t} \lambda_j(s,r)\right) \tag{4.77}$$

Multiplying the amount of operational capacity installed in year r, $\bar{x}_{o,j}(t,r)$, by its age efficiency in year t, we obtain the effective production capacity in year t of capacity installed in year r

$$\bar{x}_j(t,r) = \eta_{1,j}(t,r)\bar{x}_{o,j}(t,r). \tag{4.78}$$

Finally, summing up the contribution of each vintage of capacity to output in year t, weighted by the capacity utilization rate of each vintage, the output of sector j, $x_j(t)$, is obtained

$$x_j(t) = \sum_r \eta_{2,j}(t,r)\bar{x}_j(t,r) \tag{4.79}$$

The technology available at a given time is embodied in the productive capacity installed at that time, $\bar{x}_{in,j}(r)$. This technology changes over time, so the productive capacity installed in year r embodies the technology available at that time, denoted by $\boldsymbol{a}_{\cdot j}(\boldsymbol{r}) = [a_{1,j}(r), a_{2,j}(r), \ldots, a_{n,j}(r),]^{\top}$. The production capacity in t is the sum of all existing capital stocks installed over time. Therefore, the "average technology" in sector j will be the average of the technologies embodied in still-existing capital stocks, adjusted for efficiency, weighted by the intensity of use:

$$a_{ij}(t) = \frac{\sum_r a_{ij}(r)\eta_j(t,r)\bar{x}_o(t,r,j)}{\sum_r \eta_j(t,r)\bar{x}_o(t,r,j)} \tag{4.80}$$

where $\eta_j(,.,) = \eta_{1,j}(,.,)\eta_{2,j}(,.,)$.

It is important to note that in this model, the mortality profile $\lambda_i(r,t)$ and the efficiency profile $\eta_{1,i}(t,r)$ are independent of each other. This is in contrast to standard economics models where these two profiles are interwoven (see (B.7) in Sect. B.2). In reality, capital goods are often discarded and subjected to an end-of-life (EoL) process even when they are still functioning with a positive efficiency. This can be seen in numerous examples such as cars, home appliances, clothes, and buildings, which are often disposed of even though they are still functioning well. The independent occurrence of the mortality profile from the efficiency profile may reflect this reality. Another distinguishing feature of this model is that the mortality and efficiency profile are specific to when the capital good was installed, rather than simply being a function of time that has elapsed since its installation.

The model is closed by the balance of supply and demand for the economy as a whole, expressed as follows:

$$\boldsymbol{x}(t) = \boldsymbol{A}(t)\boldsymbol{x}(t) + \boldsymbol{B}_{in}\bar{\boldsymbol{x}}_{in}(t+1) + \boldsymbol{B}_{out}\bar{\boldsymbol{x}}_{out}(t) + \boldsymbol{y}^*(t) \tag{4.81}$$

This equation is to be solved for $\boldsymbol{x}(t)$, $\bar{\boldsymbol{x}}_I(t)$, $I \in \{in, out, o\}$, and $\boldsymbol{A}(t)$. We note that, besides $\boldsymbol{B}_{in}$, $\boldsymbol{B}_{out}$, and $\boldsymbol{y}^*(t)$, the efficiency and utilization ratios, $\eta_{1,j}$ and $\eta_{2,j}$, are exogenously given, which simplifies the model considerably. The mechanism of the model (and the way it is solved) can briefly be stated as follows.

Given an initial production capacity, (4.76) determines which production capacity needs to be retired. The retirement of production capacity reduces productive capacity and output, which has to be replaced by new capacity, resulting in a final demand for replacement of $\boldsymbol{B}_{out}\bar{\boldsymbol{x}}_{out}(t)$. The retirement and new installation of production capacity may also lead to a change in the $\boldsymbol{A}$ matrix via (4.80), as the new production capacity may embody different technology from the retiring one. Since the balance (4.81) has to be satisfied for exogenously given $\boldsymbol{y}^*$, new capacity $\bar{\boldsymbol{x}}_{in}$ may also need to be installed, generating investment demand of $\boldsymbol{B}_{in}\bar{\boldsymbol{x}}_{in}(t+1)$, resulting in new levels of output, with $\boldsymbol{A}$ possibly embodying changing technology. The entire system must be iterated every year until conversion is achieved.

A crucial assumption in solving the model is that the utilization rates, $\eta_2(t,r)$, are exogenously given, which eliminates one of the most challenging issues of the Leontief dynamic model (as noted in [56, 59]). This assumption is necessary to maintain the model's manageability. Although age-cohort-based accounting of capital stocks has a long tradition in MFA, it is rarely applied in the EEIO. Despite the strong assumption, this model offers a new opportunity to integrate LCA, MFA, and EEIO to identify potential synergies.

4.5 Price Determination in IO

In Sect. 3.3.4, we discussed the aspect of cost and price within a simple framework and derived a cost and price version of the Leontief model. In this section, we will further elaborate on the model.

4.5.1 *Primary Factors of Production*

To begin, let us recall from Sect. 3.3.4 that the value of products in an economy can be determined by the amount of "primary factors of production" embodied in them. In the example we discussed in Chap. 3, there was a single primary (exogenous) input: homogeneous labor measured in hours of work. The number of laborers is exogenously given and is a primary factor of production, while the rate of its utilization—the hours of work—can be changed.

In reality, however, there are many different types of labor. Furthermore, labor is not the only primary factor of production. Even in the simple agriculture-aquaculture example, land (rice paddy, wheat field, and dyke), pond, and infrastructure (road and water system) occur as primary factors indispensable for production. Suppose that these factors are privately owned (by landowners or "capitalists"). In that case, their owners have to be compensated for the provision of the factors, just as laborers need to be compensated for their service. Even when these factors are publicly owned, their provision incurs costs (such as maintenance and repair) that have to be compensated. In economics, these tangible items are called "capital stock", and the services they provide are termed "capital services" (please see Sect. B.2 for further details). We follow these conventions in economics.

4.5.2 *The IO Model of Cost and Price*

Assume there are two types of primary factors of production: labor services (ℓ) and capital services (K), with the cost per unit of service given by p^ℓ for labor and p^K for capital. The cost of primary factors per output in sector j is given by

$$v_j = p_j^\ell a_{\ell,j} + p_j^K a_{K,j} \tag{4.82}$$

where $a_{\ell,j} = \ell_j/x_j$ and $a_{K,j} = K_j/x_j$. Denoting by $\boldsymbol{p} = [p_i]$ the vector of product prices and by $\boldsymbol{v} = [v_i]$ the vector of value added ratios, the cost balance equation ((3.89) in Sect. 3.3) becomes

$$\boldsymbol{p} = \boldsymbol{p}\boldsymbol{A} + \boldsymbol{v} \tag{4.83}$$

or

$$\boldsymbol{v} = \boldsymbol{p}(\boldsymbol{I} - \boldsymbol{A}) \tag{4.84}$$

Solving this equation for $\boldsymbol{p}$ gives

$$\boldsymbol{p} = \boldsymbol{v}(\boldsymbol{I} - \boldsymbol{A})^{-1} = \boldsymbol{v}\boldsymbol{L} \tag{4.85}$$

From (4.83) and (4.84), we obtain

$$\begin{aligned} \boldsymbol{p}\boldsymbol{y} &= \boldsymbol{v}(\boldsymbol{I} - \boldsymbol{A})^{-1}\boldsymbol{y} = \boldsymbol{v}\boldsymbol{x} \\ \boldsymbol{v}\boldsymbol{x} &= \boldsymbol{p}(\boldsymbol{I} - \boldsymbol{A})(\boldsymbol{I} - \boldsymbol{A})^{-1}\boldsymbol{y} = \boldsymbol{p}\boldsymbol{y} \end{aligned} \tag{4.86}$$

which gives the accounting identity of the economy, where the income (the total value added produced) given by $\boldsymbol{v}\boldsymbol{x}$ is equal to the final expenditure $\boldsymbol{p}\boldsymbol{y}$.

4.5.2.1 *The Duality Between the Quantity and Price IO Models

We can rewrite the cost balance equation (4.84) as

$$(\boldsymbol{I} - \boldsymbol{A}^{\top})\boldsymbol{p}^{\top} = \boldsymbol{v}^{\top} \tag{4.87}$$

Recalling the quantity balance equation

$$(\boldsymbol{I} - \boldsymbol{A})\boldsymbol{x} = \boldsymbol{y}, \tag{4.88}$$

we notice that these equations are almost identical, except for the occurrence of the transposed $\boldsymbol{A}$ matrix in (4.87). Using these equations, we can create two optimization problems

$$\max_{\boldsymbol{x}} \boldsymbol{v}\boldsymbol{x} \quad \text{subject to} \quad (\boldsymbol{I} - \boldsymbol{A})\boldsymbol{x} \leq \boldsymbol{y} \quad \text{and } \boldsymbol{x} \geq \boldsymbol{0} \tag{4.89}$$

$$\min_{\boldsymbol{p}} \boldsymbol{p}\boldsymbol{y} \quad \text{subject to} \quad (\boldsymbol{I} - \boldsymbol{A}^{\top})\boldsymbol{p}^{\top} \geq \boldsymbol{v}^{\top} \quad \text{and } \boldsymbol{p} \geq \boldsymbol{0} \tag{4.90}$$

The first problem, (4.89), is called the primal problem, and the second problem, (4.90), is called the dual problem. These are examples of linear programming (LP) problems.

The duality theorem of LP states that if the primal problem has an optimal solution $\boldsymbol{x}^*$, then the dual problem also has an optimal solution $\boldsymbol{v}^*$ such that the following holds

$$\boldsymbol{v}\boldsymbol{x}^* = \boldsymbol{p}^*\boldsymbol{y} \tag{4.91}$$

Here, $\boldsymbol{p}^*$ is the optimal solution of the dual problem.

We can see that the system consisting of (4.87) and (4.88) is a special case of the LP problem, with the inequality replaced by equality. Moreover, (4.91) corresponds to (4.86). Therefore, the cost-price IO model is dual to the quantity IO model.

4.5.2.2 The Case of $\boldsymbol{p} = \boldsymbol{\iota}^\top$

In actual IO tables, each element of $\boldsymbol{p}$ is commonly provided not as a raw price but as an index representing the prices of a bundle of goods, which is normalized to unity for the base year as $\boldsymbol{p} = \boldsymbol{\iota}^\top$. In such cases, substituting this into (4.84), we get

$$\boldsymbol{v} = \boldsymbol{\iota}^\top(\boldsymbol{I} - \boldsymbol{A}) \tag{4.92}$$

Using (4.92), we can rewrite (4.93) as

$$\boldsymbol{v}(\boldsymbol{I} - \boldsymbol{A})^{-1} = \boldsymbol{\iota}^\top(\boldsymbol{I} - \boldsymbol{A})(\boldsymbol{I} - \boldsymbol{A})^{-1} = \boldsymbol{\iota}^\top \tag{4.93}$$

Here, if all the elements of $\boldsymbol{A}$ are nonnegative and $\boldsymbol{v} > 0$, then we can conclude from Corollary 3.1 that $\boldsymbol{A}$ is productive.

4.5.3 *The Impacts of a Change in Prices of Imports*

The modern economy relies heavily on fossil fuels, such as oil, gas, and coal. For geological reasons, these resources (except coal) are not distributed evenly and tend to be concentrated in certain regions. This concentration makes economies that lack these resources vulnerable to supply conditions imposed by exporting countries with a rich reservoir of fossil fuels. This was exemplified in the first oil crisis of 1973, where a 300% increase in oil prices ended the era of high growth in the Japanese economy, which had been growing at an average rate of 10% since 1955.

As of 2023, the world economy is currently facing a significant increase in fossil fuel prices, particularly for gas, due to supply disruptions caused by the war in Ukraine. Therefore, it is essential to consider the impacts of changes in import prices on the importing countries' prices.

4.5.3.1 A MRIO with Tree Regions

We will illustrate the three-region model in Sect. 4.3.1, consisting of regions a, b, and c, as an example. However, extending this model to any number of regions is straightforward.

Denoting by $\boldsymbol{p}^r$ and $\boldsymbol{v}^r$ the vector of (producers') prices and value added ratios in region r, the cost balance can be represented as

$$\begin{pmatrix}\boldsymbol{p}^a & \boldsymbol{p}^b & \boldsymbol{p}^c\end{pmatrix} = \begin{pmatrix}\boldsymbol{p}^a & \boldsymbol{p}^b & \boldsymbol{p}^c\end{pmatrix} \begin{pmatrix} \boldsymbol{A}^{aa} & \boldsymbol{A}^{ab} & \boldsymbol{A}^{ac} \\ \boldsymbol{A}^{ba} & \boldsymbol{A}^{bb} & \boldsymbol{A}^{bc} \\ \boldsymbol{A}^{ca} & \boldsymbol{A}^{cb} & \boldsymbol{A}^{cc} \end{pmatrix} + \begin{pmatrix}\boldsymbol{v}^a & \boldsymbol{v}^b & \boldsymbol{v}^c\end{pmatrix}, \tag{4.94}$$

where the solution for prices in each region is given by

$$\begin{pmatrix}\boldsymbol{p}^a & \boldsymbol{p}^b & \boldsymbol{p}^c\end{pmatrix} = \begin{pmatrix}\boldsymbol{v}^a & \boldsymbol{v}^b & \boldsymbol{v}^c\end{pmatrix} \begin{pmatrix} \boldsymbol{L}^{aa} & \boldsymbol{L}^{ab} & \boldsymbol{L}^{ac} \\ \boldsymbol{L}^{ba} & \boldsymbol{L}^{bb} & \boldsymbol{L}^{bc} \\ \boldsymbol{L}^{ca} & \boldsymbol{L}^{cb} & \boldsymbol{L}^{cc} \end{pmatrix} \tag{4.95}$$

Now, suppose that product r is a resource with limited global supply, such as natural gas, which is produced in only country c. We investigate the effects of an artificial increase in the export price of r from c by Δp_r^c on the product prices of the three countries. For simplicity, we assume away the feedback effects on p_r^c of the price changes and consider the sector r in c as exogenous.

Let $\bar{\boldsymbol{A}}^{cj}$ denote the matrix $\boldsymbol{A}^{cj}$ with the rth row deleted, $j \in \{a, b\}$, $\bar{\boldsymbol{A}}^{ic}$ the matrix $\boldsymbol{A}^{ic}$ with the rth column deleted, $i \in \{a, b\}$, and $\bar{\boldsymbol{A}}^{cc}$ the matrix $\boldsymbol{A}^{cc}$ with the rth row and rth column deleted. Similarly, let $\bar{\boldsymbol{p}}^c$ and $\bar{\boldsymbol{v}}^c$ be defined in an analogous way. When p_r^c is exogenized, (4.95) can be rewritten as

$$\begin{aligned}\begin{pmatrix}\boldsymbol{p}^a & \boldsymbol{p}^b & \bar{\boldsymbol{p}}^c\end{pmatrix} &= \begin{pmatrix}\boldsymbol{p}^a & \boldsymbol{p}^b & \bar{\boldsymbol{p}}^c\end{pmatrix} \begin{pmatrix} \boldsymbol{A}^{aa} & \boldsymbol{A}^{ab} & \bar{\boldsymbol{A}}^{ac} \\ \boldsymbol{A}^{ba} & \boldsymbol{A}^{bb} & \bar{\boldsymbol{A}}^{bc} \\ \bar{\boldsymbol{A}}^{ca} & \bar{\boldsymbol{A}}^{cb} & \bar{\boldsymbol{A}}^{cc} \end{pmatrix} \\ &+ \begin{pmatrix}\boldsymbol{v}^a & \boldsymbol{v}^b & \bar{\boldsymbol{v}}^c\end{pmatrix} + p_r^c \begin{pmatrix}\boldsymbol{A}_{r\cdot}^{ca} & \boldsymbol{A}_{r\cdot}^{cb} & \bar{\boldsymbol{A}}_{r\cdot}^{cc}\end{pmatrix}\end{aligned} \tag{4.96}$$

The effects on the product prices of a change in p_r^c can then be obtained by

$$\begin{pmatrix}\Delta\boldsymbol{p}^a & \Delta\boldsymbol{p}^b & \Delta\bar{\boldsymbol{p}}^c\end{pmatrix} = \Delta p_r^c \begin{pmatrix}\boldsymbol{A}_{r\cdot}^{ca} & \boldsymbol{A}_{r\cdot}^{cb} & \bar{\boldsymbol{A}}_{r\cdot}^{cc}\end{pmatrix} \begin{pmatrix} \boldsymbol{I} - \boldsymbol{A}^{aa} & -\boldsymbol{A}^{ab} & -\bar{\boldsymbol{A}}^{ac} \\ -\boldsymbol{A}^{ba} & \boldsymbol{I} - \boldsymbol{A}^{bb} & -\bar{\boldsymbol{A}}^{bc} \\ -\bar{\boldsymbol{A}}^{ca} & -\bar{\boldsymbol{A}}^{cb} & \boldsymbol{I} - \bar{\boldsymbol{A}}^{cc} \end{pmatrix}^{-1} \tag{4.97}$$

4.5.3.2 National IO Models with Exogenous Imports

We will now discuss modeling a change in import prices using national IO models with exogenous foreign trade, as we discussed in Sect. 4.3.2.

First, let us consider the noncompetitive import model. Denoting the $1 \times n$ vector of the prices of domestic products by $\boldsymbol{p}^d$ and the $1 \times n_m$ vector of the prices of imports by $\boldsymbol{p}^m$, the cost balance equation (4.83) is modified as

$$\boldsymbol{p}^d = \boldsymbol{p}^d \boldsymbol{A}^d + \boldsymbol{p}^m \boldsymbol{A}^m + \boldsymbol{v} \tag{4.98}$$

Here, $\boldsymbol{A}^d$ and $\boldsymbol{A}^m$ correspond to $\boldsymbol{A}^{aa}$ and $\boldsymbol{A}^{*a}$ for country a in Sect. 4.3.2. Solving for $\boldsymbol{p}^d$, we obtain

$$\boldsymbol{p}^d = (\boldsymbol{p}^m \boldsymbol{A}^m + \boldsymbol{v})(\boldsymbol{I} - \boldsymbol{A}^d)^{-1} \tag{4.99}$$

The effects of a change in $\boldsymbol{p}^m$ on $\boldsymbol{p}^d$ are given by

$$\Delta \boldsymbol{p}^m \boldsymbol{A}^m (\boldsymbol{I} - \boldsymbol{A}^d)^{-1} \tag{4.100}$$

Next, we consider modeling based on the competitive import model. Replacing $\boldsymbol{A}^d$ and $\boldsymbol{A}^m$ with $(\boldsymbol{I} - \hat{\boldsymbol{\mu}})\boldsymbol{A}$ and $\hat{\boldsymbol{\mu}}\boldsymbol{A}$, respectively, we obtain the competitive-import counterpart of (4.98)

$$\boldsymbol{p}^d = \boldsymbol{p}^d (\boldsymbol{I} - \hat{\boldsymbol{\mu}})\boldsymbol{A} + \boldsymbol{p}^m \hat{\boldsymbol{\mu}} \boldsymbol{A} + \boldsymbol{v} \tag{4.101}$$

Here, $\boldsymbol{A}$ corresponds to $\boldsymbol{A}^a$ in Sect. 4.3.2. We can solve for $\boldsymbol{p}^d$ as

$$\boldsymbol{p}^d = (\boldsymbol{p}^m \hat{\mu}\, \boldsymbol{A} + \boldsymbol{v})(\boldsymbol{I} - (\boldsymbol{I} - \hat{\mu})\boldsymbol{A})^{-1} \tag{4.102}$$

To determine the effects of a change in $\boldsymbol{p}^m$ on $\boldsymbol{p}^d$, we use the following equation

$$\Delta \boldsymbol{p}^m \hat{\boldsymbol{\mu}} \boldsymbol{A} (\boldsymbol{I} - (\boldsymbol{I} - \hat{\boldsymbol{\mu}})\boldsymbol{A})^{-1}. \tag{4.103}$$

4.5.4 ***Why Can Prices be Determined Independent of Quantities in the IO Model?*

The conventional two-dimensional schematic diagram that many of you may have seen shows the price and quantity of a product, determined by the intersection of a downward-sloping demand curve and an upward-sloping supply curve.[5] In contrast, the IO model we just discussed shows that prices and quantities are determined independently, not simultaneously. Prices are determined by $\boldsymbol{p} = \boldsymbol{v}\boldsymbol{L}$, where no quantities of products are involved, while quantities are determined by $\boldsymbol{x} = \boldsymbol{L}\boldsymbol{y}$, where no prices of products are involved.

[5] However, this diagram is rarely based on empirical data and should be viewed as a cartoon rather than an actual representation.

The independent determination of $\boldsymbol{p}$ and $\boldsymbol{x}$ in the IO model originates from the assumption of constant returns to scale (CRTS) that excludes any change in the unit costs of production when the level of output changes. Please see ([19], Sect. 4.4.3) for further details of the underlying microeconomics.

To better understand the impact of the assumption of CRTS being lifted, we can examine the quantity and price models (4.88) and (4.87) when the input coefficients $\boldsymbol{A}$ and $\boldsymbol{v}$ become dependent on the scale of production. In this case, the models become:

$$\begin{aligned}(\boldsymbol{I} - \boldsymbol{A}(\boldsymbol{x}))\boldsymbol{x} &= \boldsymbol{y} \\ \boldsymbol{p}(\boldsymbol{I} - \boldsymbol{A}(\boldsymbol{x})) &= \boldsymbol{v}(\boldsymbol{x})\end{aligned} \tag{4.104}$$

When CRTS is no longer assumed, product prices and quantities can no longer be determined independently. Additionally, the quantity model becomes nonlinear in $\boldsymbol{x}$ and requires iterative solutions.

CRTS may not hold true when there are limitations in the availability of resources, including production capacity. An example is where available production capacity consists of different technologies embodied in the capital stock of different ages, resulting in the coexistence of facilities with different unit operating costs. Reference [72] provides an example of this type of model. As output levels increase, the facility with the next cheapest operating cost will be put into operation, leading to diseconomies of scale. Those interested in this topic can read [73], which provides insight into this issue, despite having been published more than 60 years ago.

References

1. Leontief, Wassily. 1951. *The structure of American economy, 1919–1939: An empirical application of equilibrium analysis*, 2nd ed. New York: Oxford University Press.
2. Leontief, Wassily. 1985. *Input-output analysis*. Input-output economics. USA, New York: Oxford University Press.
3. Haan, Mark De, and Steven J. Keuning. 1996. Taking the environment into account: The NAMEA approach. *Review of Income and Wealth* 42 (2): 131–148.
4. Nakamura, Shinichiro. 2020. Tracking the product origins of waste for treatment using the WIO data developed by the Japanese ministry of the environment. *Environmental Science and Technology* 54 (23): 14862–14867.
5. Weisz, Helga, and Faye Duchin. 2006. Physical and monetary input-output analysis: What makes the difference? *Ecological Economics* 57: 534–541.
6. U.S. Bureau of Economic Analysis. Input-output accounts data. https://www.bea.gov/industry/input-output-accounts-data. Accessed 31 May 2023.
7. Ministry of Internal Affairs and Communications. Input-output tables for Japan. https://www.soumu.go.jp. Accessed 31 May 2023.
8. Bank of Korea. 2018 updated input-output tables. https://www.bok.or.kr/. Accessed 31 May 2023.
9. Lenzen, Manfred, Daniel Moran, Keiichiro Kanemoto, and Arne Geschke. 2013. Building Eora: A global multi-region input-output database at high country and sector resolution. *Economic Systems Research* 25 (1): 20–49.

10. United Nations Statistics Division. 2008. *International standard industrial classification of all economic activities (ISIC), Rev.4*. United Nations Publication, Revision 4 edition.
11. Miller, Ronald E., and Peter D. Blair. 2009. *Input-output analysis: Foundations and extensions*. Cambridge: Cambridge University Press.
12. Miller, Ronald E., and Peter D. Blair. 2022. *Input-output analysis foundations and extensions*, 3rd ed. Cambridge University Press.
13. Beutel, Joerg. 2017. The supply and use framework of national accounts. In *Handbook of input-output analysis*, Chap. 3, ed. T. Ten Raa.
14. Timmer, Marcel P., Erik Dietzenbacher, Bart Los, Robert Stehrer, and Gaaitzen J. de Vries. 2015. An illustrated user guide to the world input-output database: The case of global automotive production. *Review of International Economics* 23 (3): 575–605.
15. Stadler, Konstantin, Richard Wood, Tatyana Bulavskaya, Carl Johan Södersten, Moana Simas, Sarah Schmidt, Arkaitz Usubiaga, José Acosta-Fernández, Jeroen Kuenen, Martin Bruckner, Stefan Giljum, Stephan Lutter, Stefano Merciai, Jannick H. Schmidt, Michaela C. Theurl, Christoph Plutzar, Thomas Kastner, Nina Eisenmenger, Karl Heinz Erb, Arjan de Koning, and Arnold Tukker. 2018. EXIOBASE 3: Developing a time series of detailed environmentally extended multi-regional input-output tables. *Journal of Industrial Ecology* 22 (3): 502–515.
16. MITI. 1962. *Input-output analysis of the Japanese economy by Input-Output table of year 1955 (in Japanese)*. Tokyo: Sohbunsya.
17. Kazuo Takayama. A Historical Consideration on Japan's Input-Output Tables Focusing on the background to the creation of the. 1951. IO Table (in Japanese). *The Journal of Economic Science* 18: 2021.
18. Arai, Sonoe. 2018. Japan's input-output tables and primary statistics (in Japanese). *Business Journal of PAPAIOS* 26 (1): 50–65.
19. Nakamura, Shinichiro, and Yasushi Kondo. 2009. *Waste input-output analysis*. Dordrecht: Springer Science & Business Media.
20. UN Department of Economic and Social Affairs. 1999. *UN Handbook of input-output table compilation and analysis*. Number 74.
21. U.S. Energy Information Administration. Gasoline explained. https://www.nasa.gov/. Accessed 31 May 2023.
22. Heijungs, Reinout, and Sangwon Suh. 2002. *The computational structure of life cycle assessment*, vol. 11. Springer Science & Business Media.
23. Heijungs, Reinout, Yi Yang, and Hung Suck Park. 2022. A or I-A? Unifying the computational structures of process- and IO-based LCA for clarity and consistency. *Journal of Industrial Ecology* 1 (1): 1824–1836.
24. von Neumann, John. 1945. A model of general economic equilibrium. *Review of Economic Studies* 13 (1): 1–9.
25. Richard Stone. 1961. *Input-output and national accounts*. Organisation for European Economic Co-operation.
26. Eurostat. 2008. *Eurostat Manual of Supply, Use and Input-Output Tables*.
27. Isard, Walter. 1951. Interregional and regional input-output analysis: A model of a space-economy. *The Review of Economics and Statistics*, 318–328.
28. Chenery, Hollis B. 1953. Regional analysis. In *The structure and growth of the Italian economy*, ed. V. Punna, H. Chenery, and P. Clark, 98–139. Rome: US Mutual Security Agency.
29. Moses, Leon. 1955. The stability of interregional trading patterns and input-output analysis. *The American Economic Review* 45 (5): 803–826.
30. Wonnacott, Ronald J. 1961. *Canadian-American dependence: An interindustry analysis of production and prices*. Number 24. Amsterdam: North-Holland.
31. Meng, Bo, Yaxiong Zhang, and Satoshi Inomata. 2013. Compilation and applications of Ide-Jetro'S international input-output tables. *Economic Systems Research* 25 (1): 122–142.
32. Lenzen, Manfred, Arne Geschke, Muhammad Daaniyall Abd Rahman, Yanyan Xiao, Jacob Fry, Rachel Reyes, Erik Dietzenbacher, Satoshi Inomata, Keiichiro Kanemoto, Bart Los, Daniel Moran, Hagen Schulte in den Bäumen, Arnold Tukker, Terrie Walmsley, Thomas Wiedmann, Richard Wood, and Norihiko Yamano. 2017. The Global MRIO Lab-charting the world economy. *Economic Systems Research* 29 (2): 158–186.

33. Lenzen, Manfred, Arne Geschke, James West, Jacob Fry, Arunima Malik, Stefan Giljum, Llorenç Milà i Canals, Pablo Piñero, Stephan Lutter, Thomas Wiedmann, Mengyu Li, Maartje Sevenster, Janez Potočnik, Izabella Teixeira, Merlyn Van Voore, Keisuke Nansai, and Heinz Schandl. 2022. Implementing the material footprint to measure progress towards sustainable development goals 8 and 12. *Nature Sustainability* 5 (2), 157–166.
34. Aguiar, Angel, Badri Narayanan, and Robert McDougall. 2016. An overview of the GTAP 9 data base. *Journal of Global Economic Analysis* 1 (1): 181–208.
35. Huo, Jingwen, Peipei Chen, Klaus Hubacek, Heran Zheng, Jing Meng, and Dabo Guan. 2022. Full-scale, near real-time multi-regional input-output table for the global emerging economies (EMERGING). *Journal of Industrial Ecology*.
36. Yamano, Norihiko, and Colin Webb. 2018. Future development of the inter-country input-output (ICIO) database for global value chain (GVC) and environmental analyses. *Journal of Industrial Ecology* 22 (3): 487–488.
37. Wiebe, Kirsten S., Martin Bruckner, Stefan Giljum, and Christian Lutz. 2012. Calculating energy-related CO 2 emissions embodied in international trade using a global input-output model. *Economic Systems Research* 24 (2): 113–139.
38. Schulte, Simon, Arthur Jakobs, and Stefan Pauliuk. 2021. Relaxing the import proportionality assumption in multi-regional input-output modelling. *Journal of Economic Structures* 10 (1).
39. Dietzenbacher, Erik, Bart Los, Robert Stehrer, Marcel Timmer, and Gaaitzen de Vries. 2013. The Construction of World Input-Output Tables in the Wiod Project. *Economic Systems Research* 25 (1): 71–98.
40. Stadler, Konstantin, Kjartan Steen-Olsen, and Richard Wood. 2014. The 'Rest of the World' - estimating the economic structure of missing regions in global multi-regional input-output tables. *Economic Systems Research* 26 (3): 303–326.
41. Chen, Chuke, Zhihan Jiang, Nan Li, Heming Wang, Peng Wang, Zhihe Zhang, Chao Zhang, Fengmei Ma, Yuanyi Huang, Lu. Xiaohui, Jianlimin Wei, Jianchuan Qi, and Wei-qiang Chen. 2022. Advancing UN comtrade for physical trade flow analysis: Review of data quality issues and solutions. *SSRN Electronic Journal* 186 (July): 106526.
42. Industrial Ecology Virtual Laboratory. 2021. Global resource input output assessment (GLORIA) database. Technical Report September.
43. Wood, Richard, Konstantin Stadler, Tatyana Bulavskaya, Stephan Lutter, Stefan Giljum, Arjan de Koning, Jeroen Kuenen, Helmut Schütz, José Acosta-Fernández, Arkaitz Usubiaga, Moana Simas, Olga Ivanova, Jan Weinzettel, Jannick H Schmidt, Stefano Merciai, and Arnold Tukker. 2015. Global sustainability accounting-developing EXIOBASE for multi-regional footprint analysis. *Sustainability (Switzerland)* 7 (1): 138–163.
44. Jakobs, Arthur, Simon Schulte, and Stefan Pauliuk. 2021. Price variance in Hybrid-LCA leads to significant uncertainty in carbon footprints. *Frontiers in Sustainability* 2 (May): 1–13.
45. Bjelle, Eivind Lekve, Johannes Többen, Konstantin Stadler, Thomas Kastner, Michaela C Theurl, Karl Heinz Erb, Kjartan Steen Olsen, Kirsten S Wiebe, and Richard Wood. 2020. Adding country resolution to EXIOBASE: Impacts on land use embodied in trade. *Journal of Economic Structures* 9 (1).
46. Cabernard, Livia, and Stephan Pfister. 2021. A highly resolved MRIO database for analyzing environmental footprints and Green Economy Progress. *Science of the Total Environment* 755.
47. Horrowitz, Mark, and Karen Planting. 2009. *Concepts and Methods of the U.S. Input-Output Accounts*. Number 1997. Bureau of Economic Analysis.
48. UN. National Accounts: A Practical Introduction. 2003. *Studies in Methods Handbook of National Accounting Series F*, No 85, 147.
49. Frischknecht, Rolf, Hans-Jörg. Althaus, Christian Bauer, Gabor Doka, Thomas Heck, Niels Jungbluth, Daniel Kellenberger, and Thomas Nemecek. 2007. The environmental relevance of capital goods in life cycle assessments of products and services. *The International Journal of Life Cycle Assessment* 12 (1): 7–17.
50. Ministry of Internal Affairs and Communications. 2007. *2015 input-output table basic guidelines*.

51. Moriizumi, Yue, Hiroki Hondo, and Satoshi Nakano. 2015. Development and application of renewable energy-focused input-output table (in Japanese). *Journal of the Japan Institute of Energy* 94 (12): 1397–1413.
52. Miller, T. Reed, Peter Berrill, Paul Wolfram, Ranran Wang, Yookyung Kim, Xinzhu Zheng, and Edgar G Hertwich. 2019. Method for endogenizing capital in the United States Environmentally-Extended input-output model. *Journal of Industrial Ecology* 23 (6): 1410–1424.
53. Berrill, Peter, T. Reed Miller, Yasushi Kondo, and Edgar G. Hertwich. Capital in the American carbon, energy, and material footprint. *Journal of Industrial Ecology* 24 (3), 589–600.
54. Leontief, Wassily. 1953. Dynamic Analysis. In *Studies in the structure of the American economy*, 53–90, ed. W. Leontief. New York: Oxford University Press.
55. Tokoyama, K., and T. Murakami. 1972. Relative stability in two types of dynamic Leontief models. *International Economic Review* 13 (2): 408–415.
56. Steenge, Albert E. 1990. On the complete instability of empirically implemented dynamic Leontief models. *Economic Systems Research* 2 (1): 3–16.
57. MacLean, Heather L., and Lester B. Lave. 2003. Life cycle assessment of automobile/fuel options. *Environmental Science & Technology* 37: 5445–5452.
58. Jorgenson, Dale W. 1961. Stability of a dynamic Input-output system. *The Review of Economic Studies* 28 (2): 105–116.
59. Duchin, F., and D.B. Szyld. 1985. A dynamic input-output model with assured positive output. *Metroeconomica* 37 (3): 269–282.
60. Lenzen, Manfred. 1998. Primary energy and greenhouse gases embodied in Australian final consumption: An input-output analysis. *Energy Policy* 26 (6): 495–506.
61. Lenzen, M. 2001. A generalized input-output multiplier calculus for Australia. *Economic Systems Research* 13 (1): 65–92.
62. Södersten, Carl Johan H, Richard Wood, and Edgar G Hertwich. 2018. Endogenizing capital in mrio models: The implications for consumption-based accounting. *Environmental Science and Technology* 52 (22): 13250–13259.
63. Nansai, Keisuke, Rokuta Inaba, Shigemi Kagawa, and Yuichi Moriguchi. 2008. Identifying common features among household consumption patterns optimized to minimize specific environmental burdens. *Journal of Cleaner Production* 16 (4): 538–548.
64. Nakamura, Shinichiro, and Yasushi Kondo. 2002. Input-output analysis of waste management. *Journal of Industrial Ecology* 6 (1): 39–63.
65. Groth, Charlotta, and Hashmat Khan. 2010. Investment adjustment costs: An empirical assessment. *Journal of Money, Credit and Banking* 42 (8): 1469–1494.
66. Lenzen, Manfred, and Graham J. Treloar. 2006. Endogenising capital : A comparison of two methods. *Journal of Applied Input-Output Analysis* 10.
67. UN. System of National Accounts. 2008.
68. O'Mahony, Mary, and Marcel P. Timmer. 2009. Output, input and productivity measures at the industry level: The EU KLEMS database. *Economic Journal* 119 (538).
69. Södersten, Carl Johan, Richard Wood, and Thomas Wiedmann. 2020. The capital load of global material footprints. *Resources, Conservation and Recycling* 158 (January): 104811.
70. Ye, Quanliang, Edgar G. Hertwich, Maarten S. Krol, David Font Vivanco, Amanda W. Lounsbury, Xinzhu Zheng, Arjen Y. Hoekstra, Yutao Wang, and Ranran Wang. 2021. Linking the environmental pressures of China's capital development to global final consumption of the past decades and into the future. *Environmental Science and Technology* 55 (9), 6421–6429.
71. OECD. 2009. *Measuring capital: OECD manual 2009*, 2nd ed. Paris: OECD.
72. Pauliuk, Stefan, Richard Wood, and Edgar G. Hertwich. 2015. Dynamic models of fixed capital stocks and their application in industrial ecology. *Journal of Industrial Ecology* 19 (1): 104–116.
73. Salter, W.E.G. 1960. *Productivity and technical change*. Cambridge University Press.

Chapter 5
Environmentally Extended Input-Output Analysis (EEIO) and Hybrid LCA

Abstract This chapter covers a range of topics related to integrating emissions into input-output (IO) models and life cycle assessment (LCA). We begin by discussing combustion stoichiometry and greenhouse gas emissions resulting from fuel consumption. Additionally, we explore modeling emissions using exogenous and endogenous approaches, and discuss direct and embodied emissions, as well as Scopes 1, 2, and 3 of the GHG protocol. We also examine how to avoid double-counting in footprint analysis and the relationship between emissions and international trade. Moving on to LCA, we provide an overview of its basics and discuss the EEIO model from the point of view of ISO-LCA. We then delve into the hybridization of LCA and EEIO, highlighting the sectoral resolution of an IO table and its implications for LCA and HLCA models. Furthermore, we explore waste input-output analysis (WIO), including two types of by-products and the WIO table and model. The chapter also covers the basics of waste treatment processes, such as incineration. Lastly, we touch on other topics related to LCA, such as environmental life cycle costing (eLCC), social LCA, distributing environmental responsibility, and attributional and consequential LCA.

5.1 Integrating Emissions into the IO Model

We consider incorporating the emissions caused by fossil fuel combustion in the production and consumption phases into the IO model. Understanding the basics of GHG emissions from fuel combustion is crucial for developing effective strategies to reduce emissions and mitigate climate change.

5.1.1 Combustion Stoichiometry and GHG Emissions of Fuel Consumption

In this section, we provide an overview of the basics of greenhouse gas (GHG) emissions resulting from fuel combustion. To represent fossil fuels, we use the chemical

S. Nakamura, *A Practical Guide to Industrial Ecology by Input-Output Analysis*,
https://doi.org/10.1007/978-3-031-43684-0_5

formula $C_aH_bO_cS_xN_y$. Assuming complete combustion, the emissions from fuel combustion can be determined using combustion stoichiometry ([1] (p.18)).

$$\begin{aligned}&C_aH_bO_cS_xN_yCl_z + [a + 0.5(b - c) + x]O_2\\&\quad\longrightarrow aCO_2 + (b - 0.5z)H_2O + xSO_2 + y[\alpha NO + 0.5(1 - \alpha)N_2] + zHCl\end{aligned} \tag{5.1}$$

The equation shows that the combustion of fossil fuels generates several types of emissions, including carbon dioxide (CO_2), water vapor (H_2O), sulfur dioxide (SO_2), nitrogen oxide (NO), and hydrogen chloride (HCl). The parameter α represents the fraction of NO that originates from the fuel itself, while $1 - \alpha$ represents the fraction that results from the combustion air. Consider some examples, starting with coal. Coal (medium rank) is around 81.1% carbon (C), followed by 11.4% oxygen (O), 5.4% hydrogen (H), 1.2% nitrogen (N), and 0.8% sulfur (S) [2]. Coal of high and low ranks mostly differ in the C and O content, with C increasing and O declining with the rank. CO_2 forms during coal combustion when one atom of C unites with two atoms of O from the air; $a = 1$ in (5.1). Since the atomic weight of carbon is 12 and that of oxygen is 16, the atomic weight of CO_2 is $12 \times 1 + 16 \times 2 = 44$ g per mol. Complete combustion of one g of coal (medium rank) emits

$$0.81\text{g-C/g-Coal} \times \frac{44}{12}\frac{\text{g-CO}_2}{\text{g-C}} = 2.97\text{g-CO}_2/\text{g-Coal}. \tag{5.2}$$

We next turn to gasoline and diesel. The chemical formula of gasoline can be represented by isooctane C_8H_{18} ([3] p.22). The mass per mole of gasoline is $12 \times 8 + 8 = 114$ g, of which C makes 96 g. Approximating that gasoline weighs 0.75 kg/L, it contains $0.75 \times 96/114 = 0.63$ kg of C per L. Complete combustion of one L of gasoline emits

$$0.63\frac{\text{kg-C}}{\text{L-C}_8\text{H}_{18}} \times \frac{44}{12}\frac{\text{g-CO}_2}{\text{g-C}} = 2.31\text{kg-CO}_2/\text{L-C}_8\text{H}_{18}. \tag{5.3}$$

Diesel can be approximated by the formula $C_{12}H_{23}$. It contains $12 \times 12 = 144$ g-C per mole. With the density of diesel being about 0.85 kg/L, combustion of one L emits

$$0.85\frac{\text{kg-C}_{12}\text{H}_{23}}{\text{L-C}_{12}\text{H}_{23}} \times \frac{144}{167}\frac{\text{kg-C}}{\text{kg-C}_{12}\text{H}_{23}} \times \frac{44}{12}\frac{\text{g-CO}_2}{\text{g-C}} = 2.68\text{kg-CO}_2/\text{L-diesel}. \tag{5.4}$$

Since diesel has a higher heat value per mass than gasoline, the difference in emission between them is smaller in terms of heat value (see Table 5.1). Compared with coal, diesel, and gasoline, natural gas (mostly CH_4) has a distinctly smaller CO_2 emission per heat value thanks to the occurrence of four H atoms per C atom (see Table 5.1).

Table 5.1 provides information on the heat value and CO_2 emission of several fuels and wastes. These values differ from the theoretical values calculated above because of variations in composition.

Table 5.1 CO_2 emission factors of fuels and resources

Fuel and resource name	Heat value		Emission factor per heat		Emission factor per mass	
Coking coal	29.0	GJ/t	0.0899	t-CO_2/GJ	2.61	t-CO_2/t
Steam coal, lignite and anthracite	25.7	GJ/t	0.0906	t-CO_2/GJ	2.33	t-CO_2/t
Coke	29.4	GJ/t	0.1077	t-CO_2/GJ	3.17	t-CO_2/t
Blast furnace coke	29.4	GJ/t	0.1077	t-CO_2/GJ	3.17	t-CO_2/t
Coke oven gas (COG)	21.1	GJ/10^3 Nm^3	0.0403	t-CO_2/GJ	0.85	t-CO_2/10^3 Nm^3
BFG (Consumption)	3410	GJ/10^6 Nm^3	0.1077	t-CO_2/GJ	367.35	t-CO_2/10^6 Nm^3
BFG (Generation)	3410	GJ/10^6 Nm^3	0.1077	t-CO_2/GJ	367.35	t-CO_2/10^6 Nm^3
LDG (Consumption)	8410	GJ/10^6 Nm^3	0.1077	t-CO_2/GJ	905.98	t-CO_2/10^6 Nm^3
LDG (Generation)	8410	GJ/10^6 Nm^3	0.1077	t-CO_2/GJ	905.98	t-CO_2/10^6 Nm^3
Carbon in steel for LDG generation	8410	GJ/$10^6$$Nm^3$	0.1077	t-CO_2/GJ	905.98	t-CO_2/10^6 Nm^3
Crude oil	38.2	GJ/kl	0.0684	t-CO_2/GJ	2.61	t-CO_2/kl
Fuel oil A	39.1	GJ/kl	0.0693	t-CO_2/GJ	2.71	t-CO_2/kl
Fuel oils B and C	41.9	GJ/kl	0.0716	t-CO_2/GJ	3.00	t-CO_2/kl
Kerosene	36.7	GJ/kl	0.0679	t-CO_2/GJ	2.49	t-CO_2/kl
Diesel oil	37.7	GJ/kl	0.0687	t-CO_2/GJ	2.59	t-CO_2/kl
Gasoline	34.6	GJ/kl	0.0671	t-CO_2/GJ	2.32	t-CO_2/kl
Jet fuel	36.7	GJ/kl	0.0671	t-CO_2/GJ	2.46	t-CO_2/kl
Naphtha	33.6	GJ/kl	0.0666	t-CO_2/GJ	2.24	t-CO_2/kl
Petroleum-based hydrocarbon gas	44.9	GJ/10^6 Nm^3	0.0519	t-CO_2/GJ	2.33	t-CO_2/10^6 Nm^3
Hydrocarbon oil	41.7	GJ/kl	0.0762	t-CO_2/GJ	3.18	t-CO_2/kl
Petroleum coke	29.9	GJ/t	0.0930	t-CO_2/GJ	2.78	t-CO_2/t
Liquefied petroleum gas (LPG)	50.8	GJ/t	0.0598	t-CO_2/GJ	3.04	t-CO_2/t
Natural gas, LNG	54.6	GJ/t	0.0494	t-CO_2/GJ	2.70	t-CO_2/t
Mains gas	44.8	GJ/$10^6$$Nm^3$	0.0501	t-CO_2/GJ	2.24	t-CO_2/10^6 Nm^3
Black liquor	13.2	GJ/t (dry)	0.0953	t-CO_2/GJ	1.26	t-CO_2/t (dry)
Waste wood	16.3	GJ/t (dry)	0.1120	t-CO_2/GJ	1.83	t-CO_2/t (dry)
Waste tires	33.2	GJ/t	0.0523	t-CO_2/GJ	1.74	t-CO_2/t
Municipal waste	10.7	GJ/t	0.0259	t-CO_2/GJ	0.28	t-CO_2/t
Industrial waste	16.8	GJ/t	0.0419	t-CO_2/GJ	0.70	t-CO_2/t
Recycled plastic of packaging origins	48	GJ/t	0.0666	t-CO_2/GJ	3.20	t-CO_2/t
Nuclear power generation	3600	GJ/GWh	0		0.00	
Hydro and other power generation	3600	GJ/GWh	0	–	0.00	
Limestone	0	–	0.440	t-CO_2/t	0.440	t-CO_2/t

Units: "t" is 10^3 kg. *Source* 3EID National Institute for Environmental Studies, 2016

5.1.2 *Modeling the Emissions: Exogenous Versus Endogenous Approaches*

Japanese IO tables offer one of the highest sectoral resolutions in the world. The National Institute for Environmental Studies (NIES) in Japan has taken advantage of this fact and developed a database called Embodied Energy and Emission Intensity Data (3EID), which provides IO-based GHG emissions data for six gases (CO_2, CH_4, N_2O, HFCs, PFCs, and SF_6) at the resolution of 400 production sectors, starting from the 1995 IO table. The 3EID is remarkable due to its high resolution and continuous data collection spanning over a quarter of a century, which is a rare case worldwide. In this regard, we provide detailed information about the methodology and outcomes of the 3EID. According to [4], the methodology is called the *endogenous approach*, as it estimates emissions fully consistent with the IO framework.

5.1.2.1 The Endogenous Approach

Let there be n production sectors, n_f types of fuel input, and n_e types of emissions generated by fuel combustion. Let $\boldsymbol{Z}_f^*$ be the $n_f \times n$ matrix representing fuel input by sector in physical units (eg., joules). For instance, Table 5.2 shows the energy consumption in three Japanese iron & steel sectors, where the generation of blast furnace gas (BFG) from the Pig Iron sector is counted as a negative value because it refers to a competitive by-product (Section 4.1.4.1). Let $\Xi = [\xi_{ij}]$ be an $n_e \times n_f$ matrix with ξ_{ij} representing the emission of i (eg., CO_2) per unit consumption of fuel j. Table 5.1 provides the heat value and CO_2 emission for a selection of fuel and wastes. It is worth noting that, among the fuel items listed in Table 5.1 (excluding gases obtained as by-products), natural gas emits the least amount of CO_2 per unit of heat (49.4 g-CO_2/MJ), while coke emits the most (107.7 g-CO_2/MJ), with fuel oils in between (around 70 g-CO_2/MJ).

A fuel input can be used either as fuel or as feedstock; for example, petroleum products are used as feedstock in the production of plastics and lubricants. To filter out fuel inputs that are not used as fuel, we use a matrix $\boldsymbol{H} = [h_{ij}]$ of size $n_f \times n$, where $h_{ij} = 1$ if fuel input i is used as fuel in sector j, and $h_{ij} = 0$ if it is used as feedstock in sector j. We can then obtain the GHG emissions associated with fuel consumption by taking the element-wise product of $\boldsymbol{H}$ and $\boldsymbol{Z}_f^*$ (the $n_f \times n$ matrix of fuel input by sector in physical units) as

$$\Xi \boldsymbol{H} \odot \boldsymbol{Z}_f^* \tag{5.5}$$

where $\odot$ refers to the Hadamard product (element-wise product), that is, $\boldsymbol{H} \odot \boldsymbol{Z}_f^* = [h_{ij} z^*_{f\,ij}]$. Dividing this matrix by the $n \times 1$ vector of outputs $\boldsymbol{x}^*$ in physical units, we get the matrix of direct emission coefficients $\boldsymbol{F}^*$ as:

$$\boldsymbol{F}^* = \Xi \boldsymbol{H} \odot \boldsymbol{Z}_f^* \hat{x}^{*-1} = \Xi \boldsymbol{H} \odot \boldsymbol{A}_f^* \tag{5.6}$$

Table 5.2 Energy consumption of iron & steel sectors

ID	Fuel and resource	Units	Sectors		
			Pig iron	Ferro alloys	Crude steel[a]
1	Coking coal	GJ	359773420	9450142	22357779
2	Steam coal, lignite, and anthracite	GJ	0	0	0
3	Coke	GJ	118898800	9029416	2041301
4	Blast furnace coke	GJ	953337869	0	0
5	Coke oven gas (COG)	GJ	103763848	1540992	6291663
6	BFG (Consumption)	GJ	211543911	3377437	11295460
7	BFG (Generation)	GJ	−441990884	0	0
8	LDG (Consumption)	GJ	35122489	560753	1875377
9	LDG (Generation)	GJ	0	0	−73383440
10	Carbon in steel for LDG generation	GJ	−73383440	0	73383440
11	Crude oil	GJ	0	0	0
12	Fuel oil A	GJ	38162	211492	697818
13	Fuel oils B and C	GJ	11089732	5515702	1256824
14	Kerosene	GJ	0	26610	83298
15	Diesel oil	GJ	450	2288	0
16	Gasoline	GJ	0	4067	0
17	Jet fuel	GJ	0	0	0
18	Naphtha	GJ	0	0	0
19	Petroleum-based hydrocarbon gas	GJ	0	0	0
20	Hydrocarbon oil	GJ	0	0	0
21	Petroleum coke	GJ	14913701	0	0
22	Liquefied petroleum gas (LPG)	GJ	592377	79172	3853988
23	Natural gas, LNG	GJ	2787161	0	54147
24	Mains gas	GJ	8785504	1971	1878778
25	Black liquor	GJ	0	0	0
26	Waste wood	GJ	0	0	0
27	Waste tires	GJ	1802982	0	2133
28	Municipal waste	GJ	0	0	0
29	Industrial waste	GJ	0	0	0
30	Recycled plastic of packages origins	GJ	5205803	64074	327269
31	Nuclear power generation	GJ	0	0	0
32	Hydro and other power generations	GJ	0	0	0
33	Limestone	GJ	0	0	0

Units: 10^3 kg. [a]Converters. *Source* 3EID National Institute for Environmental Studies, 2016 where $\boldsymbol{A}_f^*$ is the $n_f \times n$ matrix of physical input coefficients (the masses of fuels per output in physical units). The matrix $\boldsymbol{F}^*$ gives the direct emissions in physical units per unit of output in physical units. Further details can be found in [4].

In reality, the output is usually measured not in physical units but in monetary units. Accordingly, not the physical units × physical units matrix, $\boldsymbol{Z}_f^*$, but the physical units × monetary units matrix, $\boldsymbol{Z}_f$, is used. Denote by $\boldsymbol{A}_f$ the matrix of fuel input coefficients of mixed units, with its row elements, fuel, measured in physical units and its column elements, producing sectors, in monetary units

$$\boldsymbol{A}_f = \boldsymbol{Z}_f \hat{\boldsymbol{x}}^{-1}. \tag{5.7}$$

Replacing the physical matrix $\boldsymbol{A}_f^*$ in (5.6) with $\boldsymbol{A}_f$, we obtain the direct emission matrix per output in monetary units

$$\boldsymbol{F} = \boldsymbol{\Xi} \boldsymbol{H} \odot \boldsymbol{A}_f \tag{5.8}$$

which can be converted to the physical matrix when divided by the vector of output prices

$$\boldsymbol{F}^* = \boldsymbol{F} \hat{\boldsymbol{p}}_x^{-1} \tag{5.9}$$

With the direct emission matrix $\boldsymbol{F}$ obtained, the matrix of embodied emissions, $\boldsymbol{R} = [r_{ij}]$, with r_{ij} representing the amount of i directly and indirectly emitted per unit production of product j, is given by

$$\boldsymbol{R} = \boldsymbol{F}(\boldsymbol{I} - \boldsymbol{A})^{-1} \tag{5.10}$$

which we already encountered in Sect. 3.4 (3.93), but with the emission limited to CH_4. The GHG emissions embodied in the final demand, $\boldsymbol{y}$, or the GHG footprints, are given by

$$\boldsymbol{E} = \boldsymbol{F} \boldsymbol{L} \hat{\boldsymbol{y}} \tag{5.11}$$

Henceforth, we call $\boldsymbol{R} = \boldsymbol{F}\boldsymbol{L}$ the embodied emission intensity.

The dependence of the direct emission matrix $\boldsymbol{F}$ on the input matrix $\boldsymbol{A}$ is now clear. As mentioned previously, the elements of $\boldsymbol{\Xi}$ are determined by physical laws, and for a given fuel and technology, $\boldsymbol{\Xi}$ is fixed. The matrix $\boldsymbol{H}$ is also exogenously given. Therefore, it is $\boldsymbol{A}_f$ that determines $\boldsymbol{F}$. If $\boldsymbol{A}_f$ changes, such as through a change in fuel mix, $\boldsymbol{F}$ also changes. However, since $\boldsymbol{A}_f$ is a part of the input matrix $\boldsymbol{A}$, it is unlikely that $\boldsymbol{A}_f$ changes while the rest of $\boldsymbol{A}$ remains unchanged. Changing the fuel mix would require alterations to the relevant equipment with associated changes in the inputs required for its operation and maintenance. Therefore, it is unlikely that $\boldsymbol{F}$ changes independently of $\boldsymbol{A}$.

This observation raises doubts about the validity of the frequently used practice of "structural decomposition analysis" [5–8], which factors a change in the amount of emissions into a change in $\boldsymbol{F}$ and a change in $\boldsymbol{L}$, treating them as if they were independent factors

$$\Delta \boldsymbol{E} = \Delta \boldsymbol{F}\boldsymbol{L}\boldsymbol{y} + \boldsymbol{F}\Delta \boldsymbol{L}\boldsymbol{y} + \boldsymbol{F}\boldsymbol{L}\Delta \boldsymbol{y} + \text{mixed effects} \tag{5.12}$$

Mathematically, this decomposition may be correct, but it may not be technologically correct.

5.1.2.2 The Exogenous Approach

As previously discussed, the endogenous approach derives the direct emission matrix $\boldsymbol{F}$ from information contained in IO tables, which is consistent with the $\boldsymbol{A}$ matrix. On the other hand, the exogenous approach obtains $\boldsymbol{F}$ from externally available environmental data independent of $\boldsymbol{A}$. The exogenous approach is employed in various studies, including the seminal article on embodied emissions [9], a recent study on emissions in the Korean electronics industry [10], and the World Input-Output Database (WIOD) [11]. The soundness of the resulting $\boldsymbol{E}$ matrix depends on the alignment between the definition of sectoral activity in the IO table and that of the emission source categories in the environmental data [4].

The NIES employed the endogenous approach to construct the 3EID database due to the availability of high-resolution IO tables and detailed physical data on sectoral energy consumption, which is a luxury that researchers in most countries do not have. Therefore, $\boldsymbol{F}$ is usually obtained using the exogenous approach. However, the endogenous approach is useful theoretically in clarifying the relationship between $\boldsymbol{F}$ and $\boldsymbol{A}$, even if data limitations make its full implementation difficult.

5.1.3 Direct and Embodied Emissions

We will now examine the values of direct and embodied emissions, starting with some examples taken from 3EID.

5.1.3.1 Examples Taken from 3EID

Table 5.3 presents the direct emission matrix, $\boldsymbol{F}$, and the embodied emission intensity matrix, $\boldsymbol{R}$, for a small selection of sectors taken from 3EID for the year 2011.

$\boldsymbol{F}\boldsymbol{L}$ or $\boldsymbol{F}\boldsymbol{L}^d$ and the Domestic Technology Assumption

The embodied emission intensity $\boldsymbol{R} = \boldsymbol{F}\boldsymbol{L}$ is calculated using two different approaches in 3EID: one based on $(\boldsymbol{I} - \boldsymbol{A})^{-1}$ and the other based on $(\boldsymbol{I} - (\boldsymbol{I} - \boldsymbol{\mu})\boldsymbol{A})^{-1}$, depending on how imports are treated. The former assumes that

Table 5.3 Direct and embodied GHG emission intensities

Sector	Direct	Embodied		The direct ratio
	F	FL	FL^d	$F/(FL)$
Cement	126.12	136.56	134.88	0.92
Pig iron	61.20	68.52	65.09	0.89
Electricity	24.79	28.20	26.55	0.88
Air transport	9.28	11.26	10.30	0.82
Semiconductor devices	5.92	8.20	7.42	0.72
Dairy cattle farming	6.50	10.26	8.95	0.63
Ferroalloys	10.7	16.94	14.4	0.6
Rice	3.59	6.11	5.52	0.59
Chemical fertilizer	10.25	17.46	14.83	0.59
Wheat, barley and the like	2.31	5.94	4.91	0.39
Bread	0.34	2.94	2.13	0.12
Household air-conditioners	0.26	3.51	2.49	0.07
Crude steel (converters)	2.54	42.73	39.49	0.06
Processed meat products	0.22	4.78	2.78	0.05
Nonresidential construction (nonwooden)	0.14	3.33	2.84	0.04
Residential construction (nonwooden)	0.12	3.31	2.83	0.04
Nonresidential construction (wooden)	0.06	2.34	1.94	0.03
Residential construction (wooden)	0.05	2.21	1.79	0.02
Passenger motor cars	0.06	3.82	2.94	0.02

Source NIES 2016. Units: 10^3 kg-CO_2e/million Japanese yen. $\boldsymbol{L}$ refers to $(\boldsymbol{I}-\boldsymbol{A})^{-1}$. $\boldsymbol{L}^d$ refers to $(\boldsymbol{I}-(\boldsymbol{I}-\boldsymbol{\mu})\boldsymbol{A})^{-1}$. The average exchange rate in 2015 was 0.0833 US dollar/JPN yen or 120 JPN yen/US dollar

all imports can be domestically produced using the same technology as the domestic production represented by $\boldsymbol{A}$. The latter excludes all imports and considers only domestic flows (Section 4.3.2). The former method can account for the emissions associated with imports, subject to the assumption that the same technology used domestically is used in exporting countries (i.e., the "domestic technology assumption"). This assumption is reasonable for countries where the share of domestic sources is comparable to that of imports, but it may not be appropriate for countries like Japan, which relies heavily on importing natural resources. For example, Japan stopped producing aluminum domestically in the 1970s and now imports all of its aluminum ingots. As a result, the Japanese IO tables do not include any data on the aluminum smelting process (Section 2.2.4.2). Since the aluminum smelting process is highly energy intensive (Sect. 2.2.4.2), assuming domestic technology for aluminum will result in an underestimation of the impacts.

On the other hand, the latter approach based on L^d completely neglects the emissions associated with imports but can provide accurate emissions from domestic sources . In all cases in Table 5.3, FL surpass FL^d. Both approaches have advantages and disadvantages. Resorting to a multi-regional IO (MRIO) framework would be an ideal solution, at least conceptually (see Sect. 5.1.6 for further discussion). From this point on, our focus will be on FL unless stated otherwise.

Direct versus Indirect Emissions and the Degree of Fabrication

In Table 5.3, it is notable that the emissions differ significantly among sectors and that there is a considerable difference between the direct emission F and the embodied emission FL for each sector. However, comparing the values of F or FL across sectors may not be meaningful due to the wide range of product prices. For example, the emissions from "Cement" and "Pig iron" are high because these products are relatively inexpensive compared to products such as cars, electronics, and buildings. In 2015, the price of cement and pig iron per 10^3 kg was around 7000 JPY and 37000 JPY, respectively.

On the other hand, by comparing the relative magnitudes of F and FL among sectors, we can gain valuable information about the contribution of each sector to the overall emissions in the product supply chain. Note that FL gives the total emission, including direct and indirect emissions in the entire supply chain, with the relevant sector occurring at the end of the chain.

For "Cement," "Pig iron," and "Electricity," the ratio of F to FL is around 0.9, indicating that the emissions in the supply chain of these sectors mostly occur directly, with most emissions coming from cement kilns, blast furnaces, and steam boilers. These sectors are followed by "Air transport," " Semiconductor devices," "Dairy cattle farming," " Ferroalloys," "Rice," and "Chemical fertilizer," where the ratio of F to FL is around 0.6–0.8. The high share of direct GHG emissions in " Semiconductor devices" is attributed to the emission of perfluorocarbons (PFCs) in semiconductor manufacturing processes and equipment, while that of "Rice" is attributed to CH_4 emissions from rice paddies (Sect. 2.2.1.2). In these sectors, direct (Scope 1) emissions make up the majority of total emissions.

In contrast, for all other sectors in Table 5.3, the share of direct emissions is smaller than that of indirect emissions. Except for "Wheat, barley, and the like," for which the share of direct emission is 0.39, the share of direct emissions is smaller than 0.12, indicating that emissions mostly occur upstream of the supply chain indirectly, rather than directly within their facilities.

"Crude steel" is produced from "Pig iron" by reducing C to the desired level and removing contaminants such as S, P, and N. The contribution of this process to the embodied intensity of CO_2 is only 5.8%, with the rest mostly attributable to the immediate upstream process of pig iron production. "Passenger motor cars" are characterized by a remarkably low ratio of direct to embodied emission, 0.02, indicating that emissions mostly occur in the upstream processes, such as pig iron production, rather than the manufacturing process where cars are assembled and painted.

5.1.3.2 Alternative Decomposition of Emissions?

The decomposition of embodied emissions into its direct and indirect components in Table 5.3 is based on the equation:

$$e = \underbrace{\boldsymbol{Fy}}_{\text{direct emission}} + \underbrace{\boldsymbol{F(L-I)y}}_{\text{indirect emission}} \tag{5.13}$$

In Sect. 3.4.1.2, we noted that [12] proposed an alternative decomposition

$$e = \underbrace{\boldsymbol{F(I+A)y}}_{\text{direct emission}} + \underbrace{\boldsymbol{F(L-(I+A))y}}_{\text{indirect emission}} \tag{5.14}$$

This formulation is based on the definition of "direct output associated with final consumption," which is given as $(\boldsymbol{I}+\boldsymbol{A})\boldsymbol{y}$, representing "the direct requirements from producers in order to allow the industry to produce this output" [12].

Henceforth, we use the formulation in (5.13), except when otherwise stated, aligning with the methods used in previous studies such as [4, 10, 13–15]. It should be noted that the direct emissions in (5.13) correspond to Scope 1 emissions of the GHG protocol standard.

5.1.4 *Scopes 1, 2, and 3 of the GHG Protocol*

Closely related to the discussion of direct and indirect emissions is the classification by the GHG Protocol Standard [16–18] of the emissions from an organization (company) into three scopes.

Scope 1 emissions: These are emissions generated from the reporting company's own operations that are owned or controlled by them.

Scope 2 emissions: These include emissions resulting from the generation of purchased or acquired electricity, steam, heating, or cooling that are consumed by the reporting company.

Scope 3 emissions: These encompass all indirect emissions occurring in the value chain of the reporting company, including both upstream and downstream emissions. They are not included in scope 2 emissions.

"Indirect emissions" refer to emissions that are a consequence of an organization's operations, but occur at sources not owned or controlled by the organization. Scope 2 emissions are a special category of indirect emissions. Scope 3 may address upstream emissions related to the inputs to production, downstream emissions related to the use of the products produced, and emissions related to the commuting of employees [19]. Scope 3 is optional, but it provides an opportunity to be innovative in GHG management. According to the organization environmental footprint guidelines of the European Commission [20, 21], the inclusion of upstream emissions is mandatory while the inclusion of downstream emissions is optional.

Why are Indirect Emissions Important?

The importance of IO in IE lies in its ability to quantify indirect impacts in a simple and transparent manner, economy-wide or globally. As argued by [18], there is a compelling need to consider indirect emissions because they reveal opportunities for energy savings and avoiding problem-shifting. The same applies to scope 3 emissions, which may vary among technologies, making it important to consider all impacts when evaluating alternatives. We saw in Sect. 5.1.3.1 that most GHG greenhouse gas emissions associated with final products occur upstream in processes such as material production, transportation, and energy generation. Neglecting scope 3 impacts could result in promoting technologies that do not achieve the expected emission reductions.

5.1.4.1 Estimating Scopes 1, 2, and 3 Emissions Based on IO

Consumer Perspective Referring to $\boldsymbol{y}$

We consider both the emissions associated with the final products, $\boldsymbol{y}$, and the emissions associated with the outputs used to produce the final products, $\boldsymbol{x}$. The former is known as the Scope based on the consumer perspective, while the latter is known as the Scope based on the producer perspective. From Eq. (5.8), the Scope 1 emissions of product j from the consumer perspective can be given by

$$\boldsymbol{Scope1}_y = \boldsymbol{F} \boldsymbol{y}_j \tag{5.15}$$

The Scope 2 emissions refer to the indirect emissions $\boldsymbol{F}(\boldsymbol{L} - \boldsymbol{I})$ generated by the use of purchased energy, such as electricity and steam. We define $\bar{\boldsymbol{A}}_j$ to be $\boldsymbol{A}$ with its elements in the j-th column set to zero except for those referring to purchased energy. The Scope 2 emissions of product j from the consumer perspective can then be given by ([13])

$$\boldsymbol{Scope2}_y = \boldsymbol{F} \bar{\boldsymbol{A}}_j \boldsymbol{y}_j \tag{5.16}$$

The direct emissions in (5.14) can be interpreted as the sum of Scope 1 and 2 emissions when $\boldsymbol{A}$ is replaced with $\bar{\boldsymbol{A}}_j$.

Using the Korean IO table with 381 production sectors for 2017, [10] estimated Scopes 1, 2, and 3 emissions for the Korean electronics industry. The estimation is based on the decomposition of $\boldsymbol{FLy_J}$, where J refers to 17 items of electronics ranging from semiconductors to display machinery, as follows

$$\boldsymbol{FLy_J} = \underbrace{\boldsymbol{Fy_J}}_{\text{Scope 1}} + \underbrace{\boldsymbol{F\bar{A}y_J}}_{\text{Scope 2}} + \underbrace{\boldsymbol{F(A - \bar{A} + A^2 + A^3 + \cdots)y_J}}_{\text{Scope 3}} \tag{5.17}$$

Here, $\bar{\boldsymbol{A}}$ refers to $\boldsymbol{A}$ with its elements set to zero except for those referring to purchased electricity of J. Equation (5.17) provides an IO-based decomposition of the carbon footprint (CF) of product j into Scopes 1, 2, and 3. The study found that semiconductors alone account for 41% of the electronics industry's total emissions, or 3.2% of Korea's national total. Furthermore, their indirect emissions are nearly three times their direct emissions, with electricity being the primary driver, while the metal and chemical sectors also contribute significantly.

Producer Perspective Referring to $\boldsymbol{x}$

When interpreting Scope 1, 2, and 3 emissions as referring to the activity of a company, it may be more appropriate to use output $\boldsymbol{x}$ instead of final demand $\boldsymbol{y}$. This is especially relevant for companies producing intermediate products with little final demand, such as metals and basic chemicals. Based on this perspective, [22] defined Scope 1 of sector j as

$$\boldsymbol{Scope1}_j = \boldsymbol{Fx}_j \tag{5.18}$$

where $\boldsymbol{x}_j$ is an $n \times 1$ vector with all the elements zero except for x_j in the ith row.

To encompass both Scope 2 and 3, [22] introduced Upstream Emission Footprints (UEF), defined as

$$\mathbf{UEF}_j = \boldsymbol{Scope2}_j + \boldsymbol{Scope3}_j = \boldsymbol{FLZ}_j \tag{5.19}$$

where $\boldsymbol{Z}_j$ refers to the jth column of the intermediate flow matrix. $\boldsymbol{FLZ}_j$ is termed the "embodied flows of carbon across production activities" by [18].

Using the identity

$$\boldsymbol{Z} = \boldsymbol{A\hat{x}} = \boldsymbol{AL\,\hat{y}} \tag{5.20}$$

and

$$\boldsymbol{L} = \sum_{i=0}^{\infty} \boldsymbol{A}^i, \tag{5.21}$$

it follows

$$\begin{aligned}\boldsymbol{FLZ} &= \boldsymbol{FLAL\hat{y}} = \boldsymbol{FLA(I + A + A^2 + \cdots)\hat{y}} \\ &= \boldsymbol{FL(L - I)\hat{y}} = \boldsymbol{FL(\hat{x} - \hat{y})}\end{aligned} \tag{5.22}$$

which implies

$$\text{UEF}_j = \boldsymbol{FLx}_j - \boldsymbol{FLy}_j = \boldsymbol{FLx}_j - \boldsymbol{Fx}_j \tag{5.23}$$

Double Counting in Scope 3?

Note that $\sum_j \boldsymbol{FLy}_j$ gives the total emissions generated by the production sectors in the economy. From (5.18) and (5.23), the summation of Scope 1, 2, and 3 emissions for all sectors in the economy can be expressed as

$$\sum_j (\boldsymbol{FLx}_j - \boldsymbol{Fx}_j + \boldsymbol{Fx}_j) = \sum_j \boldsymbol{FLx}_j \geq \sum_j \boldsymbol{FLy}_j \tag{5.24}$$

where the last inequality results from $\boldsymbol{x}_j \geq \boldsymbol{y}_j$. This equation implies that the sum of Scope 1, 2, and 3 emissions exceeds the total emissions in the economy. Consequently, it implies that the emissions are being double counted, with Scope 3 including the Scope 1 emissions of other sectors.

The GHG protocol acknowledges the possible occurrence of double counting (in Scope 3) and urges its avoidance when compiling national inventories but finds it "less important" for GHG risk management and voluntary reporting ([17], Chap. 4). Although the presence of double-counting may appear odd within the framework of IO, it is important to note that the Scopes 1,2, and 3 were originally intended to be used at the level of individual organizations, such as corporations, rather than at the level of sectors in the IO model where all the organizations with similar products are aggregated. The fact that Scopes 2 and 3 GHG emissions are evaluated using process-based life cycle assessment in the corporate world may make maintaining consistency with national inventories less relevant. However, double counting is a serious issue in footprint analysis, particularly when assessing the environmental footprint of materials [23], which is the topic of the next section.

5.1.5 *Avoiding Double Counting in Footprint Analysis

The quantification of the embodied impacts and carbon footprints, based on $\boldsymbol{FLy}$, is widely accepted and can be applied to any final product. However, applying this model to materials such as steel alloys, basic chemicals, or synthetic fibers is not straightforward, as materials are intermediate products that require further processing before becoming final products. Accounting for only materials purchased by final consumers would underestimate the environmental impacts of materials. On the other hand, using the model based on gross output ($\boldsymbol{FLx}$) would lead to the double-counting of emissions at each processing stage, resulting in incorrect total environmental impacts. To address this problem, [24] developed a methodology that distinguishes between the upstream and downstream effects of materials and avoids double counting issues.

The methodology defines all materials of interest as "target sectors" and distinguishes between upstream and downstream flows. Upstream flows include the materials, products, and services used to produce the target material, while downstream flows include the materials, products, and services using the target material. The distinction is illustrated using the example of pig iron and crude steel. If crude steel is selected as the target material, almost all the impacts of pig iron production are allocated to crude steel production (upstream). However, if pig iron is chosen as the target material, the impacts of crude steel production are allocated to the use of pig iron production (downstream). The downstream phase of target sectors only relates to nontarget sectors, and if target sectors supply each other, double-counting is prevented by allocating the impacts to just one of the involved target sectors. The methodology provides a consistent mathematical framework for determining the production amount to which the unitary impact calculated via the standard Leontief approach is applied, thus avoiding double-counting issues.

5.1.5.1 The Output with and Without Double-Counting

We divide output vector $\boldsymbol{x}$ exclusively and exhaustively into targeted outputs $\boldsymbol{x}_t$ and nontargeted outputs $\boldsymbol{x}_o$, and partition the matrices $\boldsymbol{A}$ and $\boldsymbol{L}$ accordingly

$$\boldsymbol{A} = \begin{pmatrix} \boldsymbol{A}_{oo} & \boldsymbol{A}_{ot} \\ \boldsymbol{A}_{to} & \boldsymbol{A}_{tt} \end{pmatrix} \tag{5.25}$$

$$\boldsymbol{L} = (\boldsymbol{I} - \boldsymbol{A})^{-1} = \begin{pmatrix} \boldsymbol{I}_0 - \boldsymbol{A}_{oo} & -\boldsymbol{A}_{ot} \\ -\boldsymbol{A}_{to} & \boldsymbol{I}_t - \boldsymbol{A}_{tt} \end{pmatrix}^{-1} = \begin{pmatrix} \boldsymbol{L}_{oo} & \boldsymbol{L}_{ot} \\ \boldsymbol{L}_{tt} & \boldsymbol{L}_{tt} \end{pmatrix} \tag{5.26}$$

From the properties of the inverse of a partitioned matrix (see Sect. A.3.2 or [25]), the following holds

$$\begin{aligned} \boldsymbol{L}_{oo} &= \left(\boldsymbol{I}_o - \boldsymbol{A}_{oo} + \boldsymbol{A}_{ot}(\boldsymbol{I}_t - \boldsymbol{A}_{tt})^{-1}\boldsymbol{A}_{to}\right)^{-1} \\ \boldsymbol{L}_{tt} &= \left(\boldsymbol{I}_t - \boldsymbol{A}_{tt} - \boldsymbol{A}_{to}(\boldsymbol{I}_o - \boldsymbol{A}_{oo})^{-1}\boldsymbol{A}_{ot}\right)^{-1} \\ \boldsymbol{L}_{to} &= \boldsymbol{L}_{tt}\boldsymbol{A}_{to}(\boldsymbol{I}_o - \boldsymbol{A}_{oo})^{-1} \\ \boldsymbol{L}_{ot} &= (\boldsymbol{I} - \boldsymbol{A}_{oo})^{-1}\boldsymbol{A}_{ot}\boldsymbol{L}_{tt} \end{aligned} \tag{5.27}$$

where $\boldsymbol{I}_t$ and $\boldsymbol{I}_o$ refer to the identity matrix of order t and o. Denote by $\boldsymbol{x}_t^{wdc}$ the fraction of $\boldsymbol{x}_t$ without double-counting, such that the following condition is satisfied

$$\boldsymbol{x}_t = \boldsymbol{L}_{tt}\boldsymbol{x}_t^{wdc} \tag{5.28}$$

Denoting by $\boldsymbol{x}_t^{dc}$ the fraction of $\boldsymbol{x}_t$ with double-counting, we have

$$\boldsymbol{x}_t = \boldsymbol{x}_t^{wdc} + \boldsymbol{x}_t^{dc} \tag{5.29}$$

The amounts of x_o used upstream, x_o^u, and downstream, x_o^d, are then given by

$$\begin{aligned} x_o^u &= L_{ot} x_t^{wdc} \\ x_o^d &= x_o - x_o^u \end{aligned} \tag{5.30}$$

From (5.28), (5.29), and (5.27) we obtain

$$x_t^{dc} = x_t - x_t^{wdc} = (I_t - L_{tt}^{-1}) x_t = (A_{tt} + A_{to}(I - A_{oo})^{-1} A_{ot}) x_t \tag{5.31}$$

Rewritten from the final demand perspective, x_t can be represented as

$$x_t = L_{tt} y_t + L_{to} y_o = L_{tt}(y_t + A_{to}(I - A_{oo})^{-1} y_o) \tag{5.32}$$

and hence from (5.28)

$$x_t^{wdc} = L_{tt}^{-1} x_t = y_t + A_{to}(I_o - (I - A_{oo})^{-1}) y_o \tag{5.33}$$

The share of x_t that is double counted, the double counting factor, k, is then given by

$$k = \hat{x}_t^{-1} x_t^{dc} \tag{5.34}$$

5.1.5.2 Upstream and Downstream Emission Allocation

Writing E_t^u for the contribution of t and o to upstream emissions associated with x_t^{wdc}, we have

$$E_t^u = E_{tt}^u + E_{ot}^u = F_t L_{tt} x_t^{wdc} + F_o L_{ot} x_t^{wdc} \tag{5.35}$$

The upstream emissions can be further divided into three components

$$\begin{aligned} \text{Direct emissions of } t\text{:} \quad E_t^{u1} &= F_t x_t^{wdc} \\ \text{Indirect emissions of } t\text{:} \quad E_t^{u2} &= F_t (L_{tt} - I_t) x_t^{wdc} \\ \text{Indirect emissions of } o\text{:} \quad E_t^{u3} &= F_o (L_{ot}) x_t^{wdc} \end{aligned} \tag{5.36}$$

The emissions in the downstream phase are only related to the process emissions of o

$$E^d = F_o x_t^{wdc} \tag{5.37}$$

5.1.5.3 Scope 3 GHG Emissions of 64 Materials in Japan

The methodology described above was applied to the 2011 Japanese IO table with 393 endogenous sectors to estimate Scope 3 GHG emissions for 64 target materials, including 26 biomass items, 18 fossil fuel items, six metallic minerals, and 15 non-metallic minerals. The results show that the total GHG emissions associated with the target materials accounted for 38% of the total industrial emissions of 1.25 Gt-CO_2e, with 85% associated with the upstream phase, $\boldsymbol{E}_t^u$, and the remaining 15% associated with the downstream phase, $\boldsymbol{E}^d$.

Three target materials, crude steel (converters), petroleum refinery, and cement, were identified as major emitters, accounting for 58% of the total industrial GHG emissions. Further allocation based on Eq. (5.36) of the emissions into direct and two indirect sources revealed significant differences among these three materials, indicating the need to consider different options for mitigating emissions. While upstream emissions made up 93% and 99% of the total emissions for crude steel and cement, respectively, they were only 48% for petroleum refinery, with the remaining 52% attributed to downstream emissions, mostly from the transport and power sectors. For crude steel, the majority of emissions (89%) originated from the indirect emissions of o, $\boldsymbol{E}_t^{u3}$, namely pig iron production. For cement, almost all emissions originated from the cement process, $\boldsymbol{E}_t^{u1}$.

The double counting factor $\boldsymbol{k}$ was found to be less than 10% for 41 materials and higher than 30% for only nine materials, consisting of energy materials, chemicals, and biomass.

5.1.5.4 Further Applications and Extensions

The above method has been widely used and extended, and a few of these applications are discussed below.

Alternative allocations: From "what for" to "where from"

Reference [26] extended [24] to consider the impacts of imports by replacing $\boldsymbol{L}$ with $\boldsymbol{L}^d$. Another allocation scheme (Allocation 2) was also introduced, in which the process emissions of any target material are allocated to itself and not to the target material of destination. In this allocation, the impacts are attributed to the target sectors where they are directly caused, rather than where the target material ends up. Allocation 1 (the original allocation scheme) is deemed suitable for "what for" research questions, while Allocation 2 is appropriate for "where from" research questions. The change in allocation scheme was found to have significant effects, with the emissions reported in Allocation 2 being more than 10% less than those reported in Allocation 1 for 33 of the 64 target materials.

Total environmental impact at the national level

Reference [27] assessed the total environmental impact at the national level in Japan by analyzing the production process for 213 materials. This was made possible by the combined use of the IDEAv2 database with the IO table. The study identified four phases for environmental impact generation (foreign supply chain, domestic supply chain, production process, and use), 12 environmental impact areas, and four resource categories (biomass, fossil, metal, and nonmetallic mineral) for each material. The environmental impacts related to imports/foreign supply chains of target material production, $\boldsymbol{E}_t^m$, were obtained based on

$$\boldsymbol{E}_t^m = \boldsymbol{F}\boldsymbol{L}\boldsymbol{A}^m\boldsymbol{L}_{\cdot t}^d \,\mathrm{diag}(\boldsymbol{x}_t^{wdc}) \tag{5.38}$$

where $\boldsymbol{L}_{\cdot t}^d$ refers to the column elements referring to the target sectors in $\boldsymbol{L}^d$. The study found that biomass, particularly that related to crops, livestock, and dairy products, and fossil resources were the major impact generation resource categories, followed by metal resources. The direct emissions, $\boldsymbol{E}_t^{u1}$, accounted for approximately 30% of total impact generation in 2010, while foreign supply chain emissions, $\boldsymbol{E}_t^m$, contributed to around 29% of the total impact generation. Domestic supply chain emissions, $\boldsymbol{E}_t^{u2}$, accounted for about 25% of the total impact generation, while the use phase accounted for around 16%.

Global Scope 3 Impacts Without Double Counting

Reference [28] extended the method of [24, 26] to assess the Scope 3 impacts of any target sector and target region without double-counting, using a global supply chain perspective from four different angles (production, target, final supply, and final demand). The study employed EXIOBASE3, which has 163 sectors and 49 regions, with 75 sectors being designated as target sectors and all regions as target regions. Comparisons with previous approaches revealed that the standard MRIO procedure overestimated the impacts of global material production by 20–30% due to double-counting. In contrast, assessing only the direct impacts would underestimate the impacts of global material production by 20–25% in terms of climate change and PM health impacts. These findings demonstrate the importance of correcting for double-counting when assessing the environmental impacts of material production.

Global Impacts of Capital Consumption

Another important example with application to EXIOBASE3 is [23], which addressed GHG emissions from the global production of eight structural and functional materials (plastic & rubber, wood products, other minerals, cement, other metals, aluminum, and iron & steel). A distinguishing feature of this study is to consider the impacts associated with fixed capital consumption (depreciation/replacement) based on the Lenzen-type model discussed in Sect. 4.4.3.4. The study reveals that construction and the manufacturing of machinery, vehicles, and other durable products together account for four-fifths of the carbon footprint of the eight materials. Additionally, the replacement or formation of new capital stocks was responsible for 60% of material-related emissions in 2015.

Alternative Derivation of (5.32) *based on the Hypothetical Extraction Method*

Another noteworthy aspect of [23] is the independent derivation of (5.32) under general conditions using the hypothetical extraction method (HEM). HEM is an IO-based approach that quantifies the impacts on the total output of an economy when a particular sector is not present (or extracted) [29].

Let $\boldsymbol{A_o}$ be an $n \times n$ input coefficients matrix $\boldsymbol{A}$ with the rows referring to the target sectors extracted, and $\boldsymbol{A_t}$ be $\boldsymbol{A}$ with all the elements put equal to zero except for the rows referring to the target sectors. We then have

$$\boldsymbol{A} = \boldsymbol{A_t} + \boldsymbol{A_o} \tag{5.39}$$

Similarly, we divide the final demand $\boldsymbol{y}$ into two components

$$\boldsymbol{y} = \boldsymbol{y_t} + \boldsymbol{y_o} \tag{5.40}$$

The output not involved with the production of $\boldsymbol{t}$, $\boldsymbol{x_o}$, can be given by

$$\boldsymbol{x_o} = \boldsymbol{L_o}\boldsymbol{y_o} \tag{5.41}$$

Using HEM, the output required to satisfy the intermediate and final demand for the target sectors, $\boldsymbol{x_t}$, can be expressed as

$$\boldsymbol{x_t} = \boldsymbol{L}\boldsymbol{y} - \boldsymbol{L_o}\boldsymbol{y_o} \tag{5.42}$$

By noting that $\boldsymbol{L} = \boldsymbol{L}\boldsymbol{A_t}\boldsymbol{L_o} + \boldsymbol{L_o}$[1], we can derive an alternative representation of $\boldsymbol{x_t}$, as given by

$$\boldsymbol{x_t} = \boldsymbol{L}\boldsymbol{y_t} + \boldsymbol{L}\boldsymbol{A_t}\boldsymbol{L_o}\boldsymbol{y_o} \tag{5.43}$$

which is equivalent to (5.32).

5.1.6 Emissions and International Trade

In Sect. 5.1.3, we discussed the assumption of "domestic technology," which assumes that all imports can be domestically produced with the same domestic technology $\boldsymbol{A}$. However, this assumption may be difficult to justify for cases involving imports with

[1] Proof. Noting $\boldsymbol{AL} = \boldsymbol{LA} = \boldsymbol{L} - \boldsymbol{I}$ (see (5.22)), and $\boldsymbol{A} = \boldsymbol{A_t} + \boldsymbol{A_o}$, we have

$$\begin{aligned}\boldsymbol{L} - \boldsymbol{L_o} &= \boldsymbol{L} - \boldsymbol{L}\boldsymbol{L_o} + \boldsymbol{L}\boldsymbol{L_o} - \boldsymbol{L_o} = -\boldsymbol{L}(\boldsymbol{L_o} - \boldsymbol{I}) + (\boldsymbol{L} - \boldsymbol{I})\boldsymbol{L_o}\\ &= -\boldsymbol{L}\boldsymbol{A_o}\boldsymbol{L_o} + \boldsymbol{L}\boldsymbol{A}\boldsymbol{L_o} = \boldsymbol{L}(\boldsymbol{A} - \boldsymbol{A_o})\boldsymbol{L_o} = \boldsymbol{L}\boldsymbol{A_t}\boldsymbol{L_o}\end{aligned}$$

I owe Edgar Hertwich for this proof.

minor domestic production. The share of domestic fossil and mineral resources in the total supply is very small in most OECD countries. Additionally, with the accelerating globalization of supply chains and the resulting intensification of inter-country interdependence in the last two decades, it is necessary to consider the interrelations between global production and consumption when assessing a country's environmental impacts. To provide a comprehensive description of the global economy and analyze its effects on the environment, environmentally extended multi-regional input-output (EE-MRIO) tables have emerged as a key framework [30] (Section 4.3.4). While global MRIO databases of fairly high sectoral and regional resolution have become increasingly available recently, this was not always the case. To address this situation, a simpler version of the global EE-MRIO, called the emission embodied in bilateral trade (EEBT) model, was developed as a predecessor to the global EE-MRIO model [31–33]. We will now discuss the EEBT model in more detail.

5.1.6.1 The EEBT Model

This section refers to [31–33]. Let there be n_R regions. Denote by $\boldsymbol{m}^{sr}$ the export from region s to region r, or the import by region r from region s, with $\boldsymbol{m}^{tt} = \boldsymbol{0}$ for any region t. Denote by $\boldsymbol{A}^{rr}$ the domestic input coefficients matrix of region r, and by $\boldsymbol{y}^{rr}$ the final demand in region r for its domestic output. Since the production in region r is given by

$$\boldsymbol{x}^r = \boldsymbol{A}^{rr}\boldsymbol{x}^r + \boldsymbol{y}^{rr} + \sum_s \boldsymbol{m}^{rs} \tag{5.44}$$

we have

$$\boldsymbol{x}^r = (\boldsymbol{I} - \boldsymbol{A}^{rr})^{-1}\left(\boldsymbol{y}^{rr} + \sum_s \boldsymbol{m}^{rs}\right) \tag{5.45}$$

with the production-based emissions in r, $\boldsymbol{e}^r$, given by

$$\boldsymbol{e}^r = \boldsymbol{F}^r(\boldsymbol{I} - \boldsymbol{A}^{rr})^{-1}\left(\boldsymbol{y}^{rr} + \sum_s \boldsymbol{m}^{rs}\right) \tag{5.46}$$

which can be further decomposed into the fraction attributed to domestic demand

$$\boldsymbol{e}^{rr} = \boldsymbol{F}^r(\boldsymbol{I} - \boldsymbol{A}^{rr})^{-1}\boldsymbol{y}^{rr} \tag{5.47}$$

and the EEBT from r to the rest of the world

$$\boldsymbol{e}^{r\cdot} = \boldsymbol{F}^r(\boldsymbol{I} - \boldsymbol{A}^{rr})^{-1}\sum_s \boldsymbol{m}^{rs} \tag{5.48}$$

Since $\boldsymbol{m}^{sr}$ refers to the import from region s to region r, the emission embodied in the import by region r from the rest of the world, $\boldsymbol{e}^{\cdot r}$, is given by

$$\boldsymbol{e}^{\cdot r} = \sum_{s} \boldsymbol{F}^{s}(\boldsymbol{I} - \boldsymbol{A}^{ss})^{-1}\boldsymbol{m}^{sr} \tag{5.49}$$

The consumption-based emissions $\boldsymbol{e}_C^r$ refer to the emissions induced by domestic consumption $\boldsymbol{y}_{rr}$ and import, that is,

$$\boldsymbol{e}_C^r = \boldsymbol{e}^{rr} + \boldsymbol{e}^{\cdot r} \tag{5.50}$$

According to a pioneering study [32], which estimated EEBT for 87 countries, around 22–23% of global CO_2 emissions were embodied in international trade. The study also found that countries listed under Annex B of the Kyoto Protocol were net importers of CO_2 emissions, while non-Annex B countries were net exporters.[2]

One advantage of EEBT is its relative ease of calculation, which only requires national IO tables, namely $\boldsymbol{A}^{rr}$. Depending on data availability, $\boldsymbol{A}^{rr}$ can either be the domestic input matrix of the noncompetitive import type or one based on the competitive import model with $\boldsymbol{A}^{rr} = \boldsymbol{\mu}^r \boldsymbol{A}^r$, as discussed in Sect. 4.3.2. However, theoretically, this feature is a drawback of EEBT because it does not consider the international flow of intermediate inputs, treating all trade flows as final products. It does not take into account the fact that imports are needed to produce exports. For instance, to calculate the emissions embodied in the production of a car in region r, one must first determine the production levels and emissions occurring in region r, which require imports from regions s and t. The resulting production in regions s and t also require imports from other regions, and so on, leading to an indefinite continuation of the process throughout the global production system ([31], p.16).

Full consideration of international/regional interdependence is performed by the MRIO model. A key difference between the EEBT model and the MRIO model is that the MRIO model distinguishes between trade that goes to intermediate and final consumption.

5.1.6.2 The MRIO Model

This section refers to [31]. Let $\boldsymbol{A}^{rs}$ denote the matrix of import input coefficients in region s for imports from region r, where the i, j element a_{ij}^{rs} represents the amount of product i from region r required as an input to produce a unit of product j in region s. The matrix $\boldsymbol{A}^{rs}$ is often approximated by $\boldsymbol{\mu}^{rs}\boldsymbol{A}^s$, where $\boldsymbol{A}^s$ is the matrix of competitive import input coefficients in region s and $\boldsymbol{\mu}^{rs}$ represents the share of imports of region s that comes from region r.

[2] The Annex B countries include Belarus, Canada, the EU (with Switzerland), Japan, Russian Federation, Ukraine, and the US, Brazil, China, India, Indonesia, Korea, Mexico, Thailand, Turkey, Venezuela are non Annex B countries.

The import of s from r can then be decomposed into intermediate products and final products as

$$\boldsymbol{m}^{rs} = \boldsymbol{A}^{rs}\boldsymbol{x}_s + \sum_{s \neq r} \boldsymbol{y}^{rs} \tag{5.51}$$

Substitution into (5.44) for region r yields

$$\boldsymbol{x}^r = \boldsymbol{A}^{rr}\boldsymbol{x}^r + \boldsymbol{y}^{rr} + \sum_{s \neq r} \boldsymbol{A}^{rs}\boldsymbol{x}_s + \sum_{s \neq r} \boldsymbol{y}^{rs} \tag{5.52}$$

Stacking this equation for all n_{R} regions gives

$$\begin{pmatrix} \boldsymbol{x}^1 \\ \boldsymbol{x}^2 \\ \vdots \\ \boldsymbol{x}^{n_{\mathrm{R}}} \end{pmatrix} = \begin{pmatrix} \boldsymbol{A}^{11} & \boldsymbol{A}^{12} & \dots & \boldsymbol{A}^{1n_{\mathrm{R}}} \\ \boldsymbol{A}^{21} & \boldsymbol{A}^{22} & \dots & \boldsymbol{A}^{2n_{\mathrm{R}}} \\ \vdots & \vdots & \vdots & \vdots \\ \boldsymbol{A}^{n_{\mathrm{R}}1} & \boldsymbol{A}^{n_{\mathrm{R}}2} & \dots & \boldsymbol{A}^{n_{\mathrm{R}}n_{\mathrm{R}}} \end{pmatrix} \begin{pmatrix} \boldsymbol{x}^1 \\ \boldsymbol{x}^2 \\ \vdots \\ \boldsymbol{x}^{n_{\mathrm{R}}} \end{pmatrix} + \begin{pmatrix} \sum_s \boldsymbol{y}^{1s} \\ \sum_s \boldsymbol{y}^{2s} \\ \vdots \\ \sum_s \boldsymbol{y}^{n_{\mathrm{R}}s} \end{pmatrix} \tag{5.53}$$

or

$$\boldsymbol{x}^{\star} = \boldsymbol{A}^{\star}\boldsymbol{x}^{\star} + \boldsymbol{y}^{\star} \tag{5.54}$$

Defining the final production in region r, including the domestic final demand and export, as

$$\boldsymbol{\pi}^r = \begin{pmatrix} \boldsymbol{0} \\ \vdots \\ \boldsymbol{0} \\ \sum_s \boldsymbol{y}^{rs} \\ \boldsymbol{0} \\ \vdots \\ \boldsymbol{0} \end{pmatrix} \tag{5.55}$$

The production-based emission in region r, $\boldsymbol{e}^r_P$, is given by

$$\boldsymbol{e}^r_P = \boldsymbol{F}^{\star}(\boldsymbol{I} - \boldsymbol{A}^{\star})^{-1}\boldsymbol{\pi}^r \tag{5.56}$$

Recalling that $\boldsymbol{y}^{\cdot r}$ refers to the consumption demand in r for products from all regions, the consumption-based emission, $\boldsymbol{e}^r_C$, is given by

$$\boldsymbol{e}^r_C = \boldsymbol{F}^{\star}(\boldsymbol{I} - \boldsymbol{A}^{\star})^{-1}\boldsymbol{y}^{\cdot r} \tag{5.57}$$

The EEBT and MRIO models differ in the way they allocate imports to intermediate consumption. The EEBT model considers total consumption—intermediate plus final—through the use of bilateral trade data. On the other hand, the MRIO model treats exogenous demand as always being final consumption, while the domestic and imported intermediate consumption is determined endogenously. For a more in-depth discussion on this topic, please refer to [31].

Despite the theoretical soundness of the MRIO model, its practical implementation faces a significant challenge: the need for inter-regional matrices, A^{rs}, $r \neq s$. These matrices, which capture the interdependence between different regions, are often not readily available in official IO statistics. Consequently, they need to be estimated using assumptions and modeling techniques [33].

5.2 LCA and EEIO

The origins of LCA can be traced to studies on energy use and the emissions of alternative packaging (beverage containers) in the 1960s. The term "life cycle assessment" was coined, and commercial LCA software was released by 1990. In the early 1990s, a number of (life cycle inventory) databases emerged, and the ISO standards (14040, 14041, 14042, and 14043) were released by 2000 [34]. This section briefly introduces ISO-LCA and its relationships to the EEIO model. Those interested in further details of ISO-LCA are recommended to see [34, 35].

It is worth noting that the methodology of LCA was developed independently of EEIO. The earliest examples of empirical EEIO studies can be traced back to [36], which analyzed the marine food web of a bay in the US by using regional IO tables and detailed on-site information about the bay's ecology.

Application of EEIO in LCA started in the early 1990s. For example, [37] studied CO_2 emissions using Japanese IO tables, while [12] investigated toxic discharges using US IO tables. Reference [38] constructed and published the first EEIO table with CO_2 emissions for Japan based on the IO table for 1985 at a sectoral resolution of 406 production sectors, preceding the publication of 3EID based on the 1990 IO table by NIES, Japan, in 1997. This study is one of the earliest examples of the EEIO table and the first at this level of sectoral resolution. Unfortunately, this valuable work and its accompanying data were available only in Japanese. Remarkably, the early studies of IO-based LCA started in Japan and the US, taking advantage of the availability of IO tables of the highest sectoral resolution at that time.

By the early 2000s, the EEIO model had been extensively and systematically introduced to LCA. The International Journal of life cycle Assessment established a new section devoted to IO and (Hybrid) LCA in 2003 [39]. In [40], authors with LCA backgrounds devoted a whole chapter to "Relationship with input-output analysis" and showed the formal equivalence of the EEIO and Life Cycle Inventory (LCI) models.

5.2.1 LCA Basics: What Is LCA?

Walter Klöpfer, a key contributor to the development of ISO standards for LCA, defines it as follows (ISO 140401, [35])

> *LCA studies the environmental aspects and potential impacts throughout a product's life (i.e. cradle-to-grave) from raw material acquisition through production, use and disposal. The general categories of environmental impacts needing consideration include resource use, human health, and ecological consequences.*

The main characteristics of LCA include ([34], p.12)

- *Takes a Life Cycle Perspective.* The core reason for taking this perspective is to allow the identification and prevention of the *burden shifting* between life cycle stages or processes that happens if efforts for lowering environmental impacts in one process or life cycle stage unintentionally create (possibly larger) environmental impacts in other processes or life cycle stages.
- *Covers a Broad Range of Environmental Issues: considers multiple environmental issues simultaneously* This helps to avoid the burden shifting between impacts that happens if efforts for lowering one type of environmental impact unintentionally increase other types of environmental impacts.
- *Is Quantitative* LCA can be used to compare environmental impacts of different processes and product systems.
- *Is Based on Science* The quantification of potential impacts in LCA is rooted in natural science, with models of the relationships between emission (or resource consumption) and impact based on proven causalities.

5.2.1.1 The Functional Unit

In addition to analyzing the environmental impacts of a product or process from cradle to grave, which involves considering the entire life cycle from resource extraction to end-of-life, the functional unit is another fundamental concept in LCA. The functional unit refers to a quantifiable measure of the performance of the product or process being studied and serves as the basis for comparing different alternatives [34]. By defining a consistent functional unit for all alternatives, the environmental impacts of each can be evaluated and compared on an equal basis.

For example, in the case of outdoor paints, the functional unit could be defined as the "complete coverage of 1 m^2 of an outdoor wall for 10 years in Germany in a uniform color at 99.9% capacity" ([34], p.84). Similarly, in the context of an LCA study on lithium-ion batteries for plug-in hybrid electric vehicles, [41] defined the functional unit as a 10 kWh battery capable of sustaining 3000 charge cycles at 80% maximum discharge giving at least a 200,000 km operation during the vehicle design life time. In contrast, [42] reviewed various LCA studies on electric vehicles with widely differing functional units, including "1 km driven under European average conditions," "200 km at nominal full load within an urban area," "transportation of one person for 1 km," and "production of one vehicle."

For LCA studies on food, the functional unit can be defined in various ways, such as "a kg of meat," "1 kg of energy-corrected milk also called fat- and protein-corrected milk," "the supply of food for a Spanish citizen in the year 2005, which amounts to 787 kg of food," and "a complete meal providing ca 35 g protein and 3,100 kJ" [43]. In power generation LCA, the generation of 1 MWh of electricity in a particular location is commonly used as the functional unit [44, 45].

5.2.1.2 The Structure of LCA

According to ISO 14040, an LCA must follow a structure consisting of four phases, which are described below ([34], p.61)

1. *Goal and Scope Definition*: In this phase, the overall aim of the study is specified, the functional unit is defined, and the processes to be included in the analysis are selected. The geographical and temporal boundaries of the study are also determined, as well as the relevant level of technology for the processes in the product system, in order to scope the product system.
2. *Life Cycle Inventory Analysis (LCI)*: This phase involves collecting information on the physical flows of the product system, including the input of resources, materials, semi-products, and products, and the output of emissions, waste, and valuable products. This phase is often the most time-consuming part of an LCA.
3. *Life Cycle Impact Assessment Analysis (LCIA)* Building on the life cycle inventory, the impact assessment phase translates the physical flows and interventions of the product system into impacts on the environment using knowledge and models from environmental science. The LCI results are classified into impact categories, such as acidification, climate change, and eutrophication, and converted to common units using characterization factors. The converted results are then aggregated into a potential indicator within the same category.
4. *Interpretation*: In this final phase, the results of the study are interpreted to answer the question(s) posed as part of the goal definition. This phase may involve comparing results to other LCAs, identifying key sources of environmental impact, and considering opportunities for improvement.

It is important to note that these phases should not be interpreted as a one-way process, starting from the top and ending at the bottom. Instead, they should be interpreted as an iterative approach with feedback among the phases ([35], p.12).

5.2.2 *The EEIO Model from the Point of View of ISO-LCA*

We now turn to the relationship between the EEIO model and LCA models regarding their formal representation. Common to most LCA models are the following underlying assumptions ([34], p. 188)

1. *Steady-state*: LCA models are usually not dynamic but represent the environment as a system in a steady state, i.e., all parameters that define its behavior are not changing over time.
2. *Linearity*: A linear relationship between a change in the flow and the consequent change in its potential environmental impact—doubling the amount of a flow doubles its potential impact.

The first assumption can be replaced with "static," since a static model assumes a steady-state environment where all parameters defining its behavior are fixed and not changing over time. In contrast, dynamic models involve different times (t and $t-1$, for instance) and can be represented as differential or difference equations, depending on whether time is measured continuously or discretely. Leontief's dynamic IO model (4.61) is a discrete dynamic model. Once initial conditions are given, a dynamic model can generate the path of endogenous model variables over time. On the other hand, static (steady-state) models contain only one time index. The EEIO model we discussed in Sect. 5.1 is a static model. Together with the linearity assumption, we notice a great similarity in the mathematical representation of LCA and EEIO models.

5.2.2.1 The Four Phases of ISO-LCA and EEIO

In terms of ISO-LCA, the calculation of individual GHG gas emissions using the IO-based approach as represented by (5.11) corresponds to the Life Cycle Inventory Analysis. We previously discussed the Life Cycle Impact Assessment (Section 3.1.3), where we converted the emissions of individual GHG gases into their aggregate, CO_2e, by weighting them using the relevant GWP. Similarly, the emissions of SO_2, NO_x, and HCl are converted and aggregated into the acidification potential (AP) expressed in kg SO_2e, while emissions of NO_x, NH_3, and P are converted and aggregated into eutrophication potential (EP) represented in kg PO_4^3e. These impacts are known as category indicators, and the characterization model is the algorithm used to convert (aggregate) inventory results into category indicators. The characterization model is based on knowledge of environmental science, and further details can be found in [34, 35]. If we denote the total impact factor values by $\boldsymbol{d}$ and a matrix of characterization factors by $\boldsymbol{C}$, we can obtain the LCIA model associated with the LCI model (5.11) as follows

$$\boldsymbol{d} = \boldsymbol{C}\boldsymbol{F}(\boldsymbol{I} - \boldsymbol{A})^{-1}\boldsymbol{y} \tag{5.58}$$

In the language of LCA, $\boldsymbol{F}$ is termed the intervention matrix.

The selection of functional unit and system boundaries are two critical elements of the goal and scope phase that are highly relevant to the EEIO model (5.58). The functional unit specifies $\boldsymbol{y}$, while the system boundaries determine $\boldsymbol{A}$. For example, if the functional unit were 1 kg of beef, $\boldsymbol{y}$ would be an $n \times 1$ vector with all the elements except for the element referring to beef being equal to zero. On the other hand, if the functional unit were 1 kg of beef steak, it would be necessary to consider the inputs required for cooking such as gas, electricity, water, salt, cooking oil, and spices in $\boldsymbol{y}$. In addition, if the capital goods associated with the cooking process, such as kitchenware and an oven, were to be considered, they would need to be included as well.

Let us now turn to the system boundaries. If we were to limit the geographical system boundary to the territorial boundary of a particular nation, we would select the national $\boldsymbol{A}^d$ as $\boldsymbol{A}$. However, if we wished to extend the boundaries to the entire world, we would select the global $\boldsymbol{A}^\star$, as (5.54), based on MRIO databases. It is worth noting that most official IO tables are compiled for a nation, and thus, their system boundaries are well-defined by the activities carried out within the nation's territorial boundary. Despite differing levels of sectoral (or process) resolution, all the sectors (processes) of the nation's economy are typically included in the IO table.

5.2.2.2 EEIO and Process-Based LCA: The Top-Down Versus Bottom-Up Approach

It is important to emphasize that despite their similar mathematical representation [40], EEIO and process-based LCA (the original form of LCA) have distinct characteristics when it comes to their empirical applications during the inventory phase of ISO-LCA. IO tables are regularly compiled and published as part of the national accounting system by statistical offices worldwide. Therefore, one can usually assume the availability of an IO table with well-defined system boundaries, although there may be delays in its compilation and publication until the data become available. In other words, we have a well-defined and publicly available accounting system of IO tables. For this reason, within the LCA community, the construction of an inventory based on an IO table is called the "top-down approach." In contrast, the construction of an inventory in a process-based LCA is called the "bottom-up approach" because it uses knowledge about the individual industrial processes involved in a life cycle and the physical flows connecting them.

The bottom-up approach has an advantage over the top-down approach when it comes to process resolution. The former can adopt a level of process resolution based on the availability of relevant knowledge and resources for the LCA project. In contrast, the resolution of the latter is predetermined by the sector resolution of a given IO table, which typically ranges from less than 30 to a maximum of around 500 sectors.

However, the advantage of the bottom-up approach can become a disadvantage when establishing system boundaries. The nature of being bottom-up means that it may not be clear where to stop and how wide and deep the coverage of processes should be. This choice of where to stop is known as "cut-off" in the language of LCA, and the associated criteria are called "cut-off criteria." These criteria aim to exclude insignificant inputs into the product system, as specified by ISO 14044.

However, selecting "insignificant inputs (or processes)" can be challenging because they may not be known before the actual collection of detailed data. This means that the selection has to be made based on criteria that are as yet unknown [46].

5.3 Hybrid LCA: Hybridization of LCA and EEIO

As discussed earlier, both process-based LCA (PLCA) and EEIO models have advantages and disadvantages as LCI models. PLCA can provide highly accurate results based on data from individual processes. However, it is based on an incomplete system description, resulting in truncations in inventories and underestimations of impacts. This is because not all inputs and outputs are covered by the process-based system [12, 46–48]. On the other hand, the EEIO model has its own set of problems, including low input/process resolution and product heterogeneity. These limitations render the EEIO model inadequate for detailed LCA studies when used alone. Institutional factors can also distort process representation in the EEIO model, such as the inclusion of lunch rooms and business meals. Moreover, an evaluation based on monetary value in the EEIO model can distort physical flow relationships due to price heterogeneity [12, 47–49].

It is possible to address the inability of EEIO to provide detailed comparisons by integrating detailed data on processes from PLCA. Conversely, truncation errors inherent in PLCA can be resolved by using EEIO, which has a complete system description. This integration, known as hybrid LCA (HLCA), has been developed to combine the strengths of PLCA and EEIO to obtain more specific and complete system descriptions [48–52]. Before discussing the various HLCA models in detail, it is worth considering the issue of the sectoral resolution of IO tables and its implications for LCA, using a specific example from a recent Japanese IO table.

5.3.1 The Sectoral Resolution of an IO Table and Its Implication for LCA

This section largely refers to [48]. The use of an IO table in LCA has a significant advantage in that it allows for comprehensive system boundary coverage of a country's entire economy. However, the flip side is that the level of sectoral resolution tends to be low due to the comprehensive nature of the IO table. This means that the classification of goods and industries becomes coarser in IO tables. The sectoral categories in an IO table are typically defined based on single product groups for economic activities with large monetary value or products that merit future attention, while items with small monetary value are usually combined into aggregates (Sect. 4.1.1.2). It may not be possible to conduct inventory analysis focused on a single product for a sector that encompasses multiple products.

A specific example from Japanese IO tables is helpful here. The Japanese table is chosen because, with around 400 production sectors, it has one of the world's highest resolutions, besides the US and Korean tables. The 2015 table includes sectors such as Commercial residential air conditioners, Drinking milk, Sake (Japanese rice wine), Sodium hydroxides, and Oxidized titanium. Inventory analysis can therefore be conducted for each of these individual items. However, the sector Passenger motor cars includes hybrid vehicles (HV) and electric vehicles (EV), besides gasoline-based internal combustion vehicles (ICV) (the share of diesel-based passenger cars is minor in Japan). Accordingly, even for a sector defined for a single product, such as passenger motor cars, this includes various power trains with distinctively different input requirements: An EV has neither a fuel tank nor an engine, but needs powerful motors and batteries. An analysis reflecting the differences between these power trains cannot therefore be performed solely on the basis of this IO table.

The IO tables of the EU, developed by the Statistical Office, have a significantly lower level of sectoral resolution compared to the larger IO tables of Japan, the US, and Korea, with only around 60 categories for each European country. As a result, LCAs of individual products using these categories directly are almost impossible. However, the recent development of EXIOBASE, with around 160 production sectors (Sect. 4.3.4), has greatly improved the situation in the EU. It is essential to consider the level of sectoral resolution when assessing the pros and cons of different hybrid methods.

5.3.2 *HLCA Models*

According to [49], HLCA models can be classified into three methods

1. Tiered hybrid analysis (TH)
2. IO-based hybrid analysis (IOH)
3. Integrated hybrid analysis (IH)

We now discuss each of these methods in some detail, referring to [48].

5.3.2.1 Tiered Hybrid Analysis (TH)

Tiered hybrid analysis is a straightforward and easy-to-use hybrid LCA method. Suppose we consider conducting an LCA for an electric vehicle (EV). In this case, the sectoral classification of the IO table lacks a specific category for EVs. Consequently, the production stage inventory data for the EV must be obtained through PLCA. Subsequently, inventory data for the energy consumption in the use stage and the waste management and recycling fees for the EV's end-of-life are determined, and the IO model is employed for inventory analysis of these two stages. Therefore, inventory data from critical production stages, along with the product data required for the use and end-of-life stages, are gathered, and the IO model supplements the analysis of inventory data upstream from the collected data.

The TH method can be described mathematically as follows. Let $\bar{\boldsymbol{A}} = [\bar{a}_{ij}]$ be the matrix consisting of the inventory data collected by PLCA. Each element $\bar{a}_{ij}$ represents the input or output (negative entry) of product i to or from process j. In the context of an EV, this corresponds to the materials and energy used during the production stage. In the language of LCA, the matrix $\boldsymbol{I} - \bar{\boldsymbol{A}}$ represents the technology matrix (Sect. 4.1.4.2). Let $\bar{\boldsymbol{F}}$ be the corresponding intervention (unit environmental burden) matrix. If $\bar{\boldsymbol{y}}$ represents the functional unit of assessment, e.g., one EV, the direct and indirect GHG emissions per functional unit can be calculated as

$$\bar{\boldsymbol{F}}(\boldsymbol{I} - \bar{\boldsymbol{A}})^{-1}\bar{\boldsymbol{y}} \tag{5.59}$$

We now describe the IO side of the TH method. Let $\boldsymbol{y}$ represent the functional unit corresponding to the IO table, which refers to the purchase of one EV, and the costs associated with energy consumption during its use and end-of-life management. The direct and indirect emissions per functional unit $\boldsymbol{y}$ can then be given by

$$\boldsymbol{F}(\boldsymbol{I} - \boldsymbol{A})^{-1}\boldsymbol{y}. \tag{5.60}$$

The sum of the emission from the process method and that from the IO gives the emissions from the tiered hybrid analysis, $\boldsymbol{e}_{\text{TH}}$

$$\boldsymbol{e}_{\text{TH}} = \bar{\boldsymbol{F}}(\boldsymbol{I} - \bar{\boldsymbol{A}})^{-1}\bar{\boldsymbol{y}} + \boldsymbol{F}(\boldsymbol{I} - \boldsymbol{A})^{-1}\boldsymbol{y} \tag{5.61}$$

Although the TH method is simple, it has limitations. It does not consider the interrelationship between $\bar{\boldsymbol{y}}$ and $\boldsymbol{y}$ or $\bar{\boldsymbol{A}}$ and $\boldsymbol{A}$, which means that any impact of the LCA system on the economic system cannot be considered. Additionally, because the production system for the process method is typically included in the transactions of the IO table for the same year, there is a risk of double-counting inventory, depending on how the functional unit is specified. This problem can violate the clarity of the system boundary, which is the greatest advantage of IO. To address this issue, [53–55] developed a methodology to apply Structural Path Analysis (SPA) to the EEIO model (see [56], Chap. 5–8 for more details on SPA).

5.3.2.2 IO-Based Hybrid Analysis (IOH)

IO-based hybrid analysis subdivides the sectoral classification of the IO table instead of linking inventory data obtained by the process method. By maintaining the calculation system of the IO table, the spatial system boundary remains the entire country. For example, the passenger motor vehicle sector i can be divided into three sectors: the ICV sector i_a, the HV sector i_b, and the EV sector i_c. In this case, the number of sectoral rows and columns in the IO table increases by two, and the input coefficient $\boldsymbol{A}$ is expanded as

$$A_{\text{IOH}} = \begin{pmatrix} a_{11} & \dots & a_{1,i_a} & a_{1,i_b} & a_{1,i_c} & \dots & a_{1n} \\ \vdots & & \vdots & \vdots & \vdots & & \vdots \\ a_{i_a,1} & \dots & a_{i_a,i_a} & a_{i_a,i_b} & a_{i,i_c} & \dots & a_{i_a,n} \\ a_{i_b,1} & \dots & a_{i_b,i_a} & a_{i_b,i_b} & a_{i_b,i_c} & \dots & a_{i_b,n} \\ a_{i_c,1} & \dots & a_{i_c,i_a} & a_{i_c,i_b} & a_{i_c,i_c} & \dots & a_{i_c,n} \\ \vdots & & \vdots & \vdots & \vdots & & \vdots \\ a_{n1} & \dots & a_{n,i_a} & a_{n,i_b} & a_{n,i_c} & \dots & a_{nn} \end{pmatrix} \quad (5.62)$$

Similarly, the functional unit vector $\boldsymbol{y}$ and the unit environmental burden matrix $\boldsymbol{F}$ are also subdivided for each sector, denoted as $\boldsymbol{F}_{\text{IOH}}$ and $\boldsymbol{y}_{\text{IOH}}$, respectively. The environmental burden $\boldsymbol{e}_{\text{IOH}}$ for the functional unit $\boldsymbol{y}_{\text{IOH}}$ obtained by IO-based hybrid analysis can then be given by

$$\boldsymbol{e}_{\text{IOH}} = \boldsymbol{F}_{\text{IOH}}(\boldsymbol{I} - \boldsymbol{A}_{\text{IOH}})^{-1}\boldsymbol{y}_{\text{IOH}} \quad (5.63)$$

We briefly present some examples of the IOH method.

Insulation Materials

Reference [50] examined the optimal use of residential long-life and highly-insulating technologies, distinguished by different insulation materials and window types, to minimize CO_2 emissions across the entire life cycle of all the houses in Japan. The assessment was done based on (5.63) with the matrix $\boldsymbol{A}$ taken from the Japanese IO table of 1995, augmented with detailed information about the relevant processes and floor sizes.

A New Disassembling Technology

Reference [57] investigated the effects on CO_2 emissions of introducing a new disassembling technology called active disassembling fastener (ADF) to the end-of-life disassembling process of electric and electronic appliances, specifically refrigerators. The study used the IOH method, using the matrix $\boldsymbol{A}$ derived from the Japanese IO table of 2000. The matrix was further disaggregated to incorporate different types of steel and home electric appliances. Additionally, detailed information about the relevant processes, including the production of ADF and disassembling and shredding processes, was incorporated.

Household Air Conditioners

Reference [58] examined the impact on the life cycle CO_2 emissions derived from household air conditioners of industrial technology changes, product lifetime changes, and energy efficiency improvements based on (5.63). Since the author was using a Japanese IO table for 1990, 1995, 2000, and 2005, the inventory of household air conditioners was readily available: Commercial residential air conditioners occurs as an independent sector in the Japanese IO table.

Hydrogen Energy System

Reference [59] analyzed the environmental and socio-economic effects of a hydrogen energy system over its entire life cycle using the IOH method and (5.63). The study utilized an engineering-based disaggregated version of the Japanese IO table for 2005, which identified hydrogen production, transportation, and fueling.

Carbon Footprint of Electronics

Reference [10] investigated the carbon footprint of Korean electronics for 15 electronic sectors, including semiconductors, LCD, circuit boards, computers, computer peripheral devices, phones, TV, and audio and acoustic equipment, with the relevant inventory directly taken from the Korean IO table, with 381 sectors for the year 2017.

EIO-LCA and its Application

The Green Design Institute at Carnegie Mellon University developed a system called EIO-LCA, which provides the impact calculation of a given functional unit (final demand) on a wide range of environmental burdens based on US IO tables with around 500 production sectors. An early example of studies based on this system is [60], which examined the environmental impacts of steel-reinforced concrete using the US IO table for 1987 with 519 sectors. Another study, [61], examined the total life cycle GHG emissions associated with the production, transportation, and distribution of food consumed by American households using the US IO table for 1997 with a similar number of sectors.

A Remark

All the studies cited mentioned above as examples of HIO are based on IO tables from Japan, Korea, or the US, which are known for having the highest level of sectoral detail. What is also noteworthy is that, regarding regional modeling, all of them are characterized as a two-region open model (comprising the domestic economy and the rest of word, which is considered exogenous) or "national model" (Sect. 4.3.2), which is understandable because there are no regions except these three countries where IO tables of equivalent resolution levels are available. In this regard, it is worth mentioning that [62] developed a methodology known as the Global Link IO (GLIO) model, which addresses this limitation by incorporating elements of MRIO to enhance the regional resolution of Japanese imports.

5.3.2.3 Integrated Hybrid Analysis (IH)

This section mostly refers to [49]. As previously mentioned, there are several reasons why information obtained from an IO table, particularly one with low resolution, is commonly considered less reliable than process-specific data. To address this issue, [49] proposed an analytical system where the $\boldsymbol{A}$ matrix from an IO table is interconnected with the matrix representation of the physical product system $\bar{\boldsymbol{A}}$ from PLCA only at upstream and downstream cut-offs where more accurate data are not available.

The Method

Integrated hybrid analysis tackles the issue of the absence of interaction between the LCA technical system $\bar{A}$ described by the process method and the "economic" system A described by the IO table by establishing the interconnection. The interconnection is facilitated by the upstream cut-off matrix C^{u}, which describes the inputs from A to the physical system $\bar{A}$, and conversely the downstream cut-off matrix C^d, which describes the outputs to A from $\bar{A}$. The cut-off matrices represent those inputs from A that are missing in the physical process system. By using the matrices C^{u} and C^{d} to connect $\bar{A}$ and A, respectively, the mutual interaction between the PLCA method and EEIO is incorporated, resulting in the extended input matrix A_{IH} corrected for truncations

$$A_{\mathrm{IH}} = \begin{pmatrix} \bar{A} & C^{\mathrm{d}} \\ C^{\mathrm{u}} & A \end{pmatrix} \tag{5.64}$$

According to the convention often adopted in the LCA literature, the northwest element of the partitioned matrix referring to the physical system should be denoted as $I - \bar{A}$ [40, 63]. However, we have chosen to use the current notation to ensure consistency and minimize potential confusion, aligning with our above discussion on TH. The matrix $\bar{A} = [\bar{a}_{ij}]$ shares the same feature as A, where $\bar{a}_{ij}$ represents the use of input i per unit of product j measured in physical units. However, it is essential to note that from the perspective of PLCA, the apparent formal similarity between the two formulas masks some critical differences [63]. (see Sect. 4.1.4.2 for the case with by-products).

The environmental burden e_{IH} that accompanies the functional unit $\bar{y}$ is then given by

$$e_{\mathrm{IH}} = \begin{pmatrix} \bar{F} & F \end{pmatrix} (I - A_{\mathrm{IH}})^{-1} \begin{pmatrix} \bar{y} \\ 0 \end{pmatrix} \tag{5.65}$$

Note that if the LCA technical system has already been incorporated into A, it is necessary to replace it with the matrix A^{corr}, which has duplicated portions removed from matrix A: The inputs that are already represented in $\bar{A}$ need to be removed from A.

$$A_{\mathrm{IH}} = \begin{pmatrix} \bar{A} & C^{\mathrm{d}} \\ C^{\mathrm{u}} & A^{\mathrm{corr}} \end{pmatrix}^{-1} \tag{5.66}$$

Reference [64] describes the method for generating A^{corr}. Since the above matrix inversion involves submatrices with physical and monetary units, the actual compilation of C^u and C^d needs information about product prices to harmonize the units for calculation.

We briefly present some case studies based on the IH method.

Carbon Footprint Intensities of All Processes in the Australian National Life Cycle Inventory Database

Reference [51] computed the hybrid carbon footprint intensities (CFIs) for all processes in the Australian National Life Cycle Inventory Database (AusLCI) based on (5.65) using an Australian SU table with 1284 industries.

Natural Gas Based Fuel Chains for Transportation

Reference [65] analyzed the environmental impacts of alternative fuels for transportation (natural gas, methanol (MeOH), and hydrogen, called "fuel" for short in this paragraph), using a variant of IH with the physical processes (provided by the PLCA database) divided into foreground process (f) and background process (b). Foreground processes are those "specific to a product system, while the background processes are not" ([34], p. 80). The foreground processes "relate to the scope of the examined product system that can be directly influenced by the decision maker," whereas "Raw material acquisition, material production, and the supply of energy, transportation, and so on, are termed background process" ([35], p. 102). In this study, the foreground processes include natural gas extraction, gas power plants, fuel production, H_2 liquefication, fuel tankers, fuel storage, fuel trucks & trailers, fueling stations, and compressed natural gas internal combustion (CNG-IC) vehicles, as well as H_2 and MeOH fuel cell vehicles.

Denote the input matrix referring to the foreground processes by $\boldsymbol{A}^{\text{ff}}$, the inputs into the foreground processes from background processes by $\boldsymbol{A}^{\text{bf}}$, and the background processes available in the physical system by $\boldsymbol{A}^{\text{bb}}$. Inputs into foreground processes that were not available in the physical system were obtained from the US IO Table [66] and represented by $\boldsymbol{A}^{\text{nf}}$. Finally, the input matrix for processes or sectors that do not occur in the physical system is represented by $\boldsymbol{A}^{\text{nn}}$, which was obtained from the IO table. The $\boldsymbol{A}_{\text{IH}}$ is then given by

$$\boldsymbol{A}_{IH} = \begin{pmatrix} \boldsymbol{A}^{\text{ff}} & \boldsymbol{0} & \boldsymbol{0} \\ \boldsymbol{A}^{\text{bf}} & \boldsymbol{A}^{\text{bb}} & \boldsymbol{0} \\ \boldsymbol{A}^{\text{nf}} & \boldsymbol{0} & \boldsymbol{A}^{\text{nn}} \end{pmatrix} \tag{5.67}$$

The triangular structure of the matrix assumes that the repercussions from (f) and (b) to (n), and from (n) to (b) are negligible for simplification. Here, the notation (n) refers to the sectors occurring in $\boldsymbol{A}^{\text{nn}}$. The study's findings suggest that reducing the environmental impact of transportation is not solely about fuel efficiency, low emission fuels, and onboard emissions. It also relates significantly to green manufacturing, with impacts primarily stemming from the transformation of bulk materials into car bodies, engines, and various car components ([65], p. 2804).

Large-scale Adoption of Wind Power

Using the same IH methodology represented by (5.67), [67] investigated the potential environmental impacts of the large-scale adoption of wind power. The study identified Final manufacturing and assembly of main components, Operation and maintenance, and Installation and decommissioning as foreground processes, with Supply of electricity to foreground system and All other inputs to the foreground processes as background processes. The $\boldsymbol{A}^{\mathrm{nn}}$ matrix was obtained from Eurostat and GTAP 6. The study found that the total emissions associated with wind electricity production ranged from 4% to 14% of the direct emissions of the replaced fossil-fueled power plants.

Long-Term, Wide-Scale Implementation of Electricity Generation from Renewable Sources

In their LCA of the long-term, wide-scale implementation of electricity generation from renewable sources (photovoltaic and solar thermal, wind, and hydropower) and carbon dioxide capture and storage for fossil power generation, [68] employed a modified version of (5.67),

$$\boldsymbol{A}_{IH}(t) = \begin{pmatrix} \boldsymbol{A}^{\mathrm{ff}}(t) & \boldsymbol{A}^{\mathrm{fb}}(t) & \boldsymbol{A}^{\mathrm{fn}}(t) \\ \boldsymbol{A}^{\mathrm{bf}}(t) & \boldsymbol{A}^{\mathrm{bb}}(t) & \boldsymbol{0} \\ \boldsymbol{A}^{\mathrm{nf}}(t) & \boldsymbol{0} & \boldsymbol{A}^{\mathrm{nn}} \end{pmatrix} \tag{5.68}$$

Unlike previous studies, their model explicitly considered changes through new installation and replacement of alternative power technologies in the total capacity of power generation, which was made possible by the dynamic model of [69] (Section 4.4.3.8). Another distinctive feature of their model was the inclusion of nonzero $\boldsymbol{A}^{\mathrm{fb}}$ and $\boldsymbol{A}^{\mathrm{fn}}$ matrices. The $\boldsymbol{A}^{\mathrm{nn}}$ matrix was obtained from the EXIOBASE database. The study found that large-scale implementation of wind, photovoltaic, and solar thermal technologies has the potential to significantly reduce pollution-related environmental impacts of electricity production, such as greenhouse gas emissions, freshwater ecotoxicity, eutrophication, and particulate matter exposure.

5.4 Waste Input-Output Analysis (WIO)

LCA measures the environmental impacts of a product through its entire life cycle phases, from resource extraction and fabrication to use and end-of-life (EoL) phases, which include waste treatment, final disposal, or recycling. The EEIO model we have discussed so far explicitly considers the resource extraction and fabrication phases in the input matrix $\boldsymbol{A}$, while the use phase can easily be incorporated into $\boldsymbol{y}$. However, the EoL phase remains not explicitly accounted for in the model.

In reality, any physical production process generates waste because achieving a 100% yield rate is generally impossible. Waste needs to be treated in compliance with relevant regulations before it is discharged into the environment. This distinguishes waste from emissions such as CO_2 and CH_4, which are mostly emitted into the environment without treatment.[3]

[3] However, technology options, such as carbon capture and sequestration (CCS), can be used to capture CO_2 before it is emitted into the atmosphere.

The incorporation of waste treatment in EEIO can be traced back to the seminal article by Leontief [70], which was further elaborated by Duchin [71]. However, these works primarily focused on the treatment of specific pollutants, such as SO_X, NO_X, or hazardous substances in wastewater. They did not model the actual waste treatment processes such as the drying, shredding, sorting, incinerating, landfill, and recycling of primary and secondary waste. Furthermore, these early works primarily remained conceptual in nature and lacked actual data due to the difficulty of obtaining the needed data at that time.

With increasing awareness of the need for sustainable waste management, including recycling and reuse, a new variant of EEIO called the Waste Input-Output (WIO) model was developed [72, 73]. This section briefly introduces the WIO model and discusses its recent applications and extensions. For more details on the WIO model, readers can refer to [48, 74], and an excellent recent review on the application of IO to waste management, including WIO, is provided by [75].

5.4.1 Two Types of By-Products and the WIO Table

In Sect. 4.1.4.1, we classified by-products into two categories, competitive and noncompetitive. A by-product is considered competitive when it has a principal producer. Power from a waste incineration plant and sulfur from a desulfurization process are typical examples of competitive by-products. On the other hand, a by-product is considered noncompetitive if there is no principal producer. Typical examples of noncompetitive by-products include ash and dust from waste incineration plants and slag from metal smelting processes.

It is unlikely for a competitive by-product to be discarded as waste since it can be used as a substitute for its principal counterpart. In contrast, many noncompetitive by-products, such as ash, dust, slag, sludge, animal waste, carcasses, and food residues, are designated as industrial waste, but they can also be recycled. Durable products like cars, ships, buildings, home appliances, and clothes are discarded as waste after their end of life (with the exception of antiques), but they can also be recycled. Kitchen waste generated in a lunch facility can be referred to as a noncompetitive by-product. On the other hand, kitchen waste generated from home cooking, such as egg shells and fish bones, is usually not called a by-product.

The distinction between a noncompetitive by-product and waste is often unclear since the definition of waste is fuzzy and depends on several factors, including some that are economic and institutional ones ([74], p. 90). Therefore, we choose to use the terms noncompetitive by-product and waste interchangeably instead of attempting to draw a clear line between them.

Chapter 3 covered the integration of the flow of waste (noncompetitive by-products) into EEIO based on WIO. However, we used a highly simplified framework that assumed landfill (open dumping) as the only option for treating a small number of waste types. In Sect. 4.1.4.2, we discussed the integration of competitive by-products into IO as a negative input of the principal counterpart into the sector that generates the by-product, using Stone's negative entry method. This section focuses on integrating the generation, treatment, and recycling of noncompetitive by-products and waste into the EEIO framework based on WIO, within a more general setting.

5.4.2 *The WIO Table and Model*

5.4.2.1 The WIO Table

The Accounting Framework of WIO

Table 5.4 gives an integrated account based on WIO of the flow of products, by-products, and waste among production, waste treatment, and final demand sectors. Let I represent the production sectors, and II represent the waste treatment sectors. The matrix $\boldsymbol{Z}^{\mathrm{I}}$, with dimensions $n_{\mathrm{I}} \times n_{\mathrm{I}}$, refers to the flow of products used by the production sectors. Similarly, the matrix $\boldsymbol{Z}^{\mathrm{II}}$, with dimensions $n_{\mathrm{I}} \times n_{\mathrm{II}}$, represents the flow of products used by the waste treatment sectors. Likewise, the matrix $\boldsymbol{W}^{\mathrm{I}+}$, with dimensions $n_w \times n_{\mathrm{I}}$, represents the waste generated by the production sectors. On the other hand, the matrix $\boldsymbol{W}^{\mathrm{II}+}$, with dimensions $n_w \times n_{\mathrm{II}}$, represents the waste generated by the waste treatment sectors. Additionally, $\boldsymbol{W}^{\mathrm{I}-}$ refers to the flow of waste used by the production sectors, while $\boldsymbol{W}^{\mathrm{II}-}$ refers to the flow of waste used by the waste treatment sectors.

Let $\boldsymbol{x}^{\mathrm{II}} = [x_k^{\mathrm{II}}]$ be the $n_{\mathrm{II}} \times 1$ vector of the activity of waste treatment sectors, where x_k^{II} is the amount of waste treated by treatment sector k. We use a broad definition of waste treatment, which includes open dumping and disposal without any treatment, meaning that any waste that is not recycled is treated in some way. Since the sum of net waste generation $\boldsymbol{\iota}^{\top}\boldsymbol{w}$ gives the total amount of waste that needs treatment, we obtain

$$\boldsymbol{\iota}_{n_{\mathrm{II}}}^{\top}\boldsymbol{x}^{\mathrm{II}} = \boldsymbol{\iota}_{n_w}^{\top}\boldsymbol{w} \tag{5.69}$$

Denote by $\boldsymbol{S} = [s_{kl}]$ the $n_{\mathrm{II}} \times n_w$ matrix, with s_{kl} referring to the share of waste l treated by treatment k, with

$$\boldsymbol{\iota}_{n^{\mathrm{II}}}\boldsymbol{S} = \boldsymbol{\iota}_{n_w} \tag{5.70}$$

We then have

$$\boldsymbol{x}^{\mathrm{II}} = \boldsymbol{S}\boldsymbol{w} \tag{5.71}$$

The WIO table 2011 Developed by the Japanese MOE

To date, the MOE-WIO table for 2011, developed by the Japanese Ministry of the Environment, is the only publicly released WIO table. This table consists of 81

Table 5.4 The integrated account of products and waste by WIO

		Production	Waste treatment	Final demand	Output
		1 | 2 | ⋯ | n_{I}	1 | 2 | ⋯ | n_{II}		
Products	1	$\boldsymbol{Z}^{\mathrm{I}}$	$\boldsymbol{Z}^{\mathrm{II}}$	$\boldsymbol{y}^{\mathrm{I}}$	$\boldsymbol{x}^{\mathrm{I}}$
	2				
	⋮				
	n_{I}				
	Waste generation				
Waste	1	$\boldsymbol{W}^{\mathrm{I}+}$	$\boldsymbol{W}^{\mathrm{II}+}$	$\boldsymbol{w}^{y+}$	$\boldsymbol{w}^{+}$
	2				
	⋮				
	n_w				
	Waste use				
Waste	1	$\boldsymbol{W}^{\mathrm{I}-}$	$\boldsymbol{W}^{\mathrm{II}-}$	$\boldsymbol{w}^{y-}$	$\boldsymbol{w}^{-}$
	2				
	⋮				
	n_w				
	Net waste generation				
Waste	1	$\boldsymbol{W}^{\mathrm{I}} = \boldsymbol{W}^{\mathrm{I}+} - \boldsymbol{W}^{\mathrm{I}-}$	$\boldsymbol{W}^{\mathrm{II}} = \boldsymbol{W}^{\mathrm{II}+} - \boldsymbol{W}^{\mathrm{II}-}$	$\boldsymbol{w}^{y} = \boldsymbol{w}^{y+} - \boldsymbol{w}^{y-}$	$\boldsymbol{w} = \boldsymbol{w}^{+} - \boldsymbol{w}^{-}$
	2				
	⋮				
	n_w				

The final demand sectors are aggregated into a single sector. The waste generation and utilization are represented by several matrices. For waste k, producing sector j, and waste treatment sector l, we have

1. Waste generation in production sectors: $\boldsymbol{W}^{\mathrm{I}+} = [w_{kj}^{\mathrm{I}+}]$
2. Waste generation in waste treatment sectors: $\boldsymbol{W}^{\mathrm{II}+} = [w_{kl}^{\mathrm{II}+}]$
3. Waste generation from the final demand sector: $\boldsymbol{w}^{y+} = [w_k^{y+}]$
4. Waste utilization in production sectors: $\boldsymbol{W}^{\mathrm{I}-} = [w_{kj}^{\mathrm{I}-}]$.
5. Waste utilization in waste treatment sectors: $\boldsymbol{W}^{\mathrm{II}-} = [w_{kl}^{\mathrm{II}-}]$
6. Waste utilization in the final demand sector: $\boldsymbol{w}^{y-} = [w_k^{y-}]$

Power generation as a by-product of waste incineration is represented as a negative entry in the column of the incineration sector in $\boldsymbol{Z}^{\mathrm{II}}$, while the generation of ash from the sector is recorded in $\boldsymbol{W}^{\mathrm{II}+}$. The use of ash in cement production is recorded in the column of the cement sector in $\boldsymbol{W}^{\mathrm{I}-}$. Restaurant food waste is recorded in $\boldsymbol{W}^{\mathrm{I}+}$, while household food waste is recorded in $\boldsymbol{w}^{y+}$. End-of-life (EoL) durable products, such as end-of-life vehicles (ELV) and waste electronics, are recorded in $\boldsymbol{w}^{y+}$, while the generation of secondary wastes, such as metal scrap, waste plastics, and shredding residues, arising from dismantling, shredding, and sorting processes, are recorded in the corresponding column(s) of $\boldsymbol{W}^{\mathrm{II}+}$. If metal scrap is recycled in metal sectors, such as steel production based on an electric arc furnace (EAF), it is recorded in the corresponding column(s) of $\boldsymbol{W}^{\mathrm{II}-}$. Waste recycling mostly occurs in $\boldsymbol{W}^{\mathrm{I}-}$. A negative entry in $\boldsymbol{W}^{\mathrm{I}}$ for a specific waste k and sector j indicates that sector j uses more waste k than it generates, whereas a positive entry implies that the sector generates more waste k than it uses producing sectors, ten waste treatment sectors , 48 primary wastes, and 51 secondary wastes (See Table S4 of [72]. The MOE-WIO are available at [76]). However, the level of sectoral resolution is relatively low for a Japanese IO table, due to limited data availability on waste flows.

The waste treatment sectors convert the primary waste into secondary waste. For example, "Cinders" generated from production sectors, mostly from "Electricity," are dehydrated and converted into "Cinders (d)," which is either recycled in cement and fertilizer production or landfilled when not recycled, together with "Cinders (i)" from waste incineration facilities. Most secondary inorganic waste is disposed of in landfills when not recycled, as indicated by the share of landfill being one for the majority of these waste types. In contrast, incineration is the primary treatment method for most secondary organic waste, paper, textiles, plastics, and wood, when not recycled.

Table 5.5 presents the share of primary waste items that end up in a landfill without treatment (column (a)) and with treatment in the secondary form (column (d)), with column (e) providing the overall share of primary waste disposed of in landfills. Only "Cinders (i)" and "Other nonflammable waste" have landfill rates exceeding 50%, whereas "Cinders," "Waste plastics," "Waste fiber," "Waste rubber," "Glass etc., scrap," "Dust," and "Glass bottles" have landfill rates between 10 and 30%. For all other waste items listed in Table 5.5, the share of landfill is below 10%. On average, around 2.8% of primary waste items end up in landfills among the 39 waste items considered, excluding items with null or negligible landfill rates.

5.4.2.2 The WIO Model

Dividing the net waste flows in $\boldsymbol{W}^{\mathrm{I}}$ and $\boldsymbol{W}^{\mathrm{II}}$ by $\boldsymbol{x}^{\mathrm{I}}$ and $\boldsymbol{x}^{\mathrm{II}}$, respectively, gives the amounts of net waste generation per production and waste treatment

$$\boldsymbol{G}^{\mathrm{I}} = \boldsymbol{W}^{\mathrm{I}}\,\mathbf{diag}(\boldsymbol{x}^{\mathrm{I}})^{-1}, \quad \boldsymbol{G}^{\mathrm{II}} = \boldsymbol{W}^{\mathrm{II}}\,\mathbf{diag}(\boldsymbol{x}^{\mathrm{II}})^{-1} \tag{5.72}$$

Let there be n_y categories of final demand. The $n^{\mathrm{I}} \times n_y$ matrix of final demand for products is denoted as $\boldsymbol{Y}^{\mathrm{I}}$, while the $n_w \times n_y$ matrix of waste discard from the demand is denoted as $\boldsymbol{W}^y$. We define $\boldsymbol{y}^{\mathrm{I}} = \boldsymbol{Y}^{\mathrm{I}}\boldsymbol{\iota}_{n_y}$ and $\boldsymbol{w}^y = \boldsymbol{W}^y\boldsymbol{\iota}_{n_y}$. The balance of products and waste then imply

$$\begin{aligned} \boldsymbol{x}^{\mathrm{I}} &= \boldsymbol{A}^{\mathrm{I}}\boldsymbol{x}^{\mathrm{I}} + \boldsymbol{A}^{\mathrm{II}}\boldsymbol{x}^{\mathrm{II}} + \boldsymbol{y}^{\mathrm{I}} \\ \boldsymbol{w} &= \boldsymbol{G}^{\mathrm{I}}\boldsymbol{x}^{\mathrm{I}} + \boldsymbol{G}^{\mathrm{II}}\boldsymbol{x}^{\mathrm{II}} + \boldsymbol{w}^y \end{aligned} \tag{5.73}$$

where $\boldsymbol{A}^{\mathrm{I}}$ and $\boldsymbol{A}^{\mathrm{II}}$ refer to the standard matrices of input coefficients

$$\boldsymbol{A}^{\mathrm{I}} = \boldsymbol{Z}^{\mathrm{I}}\,\mathbf{diag}(\boldsymbol{x}^{\mathrm{I}})^{-1}, \quad \boldsymbol{A}^{\mathrm{II}} = \boldsymbol{Z}^{\mathrm{II}}\,\mathbf{diag}(\boldsymbol{x}^{\mathrm{II}})^{-1} \tag{5.74}$$

Table 5.5 The fraction of waste landfilled after treatment: Japanese WIO 2011

Primary	(*a*)	(*b*)	(*c*)	(*d*)	(*e*)
Waste	Output	Landfilled	Treatment	Landfilled	$\frac{a \times b + d}{a}$
Cinders (i)	4,669		none	3,332	0.714
Cinders	1,836	0.245	(d)	52	0.274
Sludge	167,070	0.006	(d)	2,197	0.020
Waste oil	3,151	0.011	(f)	31	0.021
Waste acid	2,685	0.000	(c)	64	0.024
Waste alkali	2,058	0.004	(c)	39	0.023
Waste plastics	5,829	0.050	(s)	718	0.174
Waste paper	1,127	0.009	(s)	20	0.026
Wood waste	6,205	0.010	(s)	136	0.032
Waste fiber	79	0.035	(s)	23	0.320
Animal & plant residue	2,791	0.005	(d)	44	0.021
Animal solid waste	94	0.008	(s)	2	0.029
Waste rubber	32	0.065	(s)	4	0.190
Iron scrap	6,903	0.010	(s)	202	0.039
Copper scrap	128	0.010	(s)	2	0.028
Aluminum scrap	152	0.010	(s)	3	0.028
Lead scrap	18	0.010	(s)	0	0.028
Zinc scrap	36	0.010	(s)	1	0.028
Other NF metal scrap	5	0.010	(s)	0	0.028
Glass etc. scrap	6,278	0.089	(s)	906	0.233
Slag	15,621	0.059	(s)	430	0.086
Construction waste	59,445	0.014	(s)	1,142	0.033
Livestock excreta	84,556	0.000	(c)	35	0.000
Livestock corpses	168	0.008	(s)	4	0.029
Dust	16,133	0.128	(d)	274	0.145
O. flam.	14,491	0.007	(s)	514	0.042
OA paper	54	0.011	(s)	3	0.060
Newspaper etc.	5,157	0.011	(s)	256	0.060
Paper drink box	101	0.011	(s)	5	0.060
Paper C&P	523	0.011	(s)	26	0.060
Styrofoam (white)	25	0.019	(s)	0	0.021
Plastic C&P	927	0.019	(s)	1	0.021
PET bottles	223	0.018	(s)	0	0.018
Steel can	132	0.064	(s)	3	0.089
Aluminum can	126	0.064	(s)	3	0.089
Other metals	149	0.064	(s)	4	0.089
Glass bottles	585	0.062	(s)	48	0.144
Wooden bulky waste	683	0.007	(s)	24	0.042
O. nonflam.	2,232	0.107	(s)	1,181	0.636

(*a*): gross generation (Gg (kt)), (*b*): the share of (*a*) landfilled, (*c*): the process (*a*) is subjected, (*d*): the amount landfilled after (*c*) (Gg (kt)), (*e*): the total share of (*a*) landfilled
Source [15]

Rearranged in matrix form, we obtain

$$\begin{pmatrix} A^{\mathrm{I}} & A^{\mathrm{II}} \\ G^{\mathrm{I}} & G^{\mathrm{II}} \end{pmatrix} \begin{pmatrix} x^{\mathrm{I}} \\ x^{\mathrm{II}} \end{pmatrix} + \begin{pmatrix} y^{\mathrm{I}} \\ w^{y} \end{pmatrix} = \begin{pmatrix} x^{\mathrm{I}} \\ w \end{pmatrix} \tag{5.75}$$

Note that (5.75) is generally unsolvable due to the occurrence of x^{II} on the left-hand side and w on the right-hand side. This renders the equation asymmetric.

In their EEIO models with pollutants/waste water, [70, 71] solved this asymmetric problem by making the simplifying assumption that

$$x^{\mathrm{II}} = w \tag{5.76}$$

This implies that each waste is exclusively treated by a single treatment sector. For example, waste plastics are either landfilled or incinerated, but not both at the same time. This assumption simplifies the model and makes it computationally more tractable, but it may not reflect the real-world situation.

Recognizing that the "one-to-one correspondence" between wastes and waste treatment does not reflect the reality of waste treatment, [72, 73] resolved the "asymmetry problem" of (5.73) by transforming the balance of waste into the balance of waste treatment activity using the matrix S, with dimensions $n_{\mathrm{II}} \times n_w$, in (5.71), which is referred to as the allocation matrix.

$$x^{\mathrm{II}} = Sw = SG^{\mathrm{I}}x^{\mathrm{I}} + SG^{\mathrm{II}}x^{\mathrm{II}} + Sw^{y} \tag{5.77}$$

The first equality follows from (5.71).

With the second equation in (5.75) replaced by (5.77), we obtain the following symmetric model

$$\begin{pmatrix} A^{\mathrm{I}} & A^{\mathrm{II}} \\ SG^{\mathrm{I}} & SG^{\mathrm{II}} \end{pmatrix} \begin{pmatrix} x^{\mathrm{I}} \\ x^{\mathrm{II}} \end{pmatrix} + \begin{pmatrix} y^{\mathrm{I}} \\ Sw^{y} \end{pmatrix} = \begin{pmatrix} x^{\mathrm{I}} \\ x^{\mathrm{II}} \end{pmatrix} \tag{5.78}$$

which can be solved for x^{I} and x^{II} as

$$\begin{pmatrix} x^{\mathrm{I}} \\ x^{\mathrm{II}} \end{pmatrix} = \begin{pmatrix} I - A^{\mathrm{I}} & -A^{\mathrm{II}} \\ -SG^{\mathrm{I}} & I - SG^{\mathrm{II}} \end{pmatrix}^{-1} \begin{pmatrix} y^{\mathrm{I}} \\ Sw^{y} \end{pmatrix} \tag{5.79}$$

Denoting by F^{I} and F^{II} the intervention matrices corresponding to the production and wast treatment sectors, we can derive the WIO counterpart of (5.11) as follows

$$\begin{pmatrix} E^{\mathrm{I}} \\ E^{\mathrm{II}} \end{pmatrix} = \begin{pmatrix} F^{\mathrm{I}} \\ F^{\mathrm{II}} \end{pmatrix} \begin{pmatrix} I - A^{\mathrm{I}} & -A^{\mathrm{II}} \\ -SG^{\mathrm{I}} & I - SG^{\mathrm{II}} \end{pmatrix}^{-1} \begin{pmatrix} \mathrm{diag}(y^{\mathrm{I}}) \\ \mathrm{diag}(Sw^{y}) \end{pmatrix} \tag{5.80}$$

5.4.2.3 The Waste Footprint of Products

From (5.79), we can attribute the generation of waste for treatment, $\boldsymbol{w}$, to the final demand for products, $\boldsymbol{y}^{\mathrm{I}}$, and the disposal of waste from the final demand, $\boldsymbol{w}^{y}$, as follows

$$
\begin{aligned}
\boldsymbol{w} &= \left(\boldsymbol{G}^{\mathrm{I}}\ \boldsymbol{G}^{\mathrm{II}}\right)\begin{pmatrix}\boldsymbol{L}^{\mathrm{I,I}} & \boldsymbol{L}^{\mathrm{I,II}}\\ \boldsymbol{L}^{\mathrm{II,I}} & \boldsymbol{L}^{\mathrm{II,II}}\end{pmatrix}\begin{pmatrix}\boldsymbol{y}^{\mathrm{I}}\\ \boldsymbol{S}\boldsymbol{w}^{y}\end{pmatrix} + \boldsymbol{w}^{y}\\
&= \left(\boldsymbol{G}^{\mathrm{I}}\boldsymbol{L}^{\mathrm{I,I}} + \boldsymbol{G}^{\mathrm{II}}\boldsymbol{L}^{\mathrm{II,I}}\right)\boldsymbol{y}^{\mathrm{I}} + \left((\boldsymbol{G}^{\mathrm{I}}\boldsymbol{L}^{\mathrm{I,II}} + \boldsymbol{G}^{\mathrm{II}}\boldsymbol{L}^{\mathrm{II,II}})\boldsymbol{S} + \boldsymbol{I}\right)\boldsymbol{w}^{y}
\end{aligned}
\tag{5.81}
$$

The right-hand side of (5.81) can be interpreted as follows. The first term on the right-hand side refers to the amount of waste attributed to the final demand for products, which represents the waste footprint of final products. The second term represents the amount of waste attributed to the discard of waste in the final demand sectors. Its first component, $(\cdot)\boldsymbol{S}\boldsymbol{w}^{y}$, refers to the amount of waste directly and indirectly generated in treating $\boldsymbol{w}^{y}$.

These terms can be further divided into the direct and indirect components as

$$
\boldsymbol{w} = \underbrace{\boldsymbol{G}^{\mathrm{I}}\boldsymbol{y}^{\mathrm{I}} + \boldsymbol{w}^{y}}_{\text{direct discharge}} + \underbrace{\left(\boldsymbol{G}^{\mathrm{I}}(\boldsymbol{L}^{\mathrm{I,I}} - \boldsymbol{I}) + \boldsymbol{G}^{\mathrm{II}}\boldsymbol{L}^{\mathrm{II,I}}\right)\boldsymbol{y}^{\mathrm{I}} + \left(\boldsymbol{G}^{\mathrm{I}}\boldsymbol{L}^{\mathrm{I,II}} + \boldsymbol{G}^{\mathrm{II}}\boldsymbol{L}^{\mathrm{II,II}}\right)\boldsymbol{S}\boldsymbol{w}^{y}}_{\text{indirect discharge}}
\tag{5.82}
$$

The indirect discharge can be further decomposed into the discharge of primary waste from production sectors (the terms involving $\boldsymbol{G}^{\mathrm{I}}$) and secondary waste from waste treatment sectors (the terms involving $\boldsymbol{G}^{\mathrm{II}}$).

Denote by $\boldsymbol{Y}^{\mathrm{I}}$ and $\boldsymbol{W}^{y}$ the matrices obtained by extending $\boldsymbol{y}^{\mathrm{I}}$ and $\boldsymbol{w}^{y}$ column-wise by final demand categories, such as household consumption, fixed capital formation by private and public sectors, and export. Replacing $\boldsymbol{y}^{\mathrm{I}}$ and $\boldsymbol{w}^{y}$ in (5.81) with $\boldsymbol{Y}^{\mathrm{I}}$ and $\boldsymbol{W}^{y}$ gives the waste footprint extended by final demand categories $\boldsymbol{W}$, an $n_w \times n_y$ matrix

$$
\boldsymbol{W} = \left(\boldsymbol{G}^{\mathrm{I}}\boldsymbol{L}^{\mathrm{I,I}} + \boldsymbol{G}^{\mathrm{II}}\boldsymbol{L}^{\mathrm{II,I}}\ \ (\boldsymbol{G}^{\mathrm{I}}\boldsymbol{L}^{\mathrm{I,II}} + \boldsymbol{G}^{\mathrm{II}}\boldsymbol{L}^{\mathrm{II,II}})\boldsymbol{S} + \boldsymbol{I}\right)\begin{pmatrix}\boldsymbol{Y}^{\mathrm{I}}\\ \boldsymbol{W}^{y}\end{pmatrix}
\tag{5.83}
$$

By multiplying the waste footprint matrix extended by final demand categories, $\boldsymbol{W}$, with the allocation matrix $\boldsymbol{S}$ from the left, we can obtain the waste treatment footprint matrix extended by final demand categories

$$
\boldsymbol{S}\boldsymbol{W} = \left(\boldsymbol{S}(\boldsymbol{G}^{\mathrm{I}}\boldsymbol{L}^{\mathrm{I,I}} + \boldsymbol{G}^{\mathrm{II}}\boldsymbol{L}^{\mathrm{II,I}})\ \ \boldsymbol{S}(\boldsymbol{G}^{\mathrm{I}}\boldsymbol{L}^{\mathrm{I,II}} + \boldsymbol{G}^{\mathrm{II}}\boldsymbol{L}^{\mathrm{II,II}})\boldsymbol{S} + \boldsymbol{S}\right)\begin{pmatrix}\boldsymbol{Y}^{\mathrm{I}}\\ \boldsymbol{W}^{y}\end{pmatrix}
\tag{5.84}
$$

Column (a) of Table 5.6 shows the amount of waste classified by treatment method. Dehydration accounted for the largest share, while waste landfilled amounted to approximately 33% of the waste incinerated. These values represent the net results

Table 5.6 Waste by treatment: Japanese WIO 2011

(a) By treatment			(b) Landfilled			(c) Incinerated		
	Tg	c.s.		Tg	c.s.		Tg	c.s.
Dehydration	173.13	0.48	Cinders (i)	3.33	0.19	Kitchen waste	18.42	0.35
Shredding	96.03	0.75	Sludge (d)	2.20	0.32	O. flam.	12.86	0.60
Incineration	52.40	0.90	Dust	1.82	0.43	Sludge	7.75	0.74
Concentration	17.38	0.95	Constr. waste (s)	1.14	0.50	Newspaper etc.	3.45	0.81
Landfill	17.10	0.99	Sludge	1.07	0.56	Waste oil	1.89	0.85
Filtration	1.10	1.00	Glass etc. (s)	0.91	0.61	Waste plastics	1.73	0.88
RDF	0.80	1.00	Constr. waste	0.82	0.66	Wood waste	1.13	0.90
Composting	0.28	1.00	Slag	0.73	0.70	O. nonflam. (s)	0.85	0.92
Gasification	0.05	1.00	Waste plastics (s)	0.72	0.75	O. nonflam.	0.77	0.93
Feed conversion	0.01	1.00	Glass etc.	0.54	0.78	Plastic C&P	0.68	0.95
Total	358.29	1.00	Total	17.10	1.00	Total	52.40	1.00

Source [15]. Tg is equal to Mt, c.s: the ratio of cumulative sum, RDF: refuse derived fuel, O. flam.: other flammable waste, O. nonflam.: other nonflammable waste, Plastic C&P: plastic container & package. The items with d, f, i, and s refer to waste obtained after dehydration, filtration, intermediate treatment, and shredding, respectively

where the amounts of waste generated were offset by those recycled (see Table 5.4 and (5.71)).

Equation (5.84) allows us to decompose the net results into the gross amounts of waste generation and recycling. The results presented in Fig. 5.1 demonstrate that, in terms of gross generation, waste sent to a landfill was approximately 30% more than that of waste incineration. However, the net amount of waste sent to a landfill was smaller than that of incineration, as around 80% of the gross amount of landfill waste was recycled through public capital formation. Without this significant recycling, the amount of waste sent to a landfill would have been five to six times higher than the actual amount. Furthermore, it is noteworthy that private consumption was the largest contributor to waste dehydration and incineration.

The Waste for a Landfill Footprint of Products

Figure 5.2 depicts a breakdown of the gross generation of waste for a landfill as shown in Fig. 5.1 into individual product- and waste items based on (5.81). Specifically, it illustrates the waste for landfill footprint of products.

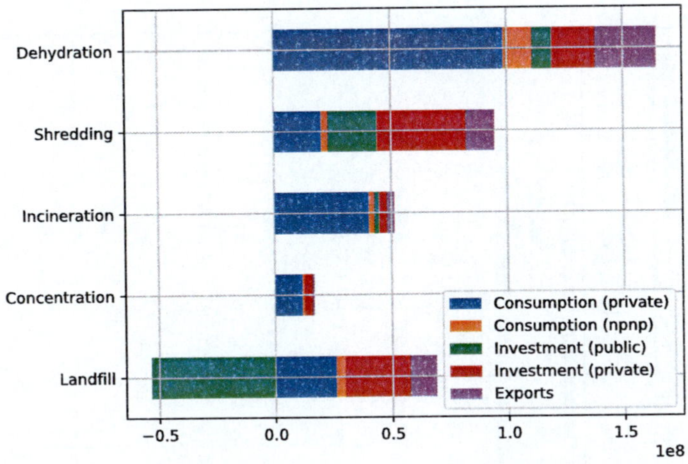

Fig. 5.1 Waste-treatment attributed to final demand categories by (5.84). Units in 100Tg (100 million tonnes). (npnp: nonprofit and nonprivate.) Fig. 1 of [15]

The recycling of "Construction waste", "Slag", and "Glass etc. scrap" is primarily carried out by " Public construction", which uses construction and demolition waste as aggregates in road construction—a key component of "Public construction". Meanwhile, "Building construction" is the primary generator of "Construction waste", "Glass etc. scrap", and "Slag", but is also the second-largest recycler of "Dust" after "Public construction." Lastly, "Civil engineering" is the largest recycler of "Sludge (d)" (dehydrated sludge), while also being the second-largest generator of "Construction waste."

"Foods" and "Livestock" are the primary generators of "Livestock excreta", of which more than 95% is recycled by "Crop cultivation", followed by "Feed production."

Service sectors such as "Medical services", " Personal services," "Real estate," and " Commerce" contribute to waste generation at a similar level to "Steel products," and their total waste generation is even larger than that of "Passenger motor cars." Interestingly, seven of the top 24 major sectors (which together make up 85% of the 13 major waste items for landfill) are in the service industry. This highlights the significant contribution of service sectors to waste generation for landfill, with their contribution to recycling being almost negligible, except for a small amount in Personal services.

The results presented here highlight the importance of considering both waste generated and waste recycled, particularly for waste items such as "Construction waste," "Sludge," "Dust," "Livestock excreta," and "Glass, etc. scrap." Without the significant recycling efforts by "Public construction," "Civil engineering," and "Crop cultivation," the landfill requirements shown in Fig. 5.2 would have been substantially different.

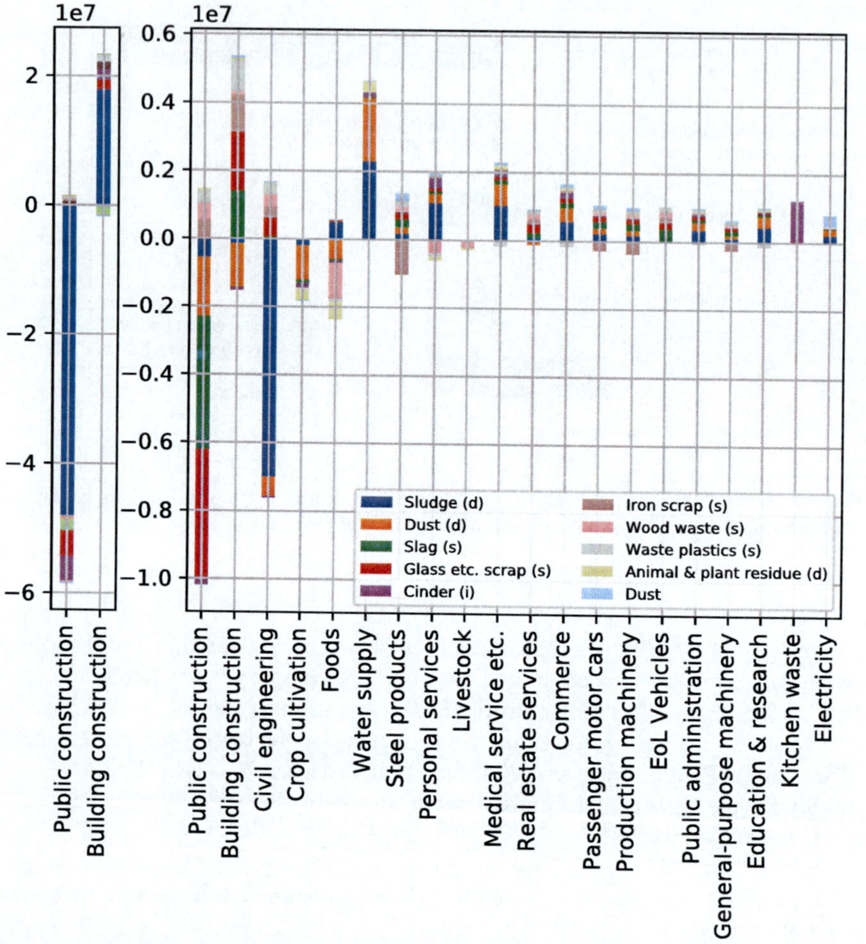

Fig. 5.2 Waste for landfill footprint of products. Units in 10 T g (10 million tonnes). The occurrence of "Kitchen waste" and "Other flammable waste" as products in the horizontal axis represents their direct discharge from private households. *Source* Fig. 2 of [15]

5.4.2.4 The SUT Extension of WIO

In the standard WIO table, waste flows are categorized based solely on treatment method, without considering the waste type. However, the WIO-SUT framework, developed by [77], addresses this limitation by allowing both waste by type and waste by treatment method to be presented together in a single table. This extension of the WIO framework results in a symmetric WIO table that does not require the conversion of waste flows into treatment flows. In terms of the WIO-SUT framework, the balance of products and waste can be represented as

$$\begin{pmatrix} A^{\mathrm{I}} & A^{\mathrm{II}} & 0 \\ 0 & 0 & S \\ G^{\mathrm{I}} & G^{\mathrm{II}} & 0 \end{pmatrix} \begin{pmatrix} x^{\mathrm{I}} \\ x^{\mathrm{II}} \\ w \end{pmatrix} + \begin{pmatrix} y^{\mathrm{I}} \\ 0 \\ w^{y} \end{pmatrix} = \begin{pmatrix} x^{\mathrm{I}} \\ x^{\mathrm{II}} \\ w \end{pmatrix} \tag{5.85}$$

Note that the matrix on the left-hand side is a square matrix of order $(n_{\mathrm{I}} + n_{\mathrm{II}} + n_w) \times (n_{\mathrm{I}} + n_{\mathrm{II}} + n_w)$. The entire system is symmetric since $(x^{\mathrm{I}}\, x^{\mathrm{II}}\, w)^{\top}$ occurs on both sides of the system. The solution to this system is given by:

$$\begin{pmatrix} x^{\mathrm{I}} \\ x^{\mathrm{II}} \\ w \end{pmatrix} = \begin{pmatrix} I - A^{\mathrm{I}} & -A^{\mathrm{II}} & 0 \\ 0 & I & -S \\ -G^{\mathrm{I}} & -G^{\mathrm{II}} & I \end{pmatrix}^{-1} \begin{pmatrix} y^{\mathrm{I}} \\ 0 \\ w^{y} \end{pmatrix} \tag{5.86}$$

The WIO-SUT model provides a solution for w in one step, together with the solution for x^{I} and x^{II}, while the original WIO requires two steps. The first step is to obtain x^{I} and x^{II} as in (5.79) and the second step is to obtain w as in (5.81). It is worth noting that despite the seemingly different forms of the two models, the Leontief inverse matrices of WIO and WIO-SUT are equivalent (see [77] for a proof.)

Case Studies

Reference [78] constructed a WIO-SUT for Australia for 2008 with 343 production sectors, ten treatment sectors (most of which involved the treatment/processing of waste materials and recycling of secondary waste), and 14 waste types. Their findings highlight the significant impact of construction waste on the Australian economy, revealing that numerous sectors contribute indirectly to the country's construction waste footprint.

Reference [79] further extended the WIO-SUT to include eight regions in Australia and constructed the first MR-WIO-SUT model. The waste footprint calculation demonstrated that the construction sector had the largest footprint across the regions. It also revealed that the total waste generated along the supply chain was two to three times higher than what households disposed of directly. Furthermore, the study estimated the impacts on overseas countries exporting products to Australia based on the domestic technology assumption, which assumes that international production processes have similar waste intensities to domestic Australian processes.

5.4.2.5 Categorizing the Methodology of WIO: IOH or IH?

We examine if the WIO is to be classified into IOH (IO-based hybrid analysis) or IH (Integrated hybrid analysis). Since the standard official IO table does not distinguish waste treatment by treatment processes and usually does not include the flows of waste (except that of competitive by-products and noncompetitive by-products items with an active market, such as waste paper and metal scraps), constructing a WIO requires (column-wise) disaggregation of the waste treatment sector by major treatment processes. On the other hand, waste flows in physical units have to be added as row elements. The need for these procedures will make WIO a method of IOH.

On the other hand, disaggregation of the waste treatment sector by individual treatment processes requires incorporating detailed engineering knowledge into the model. Different from IH, however, this incorporation (or hybridization) is not done by connecting a physical system (taken from a PLCA database) with an economic IO system via cut-off matrices as in (5.64).

Because of these features, WIO can be regarded as a mixture of IOH and IH. It takes the physical data on waste and waste management from external sources, with the latter mostly based on an engineering model. The physical system is embedded into the IO table via replacement (of data in monetary units with that in physical units) and disaggregation.

5.4.3 *The Basics of Waste Treatment Processes: Incineration*

Incineration and disposal to a landfill are the two primary methods used for treating waste that is not recycled. Incineration is the primary method of treating municipal solid waste (MSW) in Austria, Belgium, Denmark, Germany, Japan, Luxembourg, Netherlands, Norway, Singapore, Switzerland, and Taiwan [80]. In contrast, disposal to a landfill is the primary method used in other countries such as Brazil, Canada, China, Ethiopia, Russia, Spain, Turkey, the UK, Ukraine, and the US. The chemical properties of the waste being treated play a crucial role in determining the inputs and outputs of the waste treatment process, particularly in the case of incineration ([81], p. 106, [74], p.229). Therefore, it is essential to consider waste incineration processes in detail due to their widespread use and unique characteristics Fig. 5.3 shows a typical waste incineration plant in Japan. Advanced flue gas treatment enables the location of such a plant to be urban.

5.4.3.1 The Energy Content of the Waste and Its Measurement

The energy content of the waste is a crucial factor that affects the performance of a municipal solid waste (MSW) incineration plant. To measure the heating value of fuel, including waste fed into an incineration plant, the most common methods are to use either the gross heat content (higher heating value or H_H) or the net heat content (lower heating value or H_L). The primary difference between H_H and H_L is that H_H includes the energy used to vaporize water (or the latent heat of water vapor formed during the combustion of hydrocarbon fuels) [82]. The lower heating value (H_L), which is calculated using the formula

$$H_L = H_H - 24.5(w + 9h), \tag{5.87}$$

Fig. 5.3 Chuo incineration plant, Tokyo, Japan. Capacity: 0.6 Gg (600 tonnes) per day (0.6 Gg × 2 furnaces), 15 MW. Courtesy of Tokyo Twenty-three Wards Cleaning Office Association

is commonly used to determine the total energy input, as the latent heat of water vapor cannot be recovered and used [82]. Here, H_L represents the lower heating value in kJ/kg, H_H represents the higher heating value in kJ/kg, w represents the weight percentage of moisture in the fuel, and h represents the weight percentage of hydrogen in the fuel [82].

Various models are available for estimating H_H, including physical composition analysis based on the composition of plastics, paper, water, and garbage, ultimate analysis based on the elemental composition of C, H, O, N, and S, and proximate analysis based on the levels of moisture and combustible matter. We focus on the models based on the ultimate analysis as they are commonly used in designing a MSW incineration plant in Japan.

Of the ultimate analysis-based equations, Dulong's equation, Steuer's equation, and Steuer-Kestner's equation are widely used, which are given by

$$\text{Dulong's equation:}\quad H_H = 338.9C + 1433.0(H - O/8) + 94.1S \tag{5.88}$$

$$\text{Steuer's equation:}\quad H_H = 338.9(C - \frac{3}{8}O) + 238.5\frac{3}{8}O + 1443.5(H - \frac{O}{16}) + 104.6\,S \tag{5.89}$$

$$\text{Steuer-Kestner's equation:}\quad H_H = 338.9(C - \frac{3}{4}O) + 238.5\frac{3}{4}O + 1443.5\,H + 104.6\,S \tag{5.90}$$

Table 5.7 Chemical composition of waste for incineration generated by private consumption, Japan, 2011

	Kitchen waste	Other flammable waste	Newspaper etc.	Waste plastics
	Share in waste for incineration			
	0.47	0.30	0.16	0.02
	Composition: 1.0 = 100%			
C	0.107	0.289	0.315	0.364
H	0.015	0.038	0.024	0.063
N	0.007	0.012	0.003	0.003
O	0.083	0.241	0.308	0.023
Volatile Cl	0.000	0.002	0.003	0.016
Residual Cl	0.000	0.001	0.001	0.002
Ash	0.038	0.017	0.054	0.028
W	0.85	0.4	0.27	0.5
	Lower Heat Value			
kJ/kg	486.7	10336.4	8960.6	18436.9

Source The share is from [15]. The composition is from [83]

In these equations, C, O, and S refer to the % in weight of carbon, oxygen, and sulfur, respectively. The equations differ in terms of the treatment of oxygen in the combustible, with Dulong's equation assuming all oxygen is in the form of H_2O, Steuer-Kestner's equation assuming all oxygen is in the form of CO, and Steuer's equation assuming 50% of oxygen is in H_2O and the remainder is in CO.

Table 5.7 gives a real example of the chemical composition of Kitchen waste, Other flammable waste, and Newspaper etc., which occupied around 93% of waste for incineration generated by private consumption (private households) in Japan, in 2011. Based on the Steuer equation (5.89), we find that the lower heat value H_L ranges from 486.7J for Kitchen waste to 10336.4J for Other flammable waste. Since H_L is a critical factor determining the performance of an MSW incineration plant, it follows that a proper consideration of waste composition is vital for modeling the waste incineration process.

5.4.3.2 Modeling the Waste Incineration Process: A Brief Introduction

The HIWM Model: A System Engineering Model of Waste Treatment

Toshihiko Matsuto and Nobutoshi Tanaka of Hokkaido University in Sapporo, Japan, developed the HIWM (Hokkaido University Integrated Waste Management) model, an integrated system engineering model of waste management. The model is based on solid engineering information on a wide range of waste treatment processes, including waste collection, sorting, incineration (distinguished by the types of incinerator and

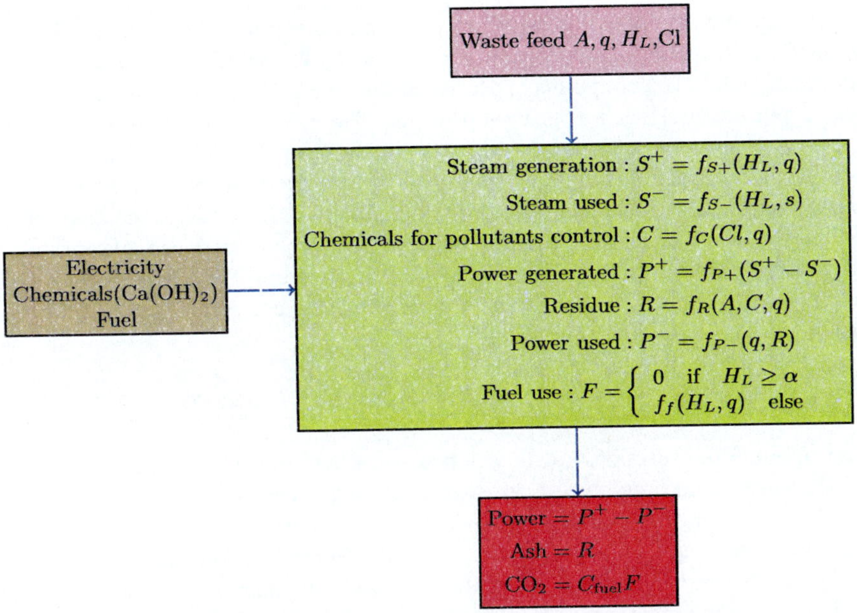

Fig. 5.4 A schematic modeling of the waste incineration process. q is the quantity of waste to be treated (10^6 g/yr) in a waste incineration plant with capacity s (10^6 g/d), A the ash content, and Cl the Cl content

the methods of heat utilization), and wastewater treatment from a landfill, and detailed data on the construction and operation of several facilities in Japan [83].

Figure 5.4 presents a schematic representation of a typical waste incineration process based on the HIWM model. The process converts waste with specific characteristics, as described by the properties outlined in Table 5.7 into ash and flue gas. To achieve this transformation, various inputs are utilized, including power, chemicals (mainly NaOH to treat flue gas containing HCl), and fuel (for incineration). In addition, if the recovered heat energy generated during the incineration process exceeds the energy consumption required by the process itself, it is possible to obtain power as an output.

Inputs and Outputs in a Waste Incineration Plant

The relationship between inputs and outputs in a waste incineration plant is complex and depends on various factors, including the characteristics of the waste feed, represented by parameters such as H_L, ash- and chlorine content, as well as the size of the plant. The system of equations inside the box in the middle of Fig. 5.4 provides a representation of the mutual interdependence of these quantities. The importance of H_L as a critical parameter in a waste incineration process is reflected in four of the seven equations. Each variable has a positive effect in all the equations, except for H_L in f_f (fuel use) and f_{S-} (steam use).

The amount of fuel required for incineration can be reduced or eliminated if the waste feed has a high energy content, represented by H_L. When $H_L \geq 4184$ (kJ/kg), the waste will have enough energy to incinerate itself. Otherwise, incinerating the waste requires fuel, such as heavy oil or gas. The effects of H_L on steam use are mixed and cannot be determined a priori. While steam use for air heating will become zero when $H_L \geq 8368$ (kJ/kg), steam use for deaerators[4] may increase with H_L. By increasing the heat generated in incineration, a higher H_L increases the quantity of steam via f_{S+} and hence the quantity of power that is generated using the steam.

The presence of volatile Cl in the waste (mostly from PVC) will produce HCl upon combustion, which needs to be reduced by using chemicals such as $Ca(OH)_2$ making use of the following reaction

$$\mathrm{Ca(OH)_2}(s) + 2\mathrm{HCl}(g) \rightarrow \mathrm{CaCl_2}(s) + 2\mathrm{H_2O}(l). \tag{5.91}$$

The output of this reaction, $CaCl_2$, will become an additional component of incineration residues, and C will have a positive effect on the residue R. The dependence of steam use on plant size s can be explained by the fact that a larger furnace needs more heat for air heating than a smaller one. The HIWM model takes into account these and other factors to provide a comprehensive understanding of waste management and incineration processes.

The aforementioned observation highlights the significance of waste properties, such as H_L, ash, and Cl content, in determining the quantitative relationship between inputs and outputs of a waste incineration process. This relationship is heavily influenced by the specific technical parameters and the type of incinerator used, as well as the associated pollution abatement and heat utilization systems. Failing to account for this critical aspect of waste incineration processes can lead to erroneous conclusions, especially when the properties of the waste feed are changing.

It is not advisable to assume a fixed input-output relationship or use black-box modeling when it comes to waste incineration processes. These methods may not be appropriate due to the critical dependence of the input-output relationship on the unique properties of the waste being incinerated, as well as the specific technical parameters and systems used in the incineration process. Instead, it is crucial to use a "white box representation" (Sect. 2.1) that explicitly considers the science and engineering principles that govern the relationship between the inputs and outputs of the waste incineration process, which takes into account the dependence of the input-output relationship on the fundamental properties of the waste.

---

[4] A device used to remove air and other dissolved gases from the feedwater to steam generating boilers.

5.4.3.3 Waste Composition, Incineration Residues, Power, and CO_2 Emissions

To illustrate the impacts of waste characteristics on the outputs of incineration processes, Table 5.8 gives the results obtained by applying the HIWM model [83] to four items of MSW in Table 5.7. We consider the case where the incinerator is of a stoker type, with a treatment capacity of 3Gg/day, and the surplus steam is entirely used for power generation. Collection and transport of waste from households to the incinerator are not considered. Our concern is the effects of waste's characteristics on the incineration process's outputs.

The Case of Separated Incineration

We first consider the case where the same quantity of each waste item is separately incinerated, with the results shown in the first four columns of Table 5.8. Of the four items of MSW, which make up around 95% of MSW incinerated in Japan, Kitchen waste is the only item with a negative net output of steam and hence power. Kitchen waste is also the only item that needs fuel for incineration because of its low H_L. For all other waste items considered in Table 5.8, incineration produces more steam than needed for operating the plant, resulting in net production of electric power for sale outside the plant.

The power output per unit varies among the waste items, with Plastic container & packaging having the highest value of around 800 Wh/kg and Kitchen waste having the lowest value of -110 Wh/kg. The difference in power output corresponds to the variation in H_L among the waste items. When these outputs are averaged with weights

Table 5.8 Incineration outputs of MSW under separated and mixed treatment

		Separated					Mixed
Items	Units	Kitchen waste	Other flammable waste	Newspaper etc.	Plastic container & package	Weighted Average	
H_L	kJ/kg	487	10336	9284	18479		5150
Share in treatment		0.47	0.30	0.16	0.02		
Steam used	kg/kg	0.255	0.620	0.579	0.940	0.496	0.587
Steam generated	kg/kg	0.142	3.006	2.700	5.373	1.816	1.497
CO_2 fossil-fuel based	kg-CO_2/kg	0.258	0.000	0.000	0.000	0.046	0.000
Electricity for sale	Wh/kg	−110	379	325	799	129	77
Incineration residue for landfill	kg/kg	0.024	0.018	0.056	0.029	0.026	0.026

Author's calculation based on the HIWM model [83] with data taken from [15]

proportional to the waste items' relative shares in incineration (shown in the second row of Table 5.8), the resulting power output is positive at 129 Wh/kg. Kitchen waste, which has the largest share in MSW for incineration, is a net consumer of electric power due to its low H_L. Nevertheless, separately incinerating the remaining waste items with high H_L can generate sufficient electric power to compensate for the net consumption of incinerating Kitchen waste. Newspaper etc. generates the largest amount of residue for landfill disposal, which is mainly due to its high ash content, supplemented by a small contribution from its Cl content.

The Case of Mixed Incineration

Next, we examine the case where the four waste items are incinerated simultaneously after being mixed in a given incineration plant. The waste items are mixed until a homogeneous composition is obtained, characterized by $H_L = 5150$ kJ/kg, 53.3% moisture content, and 2.5% ash content, and then incinerated. The last column of Table 5.8 displays the results of this mixed treatment. When compared to the results obtained from the separate treatment (shown under "Weighted Average"), the mixed treatment decreases power generation by approximately 25% from 129 Wh/kg to 77 Wh/kg, while the amount of residue remains constant and the CO_2 emission from fuel use is reduced to zero.

The difference between the results of mixed incineration and the weighted sum of separated incineration results indicates the presence of nonlinearity in the incineration process. The sum of separately incinerated waste is not equivalent to the incineration of the combined waste in terms of power output, as different incineration schemes are initiated depending on whether certain H_L threshold values are met. Depending on the H_L value, fuel for incineration or steam for heating the air may not be required, which leads to nonlinearity in the process.

To compare the net effects of mixed and separated treatment on fuel-based CO_2 emissions, we can calculate the additional emissions from power generation needed to compensate for the reduction in power output under mixed treatment, as well as the emissions due to fuel used for incineration under separated treatment. Under mixed treatment, power output decreases by 0.052 kWh per kg-waste compared with separated treatment, which means additional power needs to be produced to make up for this reduction. Assuming an average emission factor of 0.464 kg-CO_2 per kWh in Japan in 2011,[5] this results in additional emissions of around 0.024 kg-CO_2 per kg-waste. On the other hand, under separated treatment, the emissions due to fuel used for incineration would increase emissions by 0.046 kg-CO_2 per kg-waste. Therefore, mixed treatment results in lower emissions in the incineration process compared to separated treatment, even when accounting for the additional emissions from power generation needed to compensate for the reduction in power output.

Responsible for the low energy value of Kitchen waste is its high moisture content, 85%, as shown in Table 5.7. We applied the HIWM model to investigate the effects of reducing the moisture content to 75% on the power output. The model calculation shows that this reduction in moisture content would increase the H_L of Kitchen

[5] TEPCO web page.

waste from 487 kJ/kg to 2489 kJ/kg and reduce the power used in its incineration to 44 Wh/kg. The overall power output per kg-waste would increase from 129 to 172 Wh/kg under separated treatment and from 77 to 130 Wh/kg under mixed treatment, highlighting the importance of reducing the moisture content of waste for incineration to increase power output.

5.4.3.4 Consequences of Nonlinearity in Waste Incineration Processes

The nonlinearity present in waste incineration with regards to heat use and power generation has important implications for waste treatment modeling, including IO/WIO. Therefore, caution must be exercised when disaggregating mixed incineration results into separate waste items or when aggregating the results of separated incineration.

If the observed data pertains to mixed incineration, a mechanical disaggregation of inputs and outputs based on a proportionality assumption may lead to erroneous results, particularly in terms of energy use and power output. Conversely, if the observed data pertains to separated incineration, an average result weighted by the respective waste shares may not accurately reflect the results of mixed incineration.

Having a basic understanding of the relationship between waste composition and the treatment process is necessary to conduct IE studies on waste incineration. With regards to the WIO model, the nonlinearity of the results mentioned above indicates that the column elements of waste treatment sectors, such as $\boldsymbol{A}^{\mathrm{II}}$ and $\boldsymbol{G}^{\mathrm{II}}$, are dependent on waste composition. As a result, when the waste composition changes, $\boldsymbol{A}^{\mathrm{II}}$ and $\boldsymbol{G}^{\mathrm{II}}$ also change, most likely in a nonlinear fashion.

To consider the dependence on waste composition and plant specifications of $\boldsymbol{A}^{\mathrm{II}}$ and $\boldsymbol{G}^{\mathrm{II}}$, we suggest integrating a system engineering model of waste treatment, such as the HIWM model, into the WIO model.

5.4.3.5 Solving the WIO Integrated with a System Engineering Model of Waste Management: Determining $\boldsymbol{A}^{\mathrm{II}}$ and $\boldsymbol{G}^{\mathrm{II}}$

Denote by $\boldsymbol{\Psi}$ a set of (presumably nonlinear) simultaneous equations representing the relationship between waste and plant characteristics and the input-output coefficients of waste treatment sectors, $\boldsymbol{A}^{\mathrm{II}}$, $\boldsymbol{G}^{\mathrm{II}}$ (or the system of equations schematically represented by the box in the center of Fig. 5.4). Recalling that the characteristics of the waste entering each treatment process depend on how wastes for treatment, $\boldsymbol{w}$, are allocated to each process, the following general expression can be obtained

$$(\boldsymbol{A}^{\mathrm{II}}, \boldsymbol{G}^{\mathrm{II}}) = \boldsymbol{\Psi}(\boldsymbol{w}, \boldsymbol{S}; \boldsymbol{\theta}, \boldsymbol{\nu}) \tag{5.92}$$

where $\boldsymbol{\theta}$ refers to waste characteristics, such as chemical composition, and $\boldsymbol{\nu}$ to the characteristics of treatment processes, such as the pattern of heat use and pollution control. With $\boldsymbol{\Psi}$ providing the engineering model for waste treatment, we can solve the (nonlinear) WIO model for $\boldsymbol{x}$ and $\boldsymbol{w}$ iteratively following the steps illustrated in Fig. 5.5.

The iterative algorithm for solving the (nonlinear) waste input-output (WIO) model involves four steps. In Step 0, the production sectors are solved for their output and waste generation based on a given set of input-output coefficients and final demand. In Step 1, for a given set of waste allocation, the characteristics of waste treatment sectors are obtained through the engineering model. In Step 2, the model encompassing both production and waste treatment sectors is solved for the net generation of waste. In Step 3, the algorithm iterates until the difference between the current and previous estimates of waste generation is smaller than a given criterion. The algorithm terminates when the criteria is met.

5.4.4 *Further Topics in WIO and IO of Waste and Waste Management

We present further topics of IO/WIO-based approaches for waste analysis and CO_2 emissions mitigation, including an alternative method to obtain a symmetric WIO matrix, footprint analysis of products for waste, wastewater, and sludge, a Linear Programming (LP) study on treating waste plastics.

5.4.4.1 Disaggregate Wastes by Treatment

As we saw in Sect. 5.4.2.2, the balance of goods and waste (5.75) is generally unsolvable due to its asymmetry: Waste for treatment $\boldsymbol{w}$ appears on the right-hand side of the balance, while the activity level of waste treatment $\boldsymbol{x}^{\mathrm{II}}$ appears on the left-hand side. To obtain a solvable WIO model, [72] transformed the asymmetric system into a symmetric one using the allocation matrix $\boldsymbol{S}$ of order $n_{\mathrm{I}} \times n_w$, which maps wastes into treatment processes. This conversion aggregates wastes into a smaller set distinguished by treatment processes, and multiple wastes are processed simultaneously in the same treatment process. Thus, the WIO model of [72] can be categorized as an approach based on the aggregation of waste by treatment. Simultaneous processing of multiple types of waste, or mixed waste, is a common practice in waste treatment. For instance, a waste incinerator may receive waste paper, waste plastics, waste textiles, and kitchen garbage, while a landfill may accept construction debris, shredding residues, and ash.

An alternative approach to obtaining a symmetric WIO matrix is to transform the $n_w \times (n_{\mathrm{I}} + n_{\mathrm{II}})$ matrix of waste flow, $\boldsymbol{W} = (\boldsymbol{W}^{\mathrm{I}}\ \boldsymbol{W}^{\mathrm{II}})$, into an $(n_w n_{\mathrm{II}}) \times (n_{\mathrm{I}} + (n_w n_{\mathrm{II}}))$ matrix by disaggregating each of n_w wastes by the methods by which it

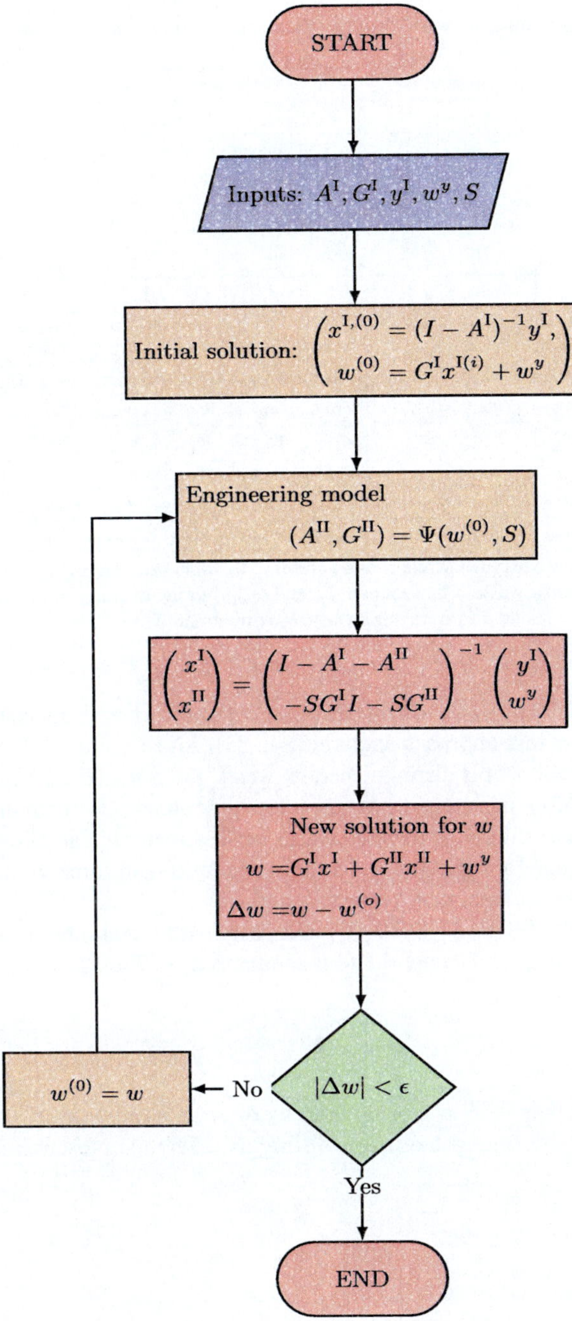

Fig. 5.5 Solving the WIO model integrating a system engineering model of waste management in an iterative fashion

Table 5.9 Disaggregating the waste flow by treatment: A simple example of three wastes and two treatment sectors

		Production sectors				Treatment sectors					
						t_1			t_2		
		1	2	···	n_{I}	waste a	waste b	waste c	waste a	waste b	waste c
	1	$z_{1,1}$	···	···	$z_{1,n_{\mathrm{I}}}$	$z_{1,a1}$	$z_{1,b1}$	$z_{1,c1}$	$z_{1,a2}$	$z_{1,b2}$	$z_{1,c2}$
	2			⋱		$z_{2,a1}$	$z_{2,b1}$	$z_{2,c1}$	$z_{2,a2}$	$z_{2,b2}$	$z_{2,c2}$
	⋮					⋮	⋮	⋮	⋮	⋮	⋮
	n_{I}	$z_{n_{\mathrm{I}},1}$	···	···	$z_{n_{\mathrm{I}},n_{\mathrm{I}}}$	$z_{n_{\mathrm{I}},a1}$	$z_{n_{\mathrm{I}},b1}$	$z_{n_{\mathrm{I}},c1}$	$z_{n_{\mathrm{I}},a2}$	$z_{n_{\mathrm{I}},b2}$	$z_{n_{\mathrm{I}},c2}$
t_1	a	$w_{a1,1}$	$w_{a1,2}$	···	$w_{a1,n_{\mathrm{I}}}$	$w_{a1,a1}$	$w_{a1,b1}$	$w_{a1,c1}$	$w_{a1,a2}$	$w_{a1,b2}$	$w_{a1,c2}$
	b	$w_{b1,1}$	$w_{b1,2}$	···	$w_{b1,n_{\mathrm{I}}}$	$w_{b1,a1}$	$w_{b1,b1}$	$w_{b1,c1}$	$w_{b1,a2}$	$w_{b1,b2}$	$w_{b1,c2}$
	c	$w_{c1,1}$	$w_{c1,2}$	···	$w_{c1,n_{\mathrm{I}}}$	$w_{c1,a1}$	$w_{c1,b1}$	$w_{c1,c1}$	$w_{c1,a2}$	$w_{c1,b2}$	$w_{c1,c2}$
t_2	a	$w_{a2,1}$	$w_{a2,2}$	···	$w_{a2,n_{\mathrm{I}}}$	$w_{a2,a1}$	$w_{a2,b1}$	$w_{a2,c1}$	$w_{a2,a2}$	$w_{a2,b2}$	$w_{a2,c2}$
	b	$w_{b2,1}$	$w_{b2,2}$	···	$w_{b2,n_{\mathrm{I}}}$	$w_{b2,a1}$	$w_{b2,b1}$	$w_{b2,c1}$	$w_{b2,a2}$	$w_{b2,b2}$	$w_{b2,c2}$
	c	$w_{c2,1}$	$w_{c2,2}$	···	$w_{c2,n_{\mathrm{I}}}$	$w_{c2,a1}$	$w_{c2,b1}$	$w_{c2,c1}$	$w_{c2,a2}$	$w_{c2,b2}$	$w_{c2,c2}$

a, b, and c refer to waste types, while t_1 and t_2 refer to treatment sectors. $z_{i,aj}$ refers to the quantity of input i used to treat waste a in treatment t_j, and $w_{ai,bj}$ to the quantity of secondary waste a to be treated by t_i that resulted from treating waste b by treatment t_j

is treated and each of n_t treatment processes by the wastes it treats.[6] Examples of studies based on this approach include [84, 85]. Since waste is distinguished by treatment and treatment is distinguished by waste, the $\boldsymbol{S}$ matrix becomes an identity matrix ([84], p.631). It is argued that "the waste type and waste treatment occur allied more straightforwardly, such as Food waste for treatment: incineration [85]."

For illustration, Table 5.9 gives a simple example with three waste types, a, b, and c, and two waste treatment sectors, t_1 and t_2: $n_w = 3$ and $n_{\mathrm{II}} = 2$. Note that the flow matrix of products and waste is symmetric, with the number of rows equal to the number of columns. Denote the four submatrices in Table 5.9 as

$$\begin{pmatrix} \mathbf{Z}^{\mathrm{I}} & \mathbf{Z}^{\spadesuit \mathrm{II}} \\ \mathbf{W}^{\spadesuit \mathrm{I}} & \mathbf{W}^{\spadesuit \mathrm{II}} \end{pmatrix} \tag{5.93}$$

It is noteworthy that $\mathbf{W}^{\spadesuit \mathrm{II}}$ is a square matrix of order $n_{\mathrm{II}} n_w \times n_{\mathrm{II}} n_w$. The disaggregated flows and the aggregated flows, distinguished by the presence/absence of $\spadesuit$, are related as

[6] Since $\boldsymbol{S}$ may contain zero elements, $n_w n_{\mathrm{II}}$ can be reduced to a smaller number excluding those with zero.

$$
\begin{aligned}
Z^{\clubsuit \mathrm{II}} \times (I_2 \otimes \iota_3) &= Z^{\mathrm{II}} \\
(\iota_2^{\top} \otimes I_3) \times W^{\clubsuit \mathrm{I}} &= W^{\mathrm{I}} \\
(\iota_2^{\top} \otimes I_3) \times W^{\clubsuit \mathrm{II}} \times (I_2 \otimes \iota_3) &= W^{\mathrm{II}}
\end{aligned}
\tag{5.94}
$$

where $\otimes$ denotes the Kronecker product.[7] We can obtain the disaggregated counterpart to x^{II}, $x^{\clubsuit \mathrm{II}}$, by

$$
x^{\clubsuit \mathrm{II}} = W^{\clubsuit \mathrm{I}} \iota_{n_{\mathrm{I}}} + W^{\clubsuit \mathrm{II}} \iota_{n_w n_{\mathrm{II}}}
\tag{5.95}
$$

and the disaggregated counterparts of As and Gs as

$$
\begin{aligned}
A^{\clubsuit \mathrm{II}} &= Z^{\clubsuit \mathrm{II}} \,\mathbf{diag}(x^{\clubsuit \mathrm{II}})^{-1} \\
G^{\clubsuit \mathrm{I}} &= W^{\clubsuit \mathrm{I}} \,\mathbf{diag}(x^{\mathrm{I}})^{-1} \\
G^{\clubsuit \mathrm{II}} &= W^{\clubsuit \mathrm{II}} \,\mathbf{diag}(x^{\clubsuit \mathrm{II}})^{-1}
\end{aligned}
\tag{5.96}
$$

where $G^{\clubsuit \mathrm{II}}$ is a square matrix of order $n_{\mathrm{II}} n_w \times n_{\mathrm{II}} n_w$. The disaggregated counterpart to (5.79) can then be given by

$$
\begin{pmatrix} x^{\mathrm{I}} \\ x^{\clubsuit \mathrm{II}} \end{pmatrix} = \begin{pmatrix} I - A^{\mathrm{I}} & -A^{\clubsuit \mathrm{II}} \\ -G^{\clubsuit \mathrm{I}} & I - G^{\clubsuit \mathrm{II}} \end{pmatrix}^{-1} \begin{pmatrix} y^{\mathrm{I}} \\ w^{\clubsuit y} \end{pmatrix}
\tag{5.97}
$$

where w^{*y} refers to w^{y} disaggregated by treatment.

In Sect. 5.4.2.4, we discussed the WIO-SUT model, which is a variant of the original WIO model. Unlike the original WIO model, the WIO-SUT model does not aggregate waste by treatment or convert waste flows into treatment flows. However, the WIO-SUT model still belongs to the aggregated approach because it does not disaggregate waste by treatment and uses an S matrix that is not an identity matrix.

Reference [84] adopts the disaggregated approach because

> *The performance of the activities of incineration and landfilling rely strongly on the incoming materials for treatment. Some materials for treatment are inert, other hazardous and each one may have a different calorific value. For example the incineration of wood has completely different heat and electricity production than the incineration of glass. Similar examples are can be done for landfills. In order to highlight these substantial differences, the CREEA approach aims to disaggregate the landfilling and incineration activity according to the input of materials for treatment.* ([86], p.37)

As we discussed in Sect. 5.4.3.3, a system engineering model is necessary to describe a waste incineration process and determine its input-output coefficients for a specific waste feed, which is usually a mixture of various waste items with different chemical compositions. Since a detailed system engineering model to describe a waste incineration process was not readily available, [84] assumed that each waste was

[7] See Appendix A.3.3 for the Kronecker product.

treated separately by a single treatment process. They then obtained the required input-output data for each waste-treatment combination primarily from the literature [87].

However, in reality, various waste items will be incinerated not separated from each other but mixed. For instance, metals and inert materials would not be deliberately subjected to incineration unless they happened to be mixed with other flammable waste items due to incomplete sorting. To obtain results that reflect the reality of mixed waste incineration, the separate incineration results for each waste item could be aggregated based on appropriate weights that represent the mixing proportions. However, as we noted in Sect. 5.4.3.3, care must be taken due to possible nonlinearity with respect to the use of heat energy.

5.4.4.2 Waste Footprint Analysis Based on IO/WIO

Compared to data on the production and consumption of goods and services (the "arteries" of the economy), data on waste and treatment (the "veins" of the economy) are relatively limited, which presents a significant obstacle to conducting IO-based waste analysis. Despite this challenge, a sizable literature has emerged on IO-based waste footprint analysis, including studies such as [78, 79, 88–93]. According to the recent review by [75] on the IO of waste and waste treatment, [88, 90–93] fall into the category of the waste-extended IO model (WEIO), where waste generation is added as satellite tables in a similar way to how resources and emissions are added in typical EEIO models. In Sect. 5.4.2.4, we discussed the WIO-based MRIO studies of [78, 79].

Despite this growing literature, recent textbooks on IO, such as [56], rarely include references to waste and waste treatment research. In the following section, we highlight several recent studies that we consider noteworthy due to their originality and resolution of methodology/data.

Waste Footprint Based on High-Resolution IO/waste Data

A High Resolution WIO for Taiwan

Reference [94] developed a high-resolution WIO model for Taiwan based on data from the Industrial Waste Control Center of the Taiwan Environmental Protection Administration. The model covers 196 types of general industrial waste and 157 types of hazardous industrial waste, which are further classified into four categories: process hazardous waste, toxic hazardous waste, characteristic-based hazardous waste, and scrap metal mixtures. The WIO model includes various treatment processes for industrial waste, such as landfill, export for treatment, sterilization, chemical treatment, composting, thermal treatment (excluding incineration), incineration, physical treatment, solidification, stabilization, washing treatment, an intermediate approach, other final disposal methods, and waste reclamation. Furthermore, wastewater treatment was divided into three processes: treatment inside factory, treatment outside factory, and joint treatment within an industrial area.

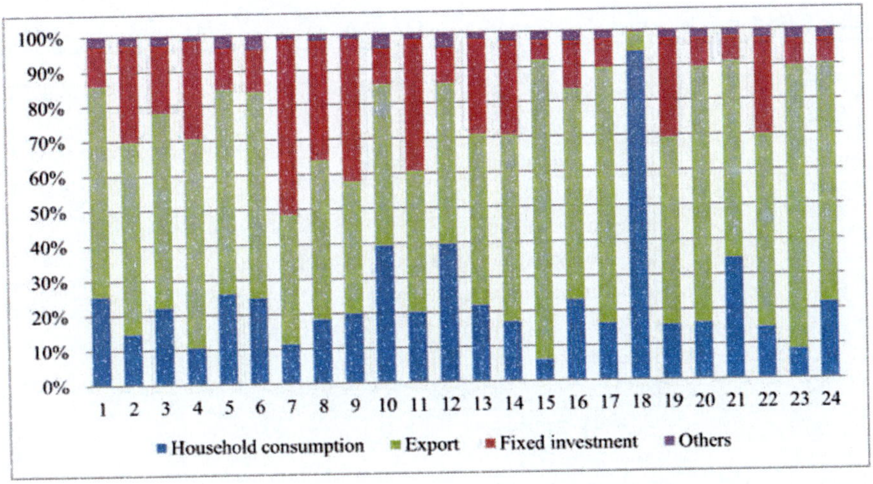

Fig. 5.6 Effects of specific final demands on different wastes, Taiwan. *Source* [94], Fig. 1. The numbers in the horizontal axis refer to waste items in Table 5.10

Figure 5.6 presents the final demand footprint of industrial wastes for 24 dominant waste items, revealing interesting insights into waste generation in Taiwan. For instance, the dominance of export was apparent for "Waste acidic etchants (w15)" and "Copper and copper compounds (w23)", where more than 90 of the waste generation of these two types of industrial wastes was attributable to export. On the other hand, "Waste lees, wine meal, and alcohol mash (w18)" had over 94% of its waste generation driven by household consumption. Other waste categories, such as "Waste-paper mixture (w10)," "Pulp and paper sludge (w12)," and "Crude solution of dimethylformamide (DMF) (w21)," were also attributable to household consumption.

Fixed investments were found to have a significant impact on the generation of certain wastes. For example, up to 50% of the amount of "Civil engineering or construction-waste mixtures (w7)" and 40% of the "Wastewater (pH > 9.0) (w9)" were attributable to fixed investments. In contrast, other parts of the final demand showed only slight influences on those wastes.

Another notable finding was that the export of "Electrical and electronic products" had the largest driving potential to generate "Coal ash (w1)," "Waste acidic etchants (w15)," "Copper and copper compounds (w23)," and "Waste liquid with a flash point less than 60 C (w24)." These results provide valuable insights into the industrial waste generation patterns in Taiwan and can inform policy decisions aimed at reducing the environmental impact of waste generation.

Solid Waste Streams in the US

Combining the US IO table with EPA's integrated waste satellite tables on 536 waste materials, [95] developed a model to estimate the commercial generation of three solid waste streams from 386 industry sectors: hazardous waste, nonhazardous

Table 5.10 Industrial wastes classification used in Fig. 5.6

1	Coal ash
2	Electric arc furnace slag
3	Inorganic sludge
4	Construction and demolition waste
5	Organic sludge
6	Waste plastic mixture
7	Civil engineering or construction-waste mixtures
8	Nonhazardous flue dust or mixtures
9	Wastewater (pH > 9.0)
10	Waste paper mixture
11	Nonhazardous mixed waste
12	Pulp and paper sludge
13	Wastewater (pH 6.0-9.0)
14	Waste foundry sand
15	Waste acidic etchants
16	Sludge mixture
17	Nonhazardous waste acid
18	Waste lees, wine meal, and alcohol mash
19	Waste acid lotion
20	Mixed waste solvents
21	Crude solution of dimethylformamide (DMF)
22	Flue dust or sludge from pollution control process in electric arc furnace
23	Copper and Copper compounds (limited to the waste catalyst, flue dust, waste water, sludge, filter, incineration fly ash or bottom ash)
24	Waste liquid with a flash point less than 60C (except that of ethanol concentration less than 24%)

waste excluding construction, and nonhazardous waste from construction. The results revealed that "Concrete" was the largest mass of waste at 341Eg (10^{18} g, million tonnes), which was more than four times greater than the next largest mass of 82E for "Paper." "Asphalt pavement" and "Waste wood" were the next largest materials at 81E and 48E, respectively, and were primarily generated by the construction sectors, such as "Highways, streets, and bridges." Even when considering the final demand perspective (i.e., the footprint), the dominance of construction sectors remained fairly stable, with "Highways, streets, and bridges" remaining the top contributor.

Waste Footprint: Three Regions Open Model, Tokyo, China-Japan

Here, we discuss a regional model of waste footprint that consists of three regions: two endogenous regions denoted as a and b, and the Rest of the World, R, which is exogenous. The balance equations of the regionally extended WIO can be expressed as

$$\begin{pmatrix} \boldsymbol{x}_a^{\mathrm{I}} \\ \boldsymbol{x}_a^{\mathrm{II}} \\ \boldsymbol{x}_b^{\mathrm{I}} \\ \boldsymbol{x}_b^{\mathrm{II}} \end{pmatrix} = \begin{pmatrix} \boldsymbol{A}_{aa}^{\mathrm{I}} & \boldsymbol{A}_{aa}^{\mathrm{II}} & \boldsymbol{A}_{ab}^{\mathrm{I}} & \boldsymbol{A}_{ab}^{\mathrm{II}} \\ \boldsymbol{S}_a\boldsymbol{G}_{aa}^{\mathrm{I}} & \boldsymbol{S}_a\boldsymbol{G}_{aa}^{\mathrm{II}} & \boldsymbol{S}_a\boldsymbol{G}_{ab}^{\mathrm{I}} & \boldsymbol{S}_a\boldsymbol{G}_{ab}^{\mathrm{II}} \\ \boldsymbol{A}_{ba}^{\mathrm{I}} & \boldsymbol{A}_{ba}^{\mathrm{II}} & \boldsymbol{A}_{bb}^{\mathrm{I}} & \boldsymbol{A}_{bb}^{\mathrm{II}} \\ \boldsymbol{S}_b\boldsymbol{G}_{ba}^{\mathrm{I}} & \boldsymbol{S}_b\boldsymbol{G}_{ba}^{\mathrm{II}} & \boldsymbol{S}_b\boldsymbol{G}_{bb}^{\mathrm{I}} & \boldsymbol{S}_b\boldsymbol{G}_{b}^{\mathrm{II}} \end{pmatrix} \begin{pmatrix} \boldsymbol{x}_a^{\mathrm{I}} \\ \boldsymbol{x}_a^{\mathrm{II}} \\ \boldsymbol{x}_b^{\mathrm{I}} \\ \boldsymbol{x}_{ba}^{\mathrm{II}} \end{pmatrix} + \begin{pmatrix} \boldsymbol{y}_{aa}^{\mathrm{I}} + \boldsymbol{y}_{ab}^{\mathrm{I}} + \boldsymbol{y}_{aR}^{\mathrm{I}} \\ \boldsymbol{S}_a\boldsymbol{w}_{aa}^{y} + \boldsymbol{S}_a\boldsymbol{w}_{ab}^{y} + \boldsymbol{S}_a\boldsymbol{w}_{aR}^{y} \\ \boldsymbol{y}_{ba}^{\mathrm{I}} + \boldsymbol{y}_{bb}^{\mathrm{I}} + \boldsymbol{y}_{bR}^{\mathrm{I}} \\ \boldsymbol{S}_b\boldsymbol{w}_{ba}^{y} + \boldsymbol{S}_b\boldsymbol{w}_{bb}^{y} + \boldsymbol{S}_b\boldsymbol{w}_{bR}^{y} \end{pmatrix} \tag{5.98}$$

The subscripts attached to matrices and vectors in (5.98) represent the regions of production, use, and waste treatment. For instance, $\boldsymbol{x}_a^{\mathrm{II}}$ and $\boldsymbol{S}_a$ represent the vector of waste treatment activities and the allocation matrix of waste to treatment in region a, respectively. $\boldsymbol{A}_{ba}^{\mathrm{II}}$ represents the matrix of inputs from b into the waste treatment sectors in a, and $\boldsymbol{w}_{ab}^{y}$ represents the export of waste from region a to b.

Reference [96] applied this model to examine the regional economic and environmental impacts of consumption activities in Tokyo (region a) on the rest of Japan (region b) and the rest of the world (region R). The results showed that Tokyo's consumption activities had a limited economic benefit for regions outside Tokyo while significantly increasing the environmental loads, particularly the utilization of landfills in those regions.

Reference [97] applied this regional model to investigate the waste embodied in bilateral trade between China (region a) and Japan (region b). This study explicitly considered the trade of waste between the two countries, represented by $\boldsymbol{w}_{ab}^{y}$. The findings showed that the final demand in Japan indirectly induced approximately 52 Tg of waste in China, while the waste impact of Chinese final demand on Japan was minimal. However, the authors noted that obtaining reliable waste data from China remained a challenge.

Global Waste Footprint Based on MRIO Models

Reference [84] addressed the lack of a global assessment of solid waste footprints by using EXIOBASE v2 to develop a WIO model of the global economy. They constructed a harmonized multiregional solid waste account that covered 48 world regions, 163 production sectors, 11 types of solid waste, and 12 waste treatment processes for the year 2007 (Table 5.11). As discussed in Sect. 4.3.4, EXIOBASE achieves its global coverage at a high sectoral resolution by estimating IO tables for countries without national tables and by disaggregating IO tables with low sectoral resolution. They used a similar approach to estimate the flow of waste, which was calculated as the balancing item between the sum of physical inputs and the sum of physical outputs ([86], p. 33; [84], p. 630). As mentioned in Sect. 5.4.4.1, what sets [84] apart from other WIO studies is its establishment of a one-to-one correspondence between waste and its treatment.

Table 5.11 Waste and treatment the physical layer of EXIOBASE

Waste items		Treatment
1	Food waste	Landfill subdivided by waste items 1–7
2	Glass waste	Composting subdivided by waste items 1 and 7
3	Metals and Inert Materials	Biogassification subdivided by waste items 1, 5, and 11
4	Oil/Hazardous waste	Incineration subdivided by waste items 1–7
5	Paper waste	Construction waste to aggregate
6	Plastic waste	Ash to clinker
7	Wood waste	Bottle reuse
8	Ash	Glass waste for recycling
9	Bottle waste	Plastic waste for recycling
10	Construction waste	Wood and paper waste for recycling
11	Sewage sludge	Other metal scrap for recycling Steel scrap for recycling

Source [84, 86]

Several important findings were obtained. The total amount of waste generated worldwide in 2007 was estimated to be approximately 3.2 Pg (10^{15} g, billion tonnes), of which 3.2 Pg was recycled or reused, 0.7 Pg was incinerated, gasified, composted, or used as aggregates, and 1.5 Pg was landfilled. Russia was found to be the largest generator of waste, followed by China, the US, the larger Western European economies, and Japan. These rankings remained relatively unchanged when a consumption-based perspective was taken. However, the waste footprint of North American and Western European countries was found to be up to 25% higher than their territorial account, whereas China's waste footprint was approximately 15% smaller than its territorial waste account.

Regional Waste Footprint Analysis at a Subnational Level Using a Multi-Regional Hybrid IO Table

Based on a regionalized version of the national hybrid SUT of EXIOBASE3 [85, 87, 98] analyzed the waste footprint at the subnational level for three Belgian regions: Brussels, Flanders, and Wallonia. They considered ten solid waste types: food, wood, textile, paper, plastic, metal, construction waste, glass, oil and hazardous waste, and ash; and six waste treatment processes, recycling, construction waste to aggregates, incineration, biogasification, composting, and landfilling. As discussed in Sect. 5.4.4.1, [85] utilized a similar approach to [84] by disaggregating waste by treatment to create a symmetric WIO model.

The study found that Flanders had the highest total waste footprint in absolute terms, while Brussels had the highest direct waste per capita, and Wallonia had the highest indirect waste and stock depletion per capita. The consumption of food products, manufactured products, and restaurants and accommodation services accounted for almost 78 ± 2% of the regional waste footprints in each region. Additionally, around 45 ± 4% of the indirect waste was generated within each region's boundaries, with 16 ± 9% in other regions and 39 ± 5% from outside Belgium, primarily from African countries and China, followed by neighboring European countries.

WIO for National Symbiosis Analysis

Reference [89] developed a WIO table for conducting national industrial symbiosis (IS) analysis in Taiwan. The WIO table included 165 production sectors, 190 types of general industrial waste, and 168 types of hazardous industrial waste. They investigated the supply and demand of by-products and found that excess by-products were either treated as industrial waste or exported for secondary material production. For instance, all zinc oxides from the EAF ash were exported because there was no facility available for their recovery in Taiwan.

Three national eco-industrial networks were identified in which producer industries were linked to receiver industries through specific by-product exchanges. These networks were: 1) dominated by fossil fuel, metal, and mineral industries, 2) dominated by agricultural and synthetic material industries, and 3) dominated by information and communications technology (ICT) and chemical industries. They also conducted a survey of the supply and demand for 23 major by-products, revealing the potential of IS both across and within individual industrial parks.

5.4.4.3 Wastewater

Proper sewage treatment is crucial for maintaining a healthy life. As UNECE Executive Secretary Olga Algayerova noted at MOP6, "Safe water and adequate sanitation are prerequisites for human dignity, gender equality, and inclusive development." Despite its significance, wastewater treatment has received limited attention in the IO community. For example, no mention is made of wastewater in [56]. In this section, we discuss WIO studies that addressed the issue of wastewater.

The First Wastewater IO (W^2IO)

Reference [99] was the first study to extend the WIO to address the issues of wastewater discharge and treatment. It created the first Wastewater IO (W^2IO) table and model, which was implemented using regional IO data for Tokyo, along with data on the discharge and treatment of wastewater in the Tokyo Metropolis. The resulting Tokyo Metropolitan W^2IO table for the year 2000 included 482 economic sectors, 11 wastewater treatment sectors, 12 types of wastewater-related waste, and 6 types of environmental load. The model was then applied to alternative wastewater treatment scenarios referring to different combinations of treatment processes and recycling of by-products. However, the model lacked detailed information about sludge generation, composition, and treatment, which is necessary to identify the key polluting sectors and explore the potential for wastewater and sludge management.

Table 5.12 Allocating waste/wastewater to treatment

Treatment	Waste/wastewater				
	Wastewater	Sludge 90	Sludge 80	Sludge 60	Ash
A2O	0.19	0	0	0	0
OD	0.43	0	0	0	0
BF	0.38	0	0	0	0
Dew 80	0	0.36	0	0	0
Dew 60	0	0.64	0	0	0
Landfill	0	0	0	0.89	1
Incineration	0	0	1	0.11	0

W²IO for Xiamen City, China

To address the lack of detailed information about sludge generation and treatment in [99, 100] developed a new W²IO model with high resolution for sludge and treatment. The model was applied to the Chinese city of Xiamen and resulted in the first W²IO for a Chinese city capable of tracing the product origins of wastewater and sludge. The model included 139 economic sectors, seven wastewater treatment sectors, and waste/wastewater items/effluents, including:

1. Wastewater: raw wastewater.
2. Sludge 90: raw sludge from the wastewater treatment processes (anaerobic-anoxic-aerobic (A2O), oxidation ditch (OD), and biological aerated filter (BF)) with water content above 90%.
3. Sludge 80: sludge with water content around 80%, generated by Dew 80 (mechanical dewatering with centrifugal dewatering equipment).
4. Sludge 60: sludge with water content less than 60%, generated by Dew 60 (plate frame pressure filtration, combining mechanical dewatering and chemical methods).
5. Ash: incineration residue of sludge.
6. Effluents: COD, NH_4N, BOD, and GHG.

Table 5.13 shows an aggregated version of the W²IO table for Xiamen City, converted to a symmetric form by use of the allocation matrix in Table 5.12; Panel C corresponds to $(SW^{I}, SW^{II}, Sw^{y})$. The final demand sectors, mostly private households, account for 70% of wastewater discharge, with the rest almost evenly discharged by the secondary and tertiary sectors. Around 63% of Sludge 80 is recycled as compost, which is used in crop cultivation. The largest fraction of wastewater is processed by OD, followed by BF, with A2O occupying the smallest share.

Figure 5.7 presents the direct wastewater discharge and wastewater footprint (W^2F) of products based on (5.81), excluding the direct discharge from the final demand sectors, for the top 20 sectors that account for around 70% of the total direct wastewater discharge from the production sectors. "Education services" and "Grain mill products" were the largest contributors to wastewater discharge in terms of both

Table 5.13 The wastewater WIO table of Xiamen city, 2012

		Panel	Production				Treat.							
			Primary	Secondary	Tertiary	Compost	A2O	OD	BF	Dew80	Dew60	Landfill	Inc	Final demand
Prod.	Primary	A	37593	533884	36901	0	0	0	0	0	0	0	0	888910
	Secondary		64679	31228532	3346075	126	2033	4529	4094	272	1517	10	974	50110665
	Tertiary		57494	9060617	9544181	0	277	638	554	0	0	0	0	8533767
	Compost		98	0	0	0	0	0	0	0	0	0	0	0
Waste	Wastewater	B	1	3024	3153	0	0	0	0	0	0	2	0	14903
	sludge 90		0	0	0	0	203604	509028	343604	0	0	0	0	0
	sludge 80		0	0	0	−30080	0	0	0	47054	0	0	0	0
	sludge 60		0	0	0	0	0	0	0	0	75723	0	0	0
	ash		0	0	0	0	0	0	0	0	0	0	5090	0
Effl.	COD		4	8896	7472	0	1326	2309	2742	0	0	2	0	50669
Treatment	A2O	C	0	562	585	0	0	0	0	0	0	0	0	2767
	OD		0	1312	1368	0	0	0	0	0	0	1	0	6466
	BF		0	1150	1199	0	0	0	0	0	0	1	0	5669
	Dew80		0	0	0	0	72809	182028	122873	0	0	0	0	0
	Dew60		0	0	0	0	130795	327000	220731	0	0	0	0	0
	landfill		0	0	0	0	0	0	0	0	67242	0	5090	0
	Incineration		0	0	0	−30080	0	0	0	47054	8481	0	0	0

Source [100]. Units: 10^3kg for sludge, ash and COD, 10^7 kg for wastewater, and 10 k RMB for the monetary flows in the "industry" rows. The latter include imports. The bottom panel (Panel C) gives the flow of waste/wastewater converted to the flow of treatment via allocation matrix in Table S2. See [100] for further details of the data

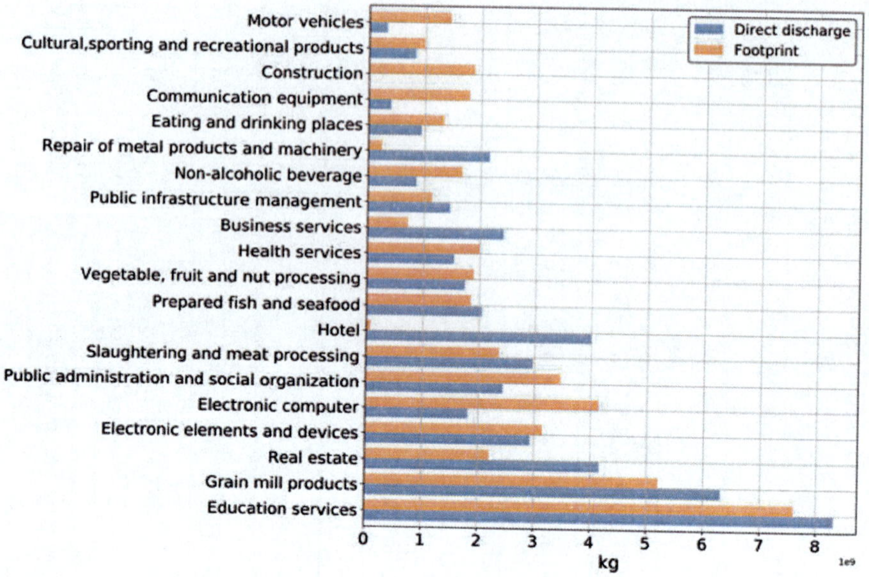

Fig. 5.7 Wastewater footprint and direct wastewater discharge by products. *Source* Fig. 1 of [100]. Units: Tg

direct discharge and W^2F. The large share of "Education services" is attributed to the high proportion of students living in dormitories where meals are provided (around 24% of the population). W^2F is significantly greater than direct discharges for sectors such as "Electronic computers," "Communication equipment," "Construction," and "Motor vehicles." Although their direct discharge may seem minor, their footprints are as significant as many food-related sectors, such as "Prepared fish and seafood" and "Vegetable, fruit, and nut processing."

To track the final demand origin of sludge, [100] introduced the sludge footprint (SF), which requires considering sector-specific features of wastewater. Different production sectors may produce wastewater with varying chemical compositions, resulting in different amounts of sludge per wastewater treated. Reference [100] used COD as a proxy for the organic content of wastewater to calculate SF. Figure 5.8 illustrates the raw sludge footprints of final demand products, SF, for the top 20 products that contribute to approximately 80% of the total raw sludge originating from production sectors. Comparing SF with W^2F (Fig. 5.7), there is a noticeable difference in the footprint ranking of sectors. Eight out of the top 20 sectors with the highest SF are food-related products, compared to five for W^2F, and food-related sectors account for approximately 48% of the total SF.

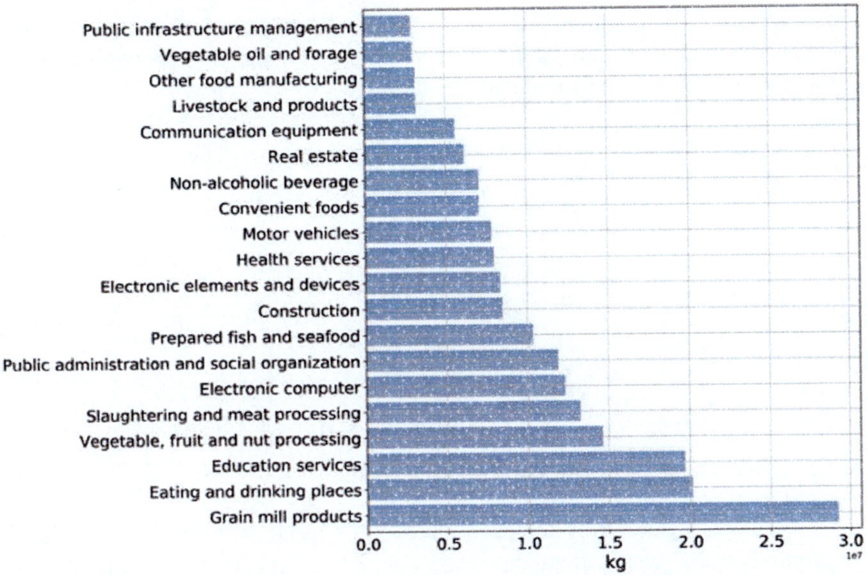

Fig. 5.8 Raw sludge footprint of the top 20 products. *Source* Fig. 2 of [100]. Units: 10Gg

5.4.4.4 Identifying the GHG Minimizing Way of Treating Waste Plastics

China is not only the largest emitter of CO_2 in the world but is also considered to be the largest contributor to marine plastic waste pollution [101]. The primary cause of this issue is the inadequate management of waste in China. In 2007, it was reported that 76% of household and business plastic waste was improperly disposed of, with 21.05% of this waste ending up in the ocean [101]. Addressing this problem requires significant improvements in waste management practices in China. To tackle this challenge, [102] proposed a method for mitigating CO_2 emissions from the treatment of waste plastics in China by exploring alternative waste treatment and recycling processes. Their approach combined Waste IO and linear programming (LP) [103–105].

Let there be $k \geq n_{\text{II}}$ waste treatment processes available. To accommodate $k - n_{\text{II}}$ alternative treatment processes, the number of columns in $\boldsymbol{A}^{\text{II}}$ and $\boldsymbol{G}^{\text{II}}$ is extended from n_{II} to k. Let $\boldsymbol{Q} = [q_{ij}]$ be a $k \times n_w$ matrix, where q_{ij} denotes the amount of waste j treated by waste treatment process i. The row sum of $\boldsymbol{Q}$ gives the total amount of waste treatment, and the column sum gives the amount of waste

$$\boldsymbol{x}^{\text{II}} = \boldsymbol{Q}\boldsymbol{\iota}_{n_w}, \quad \boldsymbol{w}^{\top} = \boldsymbol{\iota}_k^{\top}\boldsymbol{Q} \tag{5.99}$$

The WIO-LP methodology [103] is used to identify the optimal treatment strategies that minimize CO_2 emissions and marine contamination (improper discard of waste plastics). The LP is formulated as follows

$$
\begin{array}{|ll}
\text{minimize} & f_{\text{I}}x_{\text{I}} + f_{\text{II}}x_{\text{II}} \\
\text{subject to} & \\
& x_{\text{I}} = A^{\text{I}}x^{\text{I}} + A^{\text{II}}x^{\text{II}} + y^{\text{I}} \\
& w = G^{\text{I}}x^{\text{I}} + G^{\text{II}}x^{\text{II}} + h \odot w^{y} \\
& x^{\text{II}} = Q\iota_{n_w} \\
& w^{\top} = \iota_k^{\top} Q \\
& Q \odot \Upsilon = 0_{k \times n_w} \\
& x_{\text{I}} \geq 0_{n_{\text{I}}},\ x_{\text{II}} \geq 0_k,\ w \geq 0_{n_w},\ Q \geq 0_{k,n_w} \\
\text{with respect to} & x_{\text{I}},\ x_{\text{II}},\ Q
\end{array}
\tag{5.100}
$$

Here, f_{I} and f_{II} respectively represent the GHG emissions per unit of activity in the production and waste treatment sectors. $h = [h_i]$ represents the proportion of waste plastics discarded by private households that are recovered for treatment. $\odot$ denotes the Hadamard product and 0_{mn} denotes an $m \times n$ matrix of zeros. Technical, economic, and institutional constraints limit the possible combinations of waste and treatment. Some waste items can only be treated by specific processes. The $k \times n_w$ matrix $\Upsilon = [\upsilon_{ij}]$ accounts for these constraints,

The model was implemented for the Chinese WIO table for plastic waste for 2007 [106]. The table consisted of 138 plastic-producing sectors, seven types of plastic-related waste/products, and nine waste treatment processes. The seven types of plastic-related waste/products included household and business plastic waste, industrial plastic waste, imported plastic waste, plastic waste pellets, plastic residue, plastic blast furnace (BF) reductant, and refuse plastic fuel (RPF). The nine treatment processes included incineration without energy recovery, incineration with energy recovery, sanitary landfilling, RPF production, liquefaction, ammonia production, gasification (fuel), injection of plastic waste into BF as a substitute for coke, and plastic pallet production. Due to the lack of relevant data for China, the input coefficients of the treatment/recycling processes were obtained from [107].

The results found that liquefaction was the optimal option for reducing CO_2 emissions, with the potential to reduce emissions up to 9 Tg (10^{12} g, million tonnes) when the recovery rate was enhanced. "Plastic pallet production" was identified as the second-best treatment process, followed by gasification and RPF production. The study highlighted that increasing the recovery rate of discarded plastic waste could lead to reduced emissions when these processes are used. In contrast, the currently employed incineration process with energy recovery exhibited a trade-off relationship between CO_2 emissions and the reduction of improperly treated waste plastics.

5.4.4.5 Circular Economy (CE)

Few concepts in sustainability have been as popular and influential as that of the circular economy (CE) [108]. However, the concept of CE is a broad one, means many different things to different people, and its link to sustainable development is weak [109]. Leaving details of the broad concept of CE to recent review articles [108–110]), we will briefly explore the relevance of WIO in addressing issues related to CE. When we focus on the aspect of CE that aims at minimizing waste generation and resource use through reuse, refurbishment, repair of EoL products, and recycling of waste materials, the WIO framework serves as an ideal accounting and modeling platform for evaluating the environmental and economic impacts of CE implementation.

Reference [111] evaluated the impacts of extensive waste recycling, particularly construction waste, sewage sludge, and ash. The study utilized the WIO table for the Japanese economy for 1995 with 78 production sectors, three waste treatment sectors (shredding, incineration, and landfill), and 33 waste types, extended to include CO_2 emissions and employment. It was found that the quantity of waste for landfills could be reduced by around 30%, fossil-fuel based CO_2 could be reduced by 2%, with negligible negative impacts on employment (around −0.1%).

Extending the lifespan of durable products, such as electronics, through maintenance, repair, and upgrading, facilitated by advanced design strategies (design for X), is a crucial element of CE. However, it is important to consider that the extension of product lives may lead to reduced employment opportunities due to reduced demand for new products. Using the same WIO data as the above study, [103] evaluated the environmental and economic impacts of extending the product life of electronics under various CE scenarios. The results demonstrated that extending the product life could significantly reduce the environmental burden without adversely affecting economic activity and employment, provided that the decrease in expenditure for new purchases is compensated by increased expenditure on repair and maintenance.

Although the data utilized in the above WIO-based studies on CE may be outdated, the modeling techniques and approaches remain valid, providing a valuable foundation for further exploration and analysis of CE-related issues. See [112, 113] for recent reviews on the application of IO to CE.

5.5 Other Topics of LCA

We close this chapter by briefly addressing some remaining topics of LCA, including environmental life cycle-costing (an economic counterpart to LCA), social LCA, allocation of environmental responsibility, and the classification of LCA into attributional and consequential LCA.

5.5.1 *Environmental Life Cycle Costing (eLCC)*

A product with favorable environmental performance must be widely used and produced to realize its desirable effects on the economy. Otherwise, its potential to reduce environmental loads remains mostly unexploited. An important factor for it to be widely used will be that it is economically affordable for its consumers/users and profitable for its producers ([35], p. 365). This calls for the LCA of a product to be accompanied by a complementary evaluation of its affordability/profitability, namely its cost aspects [114]. The aspect of cost here should be the one that encompasses the cost associated with the whole life cycle of a product, that is, the life cycle cost [115]. The life cycle cost is generally not visible because the market price of a product does not usually reflect the cost in the use and end-of-life (EoL) phases but the costs associated with the production (and distribution) phase only. The life cycle cost, therefore, needs to be estimated just as one needs an LCA to evaluate the environmental impacts of a product. Environmental life cycle costing (eLCC) is a tool that is designed to meet this requirement (for the latest development, see [34], Chap. 15). We use the term environmental life cycle costing (eLCC) instead of life cycle costing (LCC), following [34], because the concept of LCC has a much longer history than LCA and is fairly diverse [115]; among other things, eLCC is not a method of financial or managerial accounting.

In Sect. 4.5, we discussed the IO-based cost/price model. Building on that model, we discuss in this section the cost/price counterpart of the WIO model, the WIO-price model, referring to [116]. Before turning to the WIO-price model, we make a short remark about the similarities and differences between LCA and eLCC.

5.5.1.1 LCA and eLCC: Similarities and Differences

The approach of eLCC is very similar to the standardized LCA since it is the only analysis comparable to LCA [34, 115]. However, there are two noticeable differences between the two approaches.

Firstly, eLCC is concerned with the evaluation of economic costs in monetary units, so the items of evaluation must have observable (market) prices ([34], p. 377). This means that the impacts considered in eLCC are limited to those whose costs are internalized, and impacts without internalized costs cannot be considered. This implies that the system boundary covered by eLCC will be narrower than that of LCA since the system boundary of LCA has no such limitation.

Secondly, eLCC may involve the discounting of costs to accommodate the fact that costs occurring at different points in time are not directly comparable [34, 115]. In contrast, there is no discounting in LCA.

5.5.1.2 The WIO Counterpart of Cost and Price

We will now discuss the WIO counterpart of the IO-based cost- and price model previously discussed in Sect. 4.5. For ease of exposition, we follow [116] and define the sets of natural numbers referring to sector and waste. Specifically, we define N^{I} as the set of n_{I} products, N^{II} as the set of n_{II} waste treatment sectors, N^{w} as the set of n_w waste items, and N^{y} as the set of final demand sectors. We also define $N = N^{\mathrm{I}} \cup N^{\mathrm{II}}$, $\boldsymbol{p}^{\mathrm{I}} = [p_i], i \in N^{\mathrm{I}}$, and $\boldsymbol{p}^{\mathrm{II}} = [p_i], i \in N^{\mathrm{II}}$.

We will first consider the simple case without waste recycling, where the cost categories consist of (a) the cost for the input of goods, (b) the cost of waste treatment, and (c) the cost for the input of primary factors. To correspond with this factorization of the cost categories, the unit cost equation for the production or treatment sector j can be expressed through the following equation

$$p_j = \underbrace{\sum_{i \in N^{\mathrm{I}}} a_{ij}}_{\text{(a)}} + \underbrace{\sum_{l \in N^{\mathrm{II}}} p_l \left(\sum_{k \in N^{w}} s_{lk} g_{kj} \right)}_{\text{(b)}} + \underbrace{v_j}_{\text{(c)}}, \quad j \in N^{\mathrm{I}} \cup N^{\mathrm{II}} \tag{5.101}$$

In matrix notation

$$\begin{aligned} \boldsymbol{p}^{\mathrm{I}} &= \boldsymbol{p}^{\mathrm{I}} \boldsymbol{A}^{\mathrm{I}} + \boldsymbol{p}^{\mathrm{II}} \boldsymbol{S} \boldsymbol{G}^{\mathrm{I}} + \boldsymbol{v}^{\mathrm{I}} \\ \boldsymbol{p}^{\mathrm{II}} &= \boldsymbol{p}^{\mathrm{I}} \boldsymbol{A}^{\mathrm{II}} + \boldsymbol{p}^{\mathrm{II}} \boldsymbol{S} \boldsymbol{G}^{\mathrm{II}} + \boldsymbol{v}^{\mathrm{II}} \end{aligned} \tag{5.102}$$

with the solution for $\boldsymbol{p}^{\mathrm{I}}$ and $\boldsymbol{p}^{\mathrm{I}}$ given by

$$\left(\boldsymbol{p}^{\mathrm{I}} \ \boldsymbol{p}^{\mathrm{II}}\right) = \left(\boldsymbol{v}^{\mathrm{I}} \ \boldsymbol{v}^{\mathrm{II}}\right) \begin{pmatrix} \boldsymbol{I} - \boldsymbol{A}^{\mathrm{I}} & -\boldsymbol{A}^{\mathrm{II}} \\ -\boldsymbol{S}\boldsymbol{G}^{\mathrm{I}} & \boldsymbol{I} - \boldsymbol{S}\boldsymbol{g}^{\mathrm{II}} \end{pmatrix}^{-1} \tag{5.103}$$

When waste recycling is considered, two additional cost categories should be included:

(d) The cost for the input of waste materials (recycling).
(e) The revenue from the sale of waste materials.

Incorporating these items into (5.101), we obtain:

$$\begin{aligned} p_j &= \underbrace{\sum_{i \in N_{\mathrm{I}}} p_i a_{ij}}_{\text{(a)}} + \underbrace{\sum_{l \in N_{\mathrm{II}}} p_l \sum_{k \in N_w} s_{lk}(1 - r_k) g_{kj}^{+}}_{\text{(b)}} + \underbrace{\sum_{k \in N^{w}} p_k^{\mathrm{W}} g_{kj}^{-}}_{\text{(d)}} - \underbrace{\sum_{k \in N_w} p_k^{\mathrm{W}} r_k g_{kj}^{+}}_{\text{(e)}} + \underbrace{v_j}_{\text{(c)}}, \\ &= \underbrace{\sum_{i \in N_{\mathrm{I}}} p_i a_{ij}}_{\text{(a)}} + \underbrace{\sum_{l \in N_{\mathrm{II}}} p_l \sum_{k \in N_w} s_{lk}(1 - r_k) g_{kj}^{+}}_{\text{(b)}} + \underbrace{\sum_{k \in N^{w}} p_k^{\mathrm{W}} \left(g_{kj}^{-} - r_k g_{kj}^{+} \right)}_{\text{(d) and (e)}} + \underbrace{v_j}_{\text{(c)}}, \end{aligned} \tag{5.104}$$

where p_k^{W} is the price of waste k, and $r_k \in [0, 1]$ is the rate of recycling of waste k defined as

$$r_k := \frac{\sum_{j\in N} w_{kj}^{-}}{\sum_{l\in N\cup N^y} w_{kl}^{+}} = \frac{\sum_{j\in N} g_{kj}^{-} x_j}{\sum_{j\in N} g_{kj}^{+} x_j + \sum_{l\in N^y} w_{kj}^{y}} \tag{5.105}$$

In this model, the cost of waste treatment and the sale and purchase of recovered waste materials are explicitly represented by terms (b), (c), and (d), respectively.

For a sector engaged in recycling waste k, the input cost is affected by the price of the waste, p_k^w. If p_k^w is positive, the input cost increases, while if it is negative, the input cost decreases. However, it is important to note that even when $p_k^{\mathrm{W}} \leq 0$, sector j can still reduce its cost by selling waste to other sectors, as long as the following condition is satisfied

$$p_k^{\mathrm{W}} + \sum_{l\in N\mathrm{I}} p_l s_{lk} > 0 \Longleftrightarrow | p_k^{\mathrm{W}} | = -p_k^{\mathrm{W}} < \sum_{l\in N_{\mathrm{II}}} p_l s_{lk}. \tag{5.106}$$

This condition states that if the cost of selling waste to other sectors at negative prices is lower than the cost of submitting it to waste treatment, then the sector can reduce its input cost by selling the waste.

Formulated in matrices, we have

$$\begin{aligned}(\boldsymbol{p}^{\mathrm{I}}\ \boldsymbol{p}^{\mathrm{II}}) &= (\boldsymbol{p}^{\mathrm{I}}\ \boldsymbol{p}^{\mathrm{II}}) \begin{pmatrix} \boldsymbol{A}^{\mathrm{I}} & \boldsymbol{A}^{\mathrm{II}} \\ \boldsymbol{S}(\boldsymbol{I} - \hat{\boldsymbol{r}})\boldsymbol{G}^{\mathrm{I}+} & \boldsymbol{S}(\boldsymbol{I} - \hat{\boldsymbol{r}})\boldsymbol{G}^{\mathrm{II}+} \end{pmatrix} \\ &\quad + \boldsymbol{p}^{\mathrm{W}} \left(\boldsymbol{G}^{\mathrm{I}-} - \hat{\boldsymbol{r}}\boldsymbol{G}^{\mathrm{I}+}\ \boldsymbol{G}^{\mathrm{II}-} - \hat{\boldsymbol{r}}\boldsymbol{G}^{\mathrm{II}+}\right) + (\boldsymbol{v}^{\mathrm{I}}\ \boldsymbol{v}^{\mathrm{II}}),\end{aligned} \tag{5.107}$$

which could be solved for $(\boldsymbol{p}^{\mathrm{I}}, \boldsymbol{p}^{\mathrm{II}})$ as

$$\begin{aligned}(\boldsymbol{p}^{\mathrm{I}}\ \boldsymbol{p}^{\mathrm{II}}) &= \left(\boldsymbol{p}^{\mathrm{W}} \left(\boldsymbol{G}^{\mathrm{I}-} - \hat{\boldsymbol{r}}\boldsymbol{G}^{\mathrm{I}+}\ \boldsymbol{G}^{\mathrm{II}-} - \hat{\boldsymbol{r}}\boldsymbol{G}^{\mathrm{II}+}\right) + (\boldsymbol{v}^{\mathrm{I}}\ \boldsymbol{v}^{\mathrm{II}})\right) \\ &\quad \times \begin{pmatrix} \boldsymbol{I} - \boldsymbol{A}^{\mathrm{I}} & -\boldsymbol{A}^{\mathrm{II}} \\ -\boldsymbol{S}(\boldsymbol{I} - \hat{\boldsymbol{r}})\boldsymbol{G}^{\mathrm{I}+} & \boldsymbol{I} - \boldsymbol{S}(\boldsymbol{I} - \hat{\boldsymbol{r}})\boldsymbol{G}^{\mathrm{II}+} \end{pmatrix}^{-1}, \text{ or} \qquad (5.108) \\ &= \boldsymbol{\pi} \boldsymbol{\Lambda} \qquad (5.109)\end{aligned}$$

where $\boldsymbol{\pi}$ refers to the $1 \times (n_{\mathrm{I}} + n_{\mathrm{II}})$ vector and $\boldsymbol{\Lambda}$ to the inverse matrix.

5.5.1.3 The Life Cycle Cost of a Final Product

Building on the WIO-price model derived above, we now consider the life cycle cost of a product, say product j, which includes both the use and EoL phases of the product, besides the manufacturing and distribution phase. Let $\boldsymbol{u}$ be an $m \times 1$ vector representing the amount of inputs used during the use phase of a unit of product j, such as power and fuel. Let j' be the EoL product of j, and let $\boldsymbol{S}_{\cdot j'}$ be the j'-th column of the allocation matrix referring to j'.

Using the WIO-price model (5.109), we can express the life cycle cost of j as

$$LCC_j = \pi\boldsymbol{\Lambda}\begin{pmatrix}0\\ \vdots\\ 1\\ \boldsymbol{u}\\ \vdots\\ \boldsymbol{S}_{\cdot j'}\\ \vdots\end{pmatrix} = \underbrace{\pi\boldsymbol{\Lambda}_j}_{\text{production phase}} + \underbrace{\pi\boldsymbol{\Lambda}\begin{pmatrix}\boldsymbol{0}\\ \boldsymbol{u}\\ \boldsymbol{0}\\ \boldsymbol{0}\end{pmatrix}}_{\text{use phase}} + \underbrace{\pi\boldsymbol{\Lambda}\begin{pmatrix}\boldsymbol{0}\\ \boldsymbol{0}\\ \boldsymbol{S}_{\cdot j'}\\ \boldsymbol{0}\end{pmatrix}}_{\text{EoL phase}} \tag{5.110}$$

where $\boldsymbol{\Lambda}_j$ refers to the jth column of $\boldsymbol{\Lambda}$.

Final products differ in the length of their product life; many WEEE have a life of around ten years, while many passenger vehicles have a life of around twenty years. As we notice above, when the EoL cost of a product occurs ten or longer years after its purchase, it is necessary to discount the cost to make it comparable with the manufacturing costs.

5.5.1.4 Internalizing the External Costs

Internalization of external costs can be achieved by including previously exogenous factors in $\boldsymbol{\pi}$. For example, if greenhouse gas emissions were internalized through a carbon tax with a rate of t_{GHG} dollars per kg-CO_2, $\boldsymbol{\pi}$ would be augmented with $t_{GHG} \times \boldsymbol{f}^{\mathrm{I}}$ and $t_{GHG} \times \boldsymbol{f}^{\mathrm{II}}$, where $\boldsymbol{f}$ represents the GHG emissions per activity.

Examining the possible effects of introducing a carbon tax on the eLCC of TVs and refrigerators based on WIO-LCC, [116] found that the tax could significantly reduce the comparative cost disadvantage of recycling against landfilling in general and, in particular, when combined with design for disassembling (DfD) for refrigerators.

5.5.1.5 Remarks About p^w and r

Before closing this section on WIO-LCC, it is important to note two things. Firstly, in the WIO-LCC model, $\boldsymbol{p}^w$ is not determined within the model and is an exogenous variable that needs to be provided from outside the model. In cases where scrap materials have an active market, such as metal scrap and waste paper, $\boldsymbol{p}^w$ can be obtained from observed data. However, for waste items of low economic value with no active market, it may be necessary to estimate $\boldsymbol{p}^w$ using appropriate methods (see [116] for an example).

Secondly, the rate of waste recycling $\boldsymbol{r}$ is exogenous to the WIO-price model, and needs to be determined in the WIO quantity model, as it depends on $\boldsymbol{x}$ (see (5.105)).

If there is waste recycling, the price of products cannot be determined independently of the quantity.

5.5.2 *Social LCA*

An additional extension of the life cycle assessment framework beyond the environment is the social LCA (S-LCA), which assesses the social and socio-economic aspects of products and their potential positive and negative impacts throughout their life cycle, from extraction and processing of raw materials to final disposal [117]. S-LCA, along with LCA and eLCC, is considered a vital component of life cycle sustainability assessment (LCSA), which is expressed as ([35], p. 360)

$$\text{LCSA} = \text{LCA} + \text{eLCC} + \text{S-LCA}. \tag{5.111}$$

S-LCA effectively measures the sustainability performance of products and operating systems and their contribution to the Sustainable Development Goals (SDGs) [118].

The fundamental structure of S-LCA is similar to LCA, including goal and scope, functional unit, inventory analysis, and impact assessment. However, a distinguishing feature of S-LCA is that it considers both positive and negative impacts on the product life cycle, while beneficial impacts are rare in LCA. Positive impacts are considered important to encourage performance beyond compliance with laws, international agreements, and certification standards ([117], p. 40). Unlike LCA indicators, social indicators can have a positive impact, such as knowledge-intensive jobs, total employment, trust in risk information, stakeholder involvement, and long-term control functions, which can positively impact social well-being [119].

Despite efforts to establish a harmonized S-LCA framework, including the UNEP guidelines [117, 120], the development of a general framework for S-LCA is still at an early stage of development [119, 121]. Reviewing recent studies on S-LCA, [121, 122] found a lack of a complete definition and explicit theoretical reference point. Leaving further details of S-LCA to these review articles, we briefly review recent S-LCA studies based on IO.

5.5.2.1 IO-Based S-LCA Studies

Reference [123] expanded the EXIOBASE 3 model by incorporating quantitative data related to labor issues, such as employment rate, working hours, annual salary, occupational accident rate, and unemployment rate, obtained from the International Labour Organization's database, Ilostat. They measured the impacts on human productivity caused by unemployment, quantified in terms of quality-adjusted life years (QALY) lost, for the US.

Reference [124] conducted a cradle-to-gate, IO-based S-LCA of Swedish clothing consumption to identify negative social hotspots using the Social Hotspots Database

(SHDB) and the GTAP model. The functional unit was defined as the production of one US Dollars's worth of clothing for Swedish consumption. Impact categories included child labor, injuries and fatalities, toxics and hazards, poverty wages (labor wages under two US$s per day), gender equality, labor laws (not ratified ILO conventions by sector), collective bargaining (a country lacks or does not enforce collective bargaining rights) and wage assessment. The study found that commerce in Bangladesh and business services in China were hotspots for a large number of social indicators, while some main sectors such as plant-based fiber, textiles, and apparel production were identified as negative.

In their S-LCA of the Portuguese pulp and paper sector, [125] used the World Input-Output Database (WIOD) extended with the SHDB. In addition to worker-related impacts, the study also considered local community impacts such as migrant labor, indigenous rights, and access to sanitation. Positive impacts were evaluated based on employment and remuneration indicators, including total employment, number of employees, labor compensation, and compensation of employees. The results showed that the safety of workers and the equity of remuneration levels across the supply chain are crucial aspects that need to be addressed to improve the social performance of the sector, particularly in Portugal and Spain, which are hotspot countries for social impacts.

Lastly, [119] investigated the potential social impacts that could result from the large-scale deployment of carbon capture and storage (CCS) in coal-fired power plants in OECD Europe. The functional unit was defined as 1 kWh electricity delivered to the grid, and the study included life cycle stages such as raw material extraction, resources, transport, and infrastructure. The results showed an increase in the social well-being indicator, but also highlighted potential bottlenecks such as trust in risk information, long-term control functions, and stakeholder involvement, which need to be addressed during project development and implementation.

5.5.3 Distributing Environmental Responsibility

The environmental footprint calculation allows for the attribution of environmental impacts to the final users of products, that is, consumers. With the increasing availability of global MRIO databases, it has become possible to assess the global environmental impacts of a country's final demand by considering the interdependence of global supply chains. The issue of environmental responsibility is closely related to the attribution of environmental impacts to a subject, such as consumers or countries.

For instance, suppose we have two countries, a and b, with different industrial structures. a is an affluent economy primarily based on service industries with a small manufacturing base, importing most of its products from b, which is a medium income economy focused on manufacturing and exporting to a. Since the majority of environmental impacts in these two countries occur in country b where most manufacturing takes place, under the Kyoto Protocol/Paris Agreement on GHG emissions, country b holds full responsibility for the emissions generated by the production processes within its borders. However, an MRIO-based analysis would reveal that a considerable fraction of the emissions in b could be attributed to export to a, which may require reconsideration of the fairness of responsibility sharing under the protocol.

Similarly, if we consider a and b as production sectors within a country, we can have a similar discussion about responsibility sharing among production sectors. Suppose b is engaged in producing basic materials, such as metals and basic chemicals, while a is involved in producing final products, such as cars and electronics. As discussed in Sect. 5.1.3.1, direct emissions account for a small fraction of the total emissions in a, whereas direct emissions account for the most significant share in b. Similar to the above MRIO example, it may be necessary to reconsider the fairness of attributing the entire responsibility to b, whose products are indispensable for a.

We briefly review the recent literature on distributing environmental responsibility, mostly referring to [126]. Approaches for allocating environmental responsibility can be divided into full and shared responsibility approaches. For illustration, let us consider an MRIO accounting framework involving two countries, a and b. Using the notations in Sect. 4.3, we can express the total emissions e^a and e^b, respectively, as

$$\begin{aligned} e^a &= \boldsymbol{f}^a \boldsymbol{x}^a = \boldsymbol{f}^a \left(\boldsymbol{L}^{aa}(\boldsymbol{y}^{aa} + \boldsymbol{y}^{ab}) + \boldsymbol{L}^{ab}(\boldsymbol{y}^{ba} + \boldsymbol{y}^{bb}) \right) \\ e^b &= \boldsymbol{f}^b \boldsymbol{x}^b = \boldsymbol{f}^b \left(\boldsymbol{L}^{bb}(\boldsymbol{y}^{bb} + \boldsymbol{y}^{ba}) + \boldsymbol{L}^{ba}(\boldsymbol{y}^{aa} + \boldsymbol{y}^{ab}) \right) \end{aligned} \tag{5.112}$$

We start with the full responsibility approach.

5.5.3.1 The Full Responsibility Approach

Under the full responsibility approach, environmental pressures and impacts are allocated completely to a particular group of agents. Depending on the group of agents to which the responsibility is allocated, this approach can be divided into three approaches:

Production-based (PB) responsibility The production-based responsibility (also known as producer responsibility or territorial approach) accounts for environmental pressures caused by economic processes within the country's territory (on a national scale) or within the firm's domain (on a micro level). Using (5.112), the PB of countries a and b can be represented as follows

$$\begin{aligned} \text{PB}^a &: \boldsymbol{f}^a\, \mathbf{diag}(\boldsymbol{x}^a) \\ \text{PB}^b &: \boldsymbol{f}^b\, \mathbf{diag}(\boldsymbol{x}^b) \end{aligned} \tag{5.113}$$

This corresponds to the responsibility sharing under the Kyoto Protocol/Paris Agreement, where a country is responsible for all the emissions directly generated by the production processes within its borders [127]. The production-based responsibility approach does not account for the transfer of emissions through international trade, which can lead to carbon leakage [128]. This approach also creates difficulties in assigning responsibility for emissions generated through international activities such as aviation and maritime transportation [128].

Consumption-based (CB) responsibility Consumption-based responsibility accounts for the emissions generated through international trade and minimizes the effects of carbon leakage by holding countries responsible for the emissions embodied in their trade balance [31, 127]. In terms of (5.112), the CB of a and b, $\boldsymbol{t}^a$ and $\boldsymbol{t}^b$, can be represented as

$$\begin{aligned} \boldsymbol{t}^a &= (\boldsymbol{f}^a \boldsymbol{L}^{aa} + \boldsymbol{f}^b \boldsymbol{L}^{ba})\,\mathbf{diag}(\boldsymbol{y}^{aa}) + (\boldsymbol{f}^a \boldsymbol{L}^{ab} + \boldsymbol{f}^b \boldsymbol{L}^{bb})\,\mathbf{diag}(\boldsymbol{y}^{ba}) \\ \boldsymbol{t}^b &= (\boldsymbol{f}^a \boldsymbol{L}^{ab} + \boldsymbol{f}^b \boldsymbol{L}^{bb})\,\mathbf{diag}(\boldsymbol{y}^{bb}) + (\boldsymbol{f}^a \boldsymbol{L}^{aa} + \boldsymbol{f}^b \boldsymbol{L}^{ba})\,\mathbf{diag}(\boldsymbol{y}^{ab}) \end{aligned} \tag{5.114}$$

One disadvantage of consumer responsibility is that it does not provide direct incentives to exporting countries to encourage changes in the GHG efficiency of their export industries [127]. Furthermore, this approach raises issues of fairness as production-based responsibility raised. Consuming (importing) countries provide an economic benefit, or positive impacts, to producing (exporting) countries via creating employment and income, which remain unaccounted for if a strict consumption approach is followed (this point is related to S-LCA we discussed above).

The consumption-based responsibility, however, could be "technology-adjusted" to hold countries responsible for the technology in their exports, by comparing available production alternatives for the country's exports (e.g., world averages of the environmental pressure intensity of exporting industries) with domestic technologies [129].

Income-based responsibility (IB) In the income-based (also downstream responsibility) responsibility approach, environmental responsibility is allocated according to the income generated by payments to the owners of the factors of production, which includes both capital and labor.

The calculation is based on the Ghosh model and is given by

$$\boldsymbol{i} = \hat{\boldsymbol{v}}(\boldsymbol{I} - \boldsymbol{B})^{-1}\boldsymbol{f}^\top \tag{5.115}$$

where $\boldsymbol{v}$ is the row vector of value added ratios, and $\boldsymbol{B} = [z_{ij}/x_i]$. Note that, different from $\boldsymbol{A}$, $\boldsymbol{B}$ is obtained by dividing the row elements of $\boldsymbol{Z}$ referring to the distribution of a certain product to its users by the product's output.[8] For $\boldsymbol{G} = (\boldsymbol{I} - \boldsymbol{B})^{-1}$, the IB counterpart to (5.112) is given by

[8] The Ghosh model is derived as follows. From the accounting identity of IO table in monetary units and the definition of $\boldsymbol{B}$, we have

$$\boldsymbol{x}^\top = \boldsymbol{\iota}^\top \boldsymbol{Z} + \boldsymbol{v} = \boldsymbol{\iota}^\top \hat{\boldsymbol{x}} \boldsymbol{B} + \boldsymbol{v} = \boldsymbol{x}^\top \boldsymbol{B} + \boldsymbol{v}$$

Solving for $\boldsymbol{x}^\top$ gives

$$\boldsymbol{x}^\top = \boldsymbol{v}(\boldsymbol{I} - \boldsymbol{B})^{-1}.$$

See [56, 130, 131] for further details of the Ghosh model.

$$\begin{aligned} \boldsymbol{i}^a &= \hat{\boldsymbol{v}}^a(\boldsymbol{G}^{aa}\boldsymbol{f}^{a\top} + \boldsymbol{G}^{ab}\boldsymbol{f}^{b\top}) \\ \boldsymbol{i}^b &= \hat{\boldsymbol{v}}^b(\boldsymbol{G}^{ba}\boldsymbol{f}^{a\top} + \boldsymbol{G}^{bb}\boldsymbol{f}^{b\top}) \end{aligned} \tag{5.116}$$

In contrast to the Leontief model, the Ghosh model accounts for the downstream impacts only; $\boldsymbol{B}$ provides information about how the product is distributed among sectors but not about how it is produced and which products are inputted. An example of recent studies based on IB is [132].

5.5.3.2 Shared Responsibility Approaches (VA)

Shared responsibility approaches aim to allocate environmental pressures and impacts among different groups of agents, such as producers and consumers. However, the challenge lies in determining how this allocation should be done. One proposed option is to use a predetermined sharing rule, such as the 50%–50% sharing suggested by [133]. However, such rules lack logical justification and do not necessarily reflect the actual responsibility of each group of agents.

Recently, [126] proposed a new approach to allocate environmental responsibility, the "value added-based responsibility" (VA-R) allocation, which allocates total environmental pressures occurring along an international supply chain to the participating sectors and countries according to the share of value added they generate within that specific supply chain. Denoting by $\boldsymbol{v}^a$ and $\boldsymbol{v}^b$ the ratio of value added for countries a and b, the VA-R is given by the following matrix $\boldsymbol{H} = [h_{ij}]$

$$\boldsymbol{H} = \begin{pmatrix} \hat{\boldsymbol{v}}^a & 0 \\ 0 & \hat{\boldsymbol{v}}^b \end{pmatrix} \begin{pmatrix} \boldsymbol{L}^{aa} & \boldsymbol{L}^{ab} \\ \boldsymbol{L}^{ba} & \boldsymbol{L}^{bb} \end{pmatrix} \begin{pmatrix} \boldsymbol{t}^{a\top} & \boldsymbol{0} \\ \boldsymbol{0} & \boldsymbol{t}^{b\top} \end{pmatrix} = \begin{pmatrix} \hat{\boldsymbol{v}}^a\boldsymbol{L}^{aa}\boldsymbol{t}^{a\top} & \hat{\boldsymbol{v}}^a\boldsymbol{L}^{ab}\boldsymbol{t}^{b\top} \\ \hat{\boldsymbol{v}}^b\boldsymbol{L}^{ba}\boldsymbol{t}^{a\top} & \hat{\boldsymbol{v}}^b\boldsymbol{L}^{bb}\boldsymbol{t}^{b\top} \end{pmatrix} \tag{5.117}$$

where h_{ij} refers to the origin of value added i and j to the origin of final product. The VA-R approach reallocates the consumer responsibility $\boldsymbol{t}$ given by (5.114) to all entities participating in the supply chain according to their value-added shares, holding accountable all profiteers (value generators) along the entire supply chains [126]. Recalling that income and employment (major components of value added) are positive indicators in S-LCA, the VA approach can be regarded as a scheme of responsibility sharing based on an integrated form of LCA and S-LCA; the negative environmental impacts are allocated among the groups of agents considering the associated positive social impacts.

Using the Eora database, which includes data on 189 countries and six categories of value added (compensation of employees, taxes on production, subsidies on production, net operating surplus, net mixed-income, and consumption of fixed capital), [126] applied the VA-R approach to allocate environmental responsibility for the extraction of 44 raw material categories. The study found that the use of value-added shares as an allocation principle could result in significant increases or decreases in total environmental responsibility compared to other approaches such as consumer or income responsibility. Specifically, the VA-R approach increased the responsibility of countries such as Germany or the service sector as a whole, while decreasing the responsibility of the US or the aggregated food industry.

5.5.3.3 A Summary

In general terms, the allocation schemes PB, CB, IB, and VA-R are given by

$$\begin{aligned} \text{PB:} &\quad \boldsymbol{f}\hat{\boldsymbol{x}} \\ \text{CB:} &\quad \boldsymbol{f}\boldsymbol{L}\hat{\boldsymbol{y}} \\ \text{IB:} &\quad \hat{\boldsymbol{v}}\boldsymbol{G}\boldsymbol{f}^{\top} \\ \text{VA-R:} &\quad \hat{\boldsymbol{v}}\boldsymbol{L}\,\mathbf{diag}\,(\hat{\boldsymbol{y}}\boldsymbol{L}^{\top}\boldsymbol{f}^{\top}) \end{aligned} \tag{5.118}$$

It is important to note that these allocation schemes aim to allocate the same total emission e:

$$\boldsymbol{f}\boldsymbol{x} = \boldsymbol{f}\boldsymbol{L}\boldsymbol{y} = \boldsymbol{\iota}^{\top}\hat{\boldsymbol{v}}\boldsymbol{G}\boldsymbol{f}^{\top} = \boldsymbol{\iota}^{\top}\hat{\boldsymbol{v}}\boldsymbol{L}(\hat{\boldsymbol{y}}\boldsymbol{L}^{\top}\boldsymbol{f}^{\top}) = e \tag{5.119}$$

where the third equality follows from (4.93).

5.5.4 *Attributional and Consequential LCA

According to the UNEP-SETAC Guideline on LCA [134], the methodology of LCA (inventory analysis) is divided into attributional LCA (ALCA) and consequential LCA (CLCA), distinguished by the objectives as follows

1. ALCA: The attributional approach attempts to provide information on what portion of global burdens can be associated with a product (and its life cycle).
2. CLCA: The consequential approach attempts to provide information on the environmental burdens that occur, directly or indirectly, as a consequence of a decision (usually represented by changes in demand for a product).

An ALCA aims to identify for which environmental impacts the consumption or the production of a product is accountable, whereas a CLCA to identify the environmental consequences of the consumption or the production of a product [135]. Reference ([136], p.8) provide an illustrative example of a car to explain the difference between the two approaches

> *Focusing on a car as a product as an example, ALCA would consider the share of the environmental impact over its complete life cycle of the total environmental impact of a world where the car is already present or at least hypothetically present, whereas CLCA would focus on the effects of a decision to have for example one more car on the environmental impact of the world in which the demand for that particular car was not yet occurring.*

5.5.4.1 Constraints About the Types of Models

One implication of the above definition of ALCA is the additivity restriction [134, 136]

$$\text{The total environmental impact} = \sum_i \text{the impact of product } i \tag{5.120}$$

The EEIO-based footprint calculations we discussed above fall under the category of ALCA, as they satisfy (5.120). Our discussion on environmental responsibility sharing in Sect. 5.5.3 also pertains to ALCA.

In contrast, a CLCA aims to model, in the most realistic manner, the cause-effect chain starting with a decision [136]. Unlike ALCA, CLCA does not impose constraints on the types of models used. The system boundaries in CLCA can be expanded to include processes affected by the consequences of the decision under consideration.

Let $\boldsymbol{y}_i$ be an $n \times 1$ vector representing the functional unit for the CLCA of product i. Furthermore, let $\boldsymbol{f}_i$ and $\boldsymbol{L}_i$ denote the intervention and the Leontief inverse matrix, respectively, incorporating the changes induced by $\boldsymbol{y}_i$. Since different products can have varying consequences for $\boldsymbol{f}$ and $\boldsymbol{L}$, the values of $\boldsymbol{f}_i$ and $\boldsymbol{L}_i$ may differ for different i. Additionally, the order of $\boldsymbol{f}_i$ and $\boldsymbol{L}_i$ may differ because the system boundaries can vary for different products. As a result, the condition (5.120) will not hold for a CLCA, due to the potential differences in $\boldsymbol{f}_i$ and $\boldsymbol{L}_i$ among products and the potential variability in system boundaries. The restriction (5.120) holds in ALCA because $\boldsymbol{f}_i$ and $\boldsymbol{L}_i$ remain constant for all i, and the system boundary remains the same for all products.

In Sect. 5.4.3.3, we considered the case where the sum of the impacts that result from separately incinerating waste items differs from the impacts of incinerating the mixture of all waste items simultaneously. The nonlinear WIO model incorporating the system engineering model is thus an example of CLCA: $\boldsymbol{f}$ and $\boldsymbol{L}$ change depending on the way wastes are incinerated.

5.5.4.2 Allocation in the Presence of By-Products

The division of LCA into ALCA and CLCA is closely related to the modeling of allocation in the presence of by-products or co-production. This modeling is a central divide between ALCA and CLCA [137, 138]. Allocation refers to a modeling procedure that assigns requirements specifically to the supply of a single product, even though this product is co-produced with others, without requiring a detailed understanding of the inner mechanism of a co-producing activity [139].

There are various allocation models, including system expansion (SE), partitioning (PA), and substitution (SU), among others [139]. For example, in grain farming, where 1 kg of wheat and 2 kg of straw are co-produced from the inputs of 3 kg of j and 4 kg of k, SE refers to expanding the functional unit to include both wheat

Table 5.14 Allocation models and the goal/scope of LCA

The reason for conducting LCA	ALCA	CLCA
Process-oriented	SE	SE/SU
Product-oriented	PA	SU

Source [135] Fig. 2. SE: system expansion, PA: partitioning, SU: substitution

and straw, without requiring allocation. On the other hand, under PA, the inputs are split proportionally to some common property of the wheat and straw outputs using a partition coefficient $\phi \in [0, 1]$, resulting in two partitioned processes. Finally, SU resolves co-production by assuming that secondary co-products displace some other primary production; in the grain farming example, straw is used as fuel, with 1 kg of straw substituting for μ kg of firewood consumed by private households.

In the case of WIO (5.80), this SU case can be implemented by recording the amount of straw produced in grain farming as a positive element of the wheat column of $\boldsymbol{G}^{\mathrm{I}}$, the amount of straw used for fuel as a negative element of $\boldsymbol{w}^{y}$, and the amount of firewood (or input from the forestry sector) recorded in $\boldsymbol{y}^{\mathrm{I}}$ reduced by ξ kg per kg of straw used. It is important to note that this list of modeling options is not exhaustive, and further modeling options are available [139].

In order to determine which allocation model is appropriate for ALCA and CLCA, we refer to [135] (Table 5.14). The choice of allocation model depends on whether the LCA is process-oriented or product-oriented. In a process-oriented LCA, the focus is on the wheat process itself, rather than just the wheat product. For process oriented LCAs, SE can be applied to both ALCA and CLCA, by expanding the functional unit to include both wheat and straw. In contrast, in a product-oriented LCA, we need to determine the environmental impacts of consuming only 1 kg of wheat. For product-oriented ALCA, PA is the only applicable allocation method. Since SU modeling involves changes to $\boldsymbol{y}^{\mathrm{I}}$, $\boldsymbol{w}^{y}$, and $\boldsymbol{G}^{\mathrm{I}}$, (5.120) is not satisfied. SU modeling is only applicable to CLCA.

References

1. Buekens, Alfons. 2013. *Incineration technologies*. Springer Science Business Media.
2. Vassilev, Stanislav V., Kunihiro Kitano, and Christina G. Vassileva. 1996. Some relationships between coal rank and chemical and mineral composition. *Fuel* 75 (13): 1537–1542.
3. McAllister, Sara, Jyh-Yuan Chen, and A. Carlos Fernandez-Pello. 2011. *Fundamentals of combustion processes*. Springer Science & Business Media.
4. Nansai, Keisuke. 2009. environmental input-output database building in Japan. In *Handbook of input-output economics in industrial ecology*, Chap. 31, ed. Sangwon Suh, 653–688. Springer.
5. Hoekstra, Rutger, and Jeroen C.J.M. Van Den Bergh. 2002. Structural decomposition analysis of physical flows in the economy. *Environmental and Resource Economics* 23(3): 357–378.

6. Minx, Jan C., Giovanni Baiocchi, Glen P. Peters, Christopher L. Weber, Dabo Guan, and Klaus Hubacek. 2011. A 'Carbonizing dragon" China's fast growing CO_2 emissions revisited. *Environmental Science & Technology* 45: 9144–9153.
7. Su, Bin, and B.W. Ang. 2012. Structural decomposition analysis applied to energy and emissions: Some methodological developments. *Energy Economics* 34(1): 177–188.
8. Li, Ruifei, Pei Hua, and Peter Krebs. 2022. Global trends and drivers in consumption- and income-based emissions of polycyclic aromatic hydrocarbons. *Environmental Science & Technology* 56: 131–144.
9. Lenzen, Manfred. 1998. Primary energy and greenhouse gases embodied in Australian final consumption: An input-output analysis. *Energy policy* 26 (6): 495–506.
10. Yang, Yi, Yujin Park, Timothy M Smith, Taegon Kim, and Hung-Suck Park. 2022. High-resolution environmentally extended input-output model to assess the greenhouse gas impact of electronics in South Korea. *Environmental Science & Technology.*
11. Teodora Diana Corsatea. 2019. *Soeren Lindner, Iñaki Arto, Maria Victoria Román, José Manuel Rueda-Cantuche.* A Velázquez Afonso: Antonio F Amores, and Frederik Neuwahl. World Input-Output Database Environmental Accounts.
12. Lave, Lester B., Elisa Cobas-Flores, Chris T. Hendrickson, and Francis McMichael. 1995. Using input-output analysis to estimate economy-wide discharges. *Environmental Science & Technology* 29(9): 4220–426.
13. Huang, Anny Y., Christopher L. Weber, and Scott H. Matthews. 2009. Categorization of scope 3 emissions for streamlined enterprise carbon footprinting. *Environmental Science and Technology* 43(22): 8509–8515.
14. Malik, Arunima, Manfred Lenzen, Peter J. Ralph, and Bojan Tamburic. 2015. Hybrid life-cycle assessment of algal biofuel production. *Bioresource Technology* 184: 436–443.
15. Nakamura, Shinichiro. 2020. Tracking the product origins of waste for treatment using the WIO data developed by the Japanese ministry of the environment. *Environmental Science and Technology* 54 (23): 14862–14867.
16. Wiedmann, Thomas. 2010. *Frequently asked questions about input-output output analysis.* Centre for Sustainability Accounting, York, UK: Technical report.
17. WBCSD and WRI. 2012. *The greenhouse gas protocol a corporate accounting and reporting standard.* WBCSD WRI: Technical report.
18. Hertwich, Edgar G., and Richard Wood. 2018. The growing importance of scope 3 greenhouse gas emissions from industry. *Environmental Research Letters* 13(10).
19. Bhatia, P. et al. 2011. Corporate value chain (Scope 3) standard.
20. European Commission. 2013. EC 2013 Commission recommendation of 9 April 2013 on the use of common methods to measure and communicate the life cycle environmental performance of products and organisations. *Official Journal of the European Union* 179/EU.
21. Pelletier, Nathan, Karen Allacker, Rana Pant, and Simone Manfredi. 2014. The European commission organisation environmental footprint method: Comparison with other methods, and rationales for key requirements. *International Journal of Life Cycle Assessment* 19 (2): 387–404.
22. Li, Mo, Thomas Wiedmann, and Michalis Hadjikakou. 2019. Enabling full supply chain corporate responsibility: Scope 3 emissions targets for ambitious climate change mitigation. *Environmental Science and Technology.*
23. Hertwich, Edgar G. 2021. Increased carbon footprint of materials production driven by rise in investments. *Nature Geoscience* 14(3):151–155.
24. Dente, Sébastien M.R., Chika Aoki-Suzuki, Daisuke Tanaka, and Seiji Hashimoto. 2018. Revealing the life cycle greenhouse gas emissions of materials: The Japanese case. *Resources, Conservation and Recycling* 133(September 2017): 395–403.
25. Henderson, Harold V., and Shayle R. Searle. 1981. On deriving the inverse of a sum of matrices. *SIAM review* 23(1):53–60.
26. Dente, Sébastien M.R., Chika Aoki-Suzuki, Daisuke Tanaka, Chihiro Kayo, Shinsuke Murakami, and Seiji Hashimoto. 2019. Effects of a new supply chain decomposition framework on the material life cycle greenhouse gas emissions-the Japanese case. *Resources, Conservation and Recycling* 143(January): 273–281.

27. Aoki-Suzuki, Chika, Sébastien M.R. Dente, Daisuke Tanaka, Chihiro Kayo, Shinsuke Murakami, Chiharu Fujii, Kiyotaka Tahara, and Seiji Hashimoto. 2021. Total environmental impacts of Japanese material production. *Journal of Industrial Ecology* 25(6): 1474–1485.
28. Cabernard, Livia, Stephan Pfister, and Stefanie Hellweg. 2019. A new method for analyzing sustainability performance of global supply chains and its application to material resources. *Science of the Total Environment* 684: 164–177.
29. Dietzenbacher, Erik, and Michael L. Lahr. 2013. Expanding extractions. *Economic Systems Research* 25(3): 341–360.
30. Stadler, Konstantin, Richard Wood, Tatyana Bulavskaya, Carl Johan Södersten, Moana Simas, Sarah Schmidt, Arkaitz Usubiaga, José Acosta-Fernández, Jeroen Kuenen, Martin Bruckner, Stefan Giljum, Stephan Lutter, Stefano Merciai, Jannick H. Schmidt, Michaela C Theurl, Christoph Plutzar, Thomas Kastner, Nina Eisenmenger, Karl Heinz Erb, Arjan de Koning, and Arnold Tukker. 2018. EXIOBASE 3: Developing a time series of detailed environmentally extended multi-regional input-output tables. *Journal of Industrial Ecology* 22(3): 502–515.
31. Peters, Glen P. 2008. From production-based to consumption-based national emission inventories. *Ecological Economics* 65(1): 13–23.
32. Peters, Glen, and Edgar Hertwich. 2008. CO_2 embodied in international trade with implications for global climate policy. *Environmental Science & Technology* 42: 1401–1407.
33. Peters, Glen P., and Edgar G. Hertwich. 2009. The application of multi-regional input-output analysis to industrial ecology. In *Handbook of input-output economics in industrial ecology*, vol. 1, Chap. 31, ed. Sangwon Suh, 847–863. Springer.
34. Hauschild, Michael Z., Ralph K. Rosenbaum, and Stig Irvin Olsen (eds.) *Life cycle assessment theory and practice*. Springer.
35. Klöpfer, Walter, and Birgit Grahl. 2014. *Life cycle assessment (LCA) a guide to best practice*. Wiley-VCH.
36. Isard, Walter, Kenneth Bassett, Charles Choguill, John Furtado, Ronald Izumita, John Kissin, Eliahu Romanoff, Richard Seyfarth, and Richard Tatlock. 1968. On the likage of socio-economic and ecologic systems. *Papers of the Regional Science Association* 21 (1): 79–99.
37. Moriguchi, Yuichi, and Yoshonori Kondo. 1993. Considering the global environmental impact of automobiles: Life cycle assessment of CO_2 emissions (in Japanese). *Kinzoku (Metals)* 63(6).
38. Yoshioka, Kanji, Hitoshi Hayami, Ayu Ikeda, and Masao Suga. 1993. Application of input-output tables for environmental analysis: CO_2 emissions associated with production activities and their factors (in Japanese). *Journal of Applied Input-Output Analysis* 4 (1): 37–57.
39. Suh, Sangwon, and Shinichiro Nakamura. 2007. Five years in the area of input-output and hybrid LCA. *The International Journal of Life Cycle Assessment* 12 (6): 351.
40. Heijungs, Reinout, and Sangwon Suh. 2002. *The computational structure of life cycle assessment*, vol. 11. Springer Science & Business Media.
41. Zackrisson, Mats, Lars Avellán, and Jessica Orlenius. 2010. Life cycle assessment of lithium-ion batteries for plug-in hybrid electric vehicles-Critical issues. *Journal of Cleaner Production* 18 (15): 1519–1529.
42. Dolganova, Iulia, Anne Rödl, Vanessa Bach, Martin Kaltschmitt, and Matthias Finkbeiner. 2020. A review of life cycle assessment studies of electric vehicles with a focus on resource use. *Resources* 9 (3): 1–26.
43. Basset-mens, Claudine, Niels Jungbluth, Rita Schenck, and Daniel Baumgartner. 2015. Life cycle assessment in the agri-food sector. In *Life cycle assessment in the agri-food sector*.
44. Bohnengel, Barrett and Joule Bergerson. 2014. Comparative life cycle assessment of future power system. *Environmental Science & Technology*.
45. Asdrubali, Francesco, Giorgio Baldinelli, Francesco D'Alessandro, and Flavio Scrucca. 2015. Life cycle assessment of electricity production from renewable energies: Review and results harmonization. *Renewable and Sustainable Energy Reviews* 42: 1113–1122.
46. Suh, Sangwon, and Gjalt Huppes. 2002. Missing inventory estimation tool using extended input-output analysis. *International Journal of Life Cycle Assessment* 7 (3): 134–140.
47. Suh, Sangwon, Manfred Lenzen, Graham J Treloar, Hiroki Hondo, Arpad Horvath, Gjalt Huppes, Olivier Jolliet, Uwe Klann, Wolfram Krewitt, Yuichi Moriguchi, Jesper Munksgaard,

and Gregory Norris. 2004. System boundary selection in life-cycle inventories using hybrid approaches. *Environmental Science & Technology* 38(3): 657–664.
48. Nakamura, Shinichiro, and Keisuke Nansai. 2016. Input–output and hybrid LCA. In *Special types of life cycle assessment*, ed. by Matthias Finkbeiner, 219–291. Springer.
49. Suh, Sangwon, and Gjalt Huppes. 2005. Methods for life cycle inventory of a product. *Journal of Cleaner Production* 13: 687–697.
50. Hondo, Hiroki, Yue Moriizumi, and Tomohiko Sakao. 2006. A method for technology selection considering environmental and socio-economic impacts: Input-output optimization model and its application to housing policy. *International Journal of Life Cycle Assessment* 11 (6): 383–393.
51. Yu, Man, and Thomas Wiedmann. 2018. Implementing hybrid LCA routines in an input-output virtual laboratory. *Journal of Economic Structures* 7(1).
52. Agez, Maxime, Richard Wood, Manuele Margni, Anders H Strømman, Réjean Samson, and Guillaume Majeau-Bettez. 2020. Hybridization of complete PLCA and MRIO databases for a comprehensive product system coverage. *Journal of Industrial Ecology* 24(4): 774–790.
53. Strømman, Anders Hammer, Glen P. Peters, and Edgar G. Hertwich. 2009. Approaches to correct for double counting in tiered hybrid life cycle inventories. *Journal of Cleaner Production* 17(2):248–254.
54. Strømman, Anders Hammer. 2009. Dealing with double-counting in tiered hybrid life-cycle inventories: a few comments - response. *Journal of Cleaner Production* 17 (17): 1607–1609.
55. Lenzen, Manfred. 2009. Dealing with double-counting in tiered hybrid life-cycle inventories: A few comments. *Journal of Cleaner Production* 17 (15): 1382–1384.
56. Miller, Ronald E., and Peter D. Blair. 2022. *Input-Output Analysis Foundations and Extensions*, 3rd edn. Cambridge University Press.
57. Nakamura, Shinichiro, and Eiji Yamasue. 2010. Hybrid LCA of a design for disassembly technology: Active disassembling fasteners of hydrogen storage alloys for home appliances. *Environmental Science and Technology* 44 (12): 4402–4408.
58. Nishijima, Daisuke. 2017. The role of technology, product lifetime, and energy efficiency in climate mitigation: A case study of air conditioners in Japan. *Energy Policy* 104(November 2016): 340–347.
59. Hienuki, Shunichi. 2017. Environmental and socio-economic analysis of naphtha reforming hydrogen energy using input-output tables: A case study from Japan. *Sustainability (Switzerland)* 9 (8): 1376.
60. Hendrickson, Chris, Arpad Horvath, Satish Joshi, and Lester Lave. 1998. Economic input-output models for environmental life-cycle assessment. *Environmental Science & Technology* 32 (7): 184–191.
61. Weber, Christopher L., and H. Scott Matthews. 2008. Food-miles and the relative climate impacts of food choices in the united states. *Environmental Science & Technology* 42: 3508–3513.
62. Nansai, Keisuke, Shigemi Kagawa, Yasushi Kondo, Sangwon Suh, Rokuta Inaba, and Kenichi Nakajima. 2009. Improving the completeness of product carbon footprints using a global link input-output model: The case of Japan. *Economic Systems Research* 21 (3): 267–290.
63. Heijungs, Reinout, Yi Yang, and Hung Suck Park. 2022. A or I-A? Unifying the computational structures of process- and IO-based LCA for clarity and consistency. *Journal of Industrial Ecology* 1(1): 1824–1836.
64. Suh, Sangwon. 2004. Functions, commodities and environmental impacts in an ecological-economic model. *Ecological Economics* 48: 451–467.
65. Strømman, Anders Hammer, Christian Solli, and Edgar G. Hertwich. 2006. Hybrid life-cycle assessment of natural gas based fuel chains for transportation. *Environmental Science & Technology* 40: 2797–2804.
66. Suh, Sangwon. 2003. MIET 3.0 user guide.
67. Arvesen, Anders, and Edgar G. Hertwich. 2012. Environmental implications of large-scale adoption of wind power: A scenario-based life cycle assessment (Environmental Research Letters (2011) 6 (045102)). *Environmental Research Letters* 7(3).

68. Hertwich, Edgar G., Thomas Gibon, Evert A. Bouman, Anders Arvesen, Sangwon Suh, Garvin A. Heath, Joseph D. Bergesen, Andrea Ramirez, Mabel I. Vega, and Lei Shi. 2015. Integrated life-cycle assessment of electricity-supply scenarios confirms global environmental benefit of low-carbon technologies. *Proceedings of the National Academy of Sciences* 112(20): 6277–6282.
69. Pauliuk, Stefan, Richard Wood, and Edgar G. Hertwich. 2015. Dynamic models of fixed capital stocks and their application in industrial ecology. *Journal of Industrial Ecology* 19 (1): 104–116.
70. Leontief, Wassily. 1970. Environmental repercussions and the economics structure: An input-output approach. *The Review of Economics and Statistics* 52: 262–271.
71. Duchin, Faye. 1990. The conversion of biological materials and wastes to useful products. *Structural Change and Economic Dynamics* 1: 243–261.
72. Nakamura, Shinichiro, and Yasushi Kondo. 2002. Input-output analysis of waste management. *Journal of Industrial Ecology* 6 (1): 39–63.
73. Nakamura, Shinichiro. 1999. Input-output analysis of waste cycles. In *Proceedings - 1st international symposium on environmentally conscious design and inverse manufacturing, ecodesign 1999*, 475–480.
74. Nakamura, Shinichiro, and Yasushi Kondo. 2009. *Waste input-output analysis*. Dordrecht: Springer Science & Business Media.
75. Towa, Edgar, Vanessa Zeller, and Wouter M.J. Achten. 2020. Input-output models and waste management analysis: A critical review. *Journal of Cleaner Production* 249: 119359.
76. https://doi.org/10.5281/zenodo.8088489. Accessed on 2023-06-30.
77. Lenzen, Manfred, and Christian J. Reynolds. 2014. A supply-use approach to waste input-output analysis. *Journal of Industrial Ecology*.
78. Reynolds, Christian John, Julia Piantadosi, and John Boland. 2014. A waste supply-use analysis of Australian waste flows. *Journal of Economic Structures* 3(1).
79. Fry, Jacob, Manfred Lenzen, Damien Giurco, and Stefan Pauliuk. 2016. An Australian multi-regional waste supply-use framework. *Journal of Industrial Ecology* 20 (6): 1295–1305.
80. Kaza S., L. Yao, P. Bhada-Tata, and F. Van Woerden. 2018. What a waste 2.0: a global snapshot of solid waste management to 2050. World Bank Publications.
81. McDougall, Forbes, Peter White, and Marina Franke, and Peter Hindel. 2001. *Integrated solid waste management: A life cycle inventory*. Blackwell Science.
82. U.S. Environmental Protection Agency. 2007. Methodology for thermal efficiency and energy input calculations and analysis of biomass cogeneration unit characteristics. Technical Report 5.
83. Matsuto, Toshihiko. 2005. Municipal solid waste treatment System: Analysis, planning, and evaluation, Tokyo: Gihoudou, in Japanese.
84. Tisserant, Alexandre, Stefan Pauliuk, Stefano Merciai, Jannick Schmidt, Jacob Fry, Richard Wood, and Arnold Tukker. 2017. Solid waste and the circular economy: A global analysis of waste treatment and waste footprints. *Journal of Industrial Ecology* 21 (3): 628–640.
85. Towa, Edgar, Vanessa Zeller, Stefano Merciai, and Wouter M.J. Achten. 2021. Regional waste footprint and waste treatments analysis. *Waste Management* 124: 172–184.
86. Koning, De. 2014. CREEA-Recommendation of terminology, classification, framework of waste accounts and MFA, and data collection guideline.
87. Jambeck, Jenna R., Roland Geyer, Chris Wilcox, Theodore R. Siegler, Miriam Perryman, Anthony Andrady, Ramani Narayan, and Kara Lavender Law. 2016. Physical/Hybrid supply and use tables. http://fp7desire.eu/documents/category/3-public-deliverables?download=29:physical-hybrid-supply-and-use-tablesmethodological-report report. Accessed on 2023-05-31.
88. Jensen, Christa D., Stuart Mcintyre, Max Munday, and Karen Turner. 2013. Responsibility for regional waste generation: A single-region extended input-output analysis for wales. *Regional Studies* 47 (6): 913–933.
89. Chen, Pi-Cheng., and Hwong-wen Ma. 2015. Using an industrial waste account to facilitate national level industrial symbioses by uncovering the waste exchange potential. *Journal of Industrial Ecology* 19 (6): 950–962.

90. Beylot, Antoine, Baptiste Boitier, Nicolas Lancesseur, and Jacques Villeneuve. 2016. A consumption approach to wastes from economic activities. *Waste management* 49: 505–515.
91. Salemdeeb, Ramy, Abir Al-Tabbaa, and Christian Reynolds. 2016. The UK waste input-output table: Linking waste generation to the UK economy. *Waste Management & Research* 34 (10): 1089–1094.
92. Beylot, Antoine, Baptiste Boitier, Nicolas Lancesseur, and Jacques Villeneuve. 2017. The waste footprint of French households in 2020: A comparison of scenarios of consumption growth using input-output analysis. *Journal of Industrial Ecology* 22 (2): 356–368.
93. Ruiz-Peñalver, Soraya María, Mercedes Rodríguez, and José A. Camacho. 2019. The case of Spain. A waste generation input output analysis. *Journal of Cleaner Production* 210: 1475–1482.
94. Liao, Meng-i, Pi.-cheng Chen, Hwong-wen Ma, and Shinichiro Nakamura. 2015. Identification of the driving force of waste generation using a high-resolution waste input-output table. *Journal of Cleaner Production* 94: 294-f303.
95. Meyer, David E., Mo. Li, and Wesley W. Ingwersen. 2019. Analyzing economy-scale solid waste generation using the United States environmentally-extended input-output model. *Resources, Conservation and Recycling* 157: 2020 October.
96. Tsukui, Makiko, Shigemi Kagawa, and Yasushi Kondo. 2015. Measuring the waste footprint of cities in Japan: An interregional waste input-output analysis. *Journal of Economic Structures* 4(1).
97. Tsukui, Makiko, Chen Lin, Kaiyan Ji, and Xiaoliang Lang. 2021. The true cost of trade among neighbors: The role of Japanese imports in waste generation in China. *Economic Systems Research.*
98. Merciai, Stefano, and Jannick Schmidt. 2018. Methodology for the construction of global multi-regional hybrid supply and use tables for the EXIOBASE v3 database. *Journal of Industrial Ecology* 22 (3): 516–531.
99. Lin, Chen. 2009. Hybrid input-output analysis of wastewater treatment and environmental impacts: A case study for the Tokyo Metropolis. *Ecological Economics* 68 (7): 2096–2105.
100. Xiao, Lishan, Chen Lin, and Shinichiro Nakamura. 2020. Tracing the consumption origins of wastewater and sludge for a Chinese city based on waste input-output analysis. *Environmental Science & Technology.*
101. Jambeck, Jenna R., Roland Geyer, Chris Wilcox, Theodore R. Siegler, Miriam Perryman, Anthony Andrady, Ramani Narayan, and Kara Lavender Law. 2015. Plastic waste inputs from land into the ocean. *Science* 347(6223): 768–771.
102. Lin, Chen, and Shinichiro Nakamura. 2019. Approaches to solving China's marine plastic pollution and CO_2 emission problems. *Economic Systems Research* 31 (2): 143–157.
103. Yasushi, Kondo, and Shinichiro Nakamura. 2004. Evaluating alternative life-cycle strategies for electrical appliances by the waste input-output model. *The International Journal of Life Cycle Assessment* 9: 236–246.
104. Lin, Chen. 2011. Identifying lowest-emission choices and environmental Pareto frontiers for wastewater treatment wastewater treatment input-output model based linear programming. *Journal of Industrial Ecology* 15 (3): 367–380.
105. Ohno, Hajime, Kazuyo Matsubae, Kenichi Nakajima, Yasushi Kondo, Shinichiro Nakamura, Yasuhiro Fukushima, and Tetsuya Nagasaka. 2017. Optimal recycling of steel scrap and alloying elements: Input-output based linear programming method with its application to end-of-life vehicles in Japan. *Environmental Science and Technology* 51 (22): 13086–13094.
106. Lin, Chen, Makiko Tsukui, and X. Lang. 2014. How do the waste treatment scenarios affect the environment impact of commute approaches in China? An application based on waste input-output table of China. In *The 21th international input-output conference 2014*, Kitakyusyu.
107. The Japan Containers and Packaging Recycling Association. 2007. Environmental load assessment of plastic containers and packaging recycling.
108. Figge, Frank, Andrea Thorpe, and Melissa Gutberlet. 2023. Definitions of the circular economy-circularity matters. *Ecological Economics* 208.

109. Kirchherr, Julian, Denise Reike, and Marko Hekkert. 2017. Conceptualizing the circular economy: An analysis of 114 definitions. *Resources, Conservation and Recycling* 127: 221–232.
110. Lüdeke-Freund, Florian, Stefan Gold, and Nancy M.P. Bocken. 2019. A review and typology of circular economy business model patterns. *Journal of Industrial Ecology* 23(1): 36–61.
111. Nakamura, Shinichiro, and Yasushi Kondo. 2002. Recycling, landfill consumption, and CO_2 emission: Analysis by waste input-output model. *Journal of Material Cycles and Waste Management* 4 (1): 2–11.
112. Aguilar-Hernandez, Glenn A., Carlos Pablo Sigüenza-Sanchez, Franco Donati, João FD Rodrigues, and Arnold Tukker. 2018. Assessing circularity interventions: A review of eeioa-based studies. *Journal of Economic Structures* 7(1): 1–24.
113. Donati, Franco, Glenn A. Aguilar-Hernandez, Carlos Pablo Sigüenza-Sánchez, Arjan de Koning, João FD Rodrigues, and Arnold Tukker. 2020. Modeling the circular economy in environmentally extended input-output tables: Methods, software and case study. *Resources, Conservation and Recycling* 152: 104508.
114. Rebitzer, Gerald. 2002. Integrating life cycle costing and life cycle assessment for managing costs and environmental impacts in supply chain analysis and LCA. In *Cost management in supply chains*, ed. by M. Seuring, S. Goldbach, 128–146. Physica-Verlag.
115. Hunkeler, David, Kerstin Lichtenvort, and Gerald Rebitzer, eds. 2008. *Environmental life cycle costing*. Pansacola, Florida: SETAC Press.
116. Nakamura, Shinichiro, and Yasushi Kondo. 2006. A waste input-output life-cycle cost analysis of the recycling of end-of-life electrical home appliances. *Ecological Economics* 57 (3): 494–506.
117. Andrews, Evan Stuart. 2009. *Guidelines for social life cycle assessment of products: social and socio-economic LCA guidelines complementing environmental LCA and Life Cycle Costing, contributing to the full assessment of goods and services within the context of sustainable development*. UNEP/Earthprint.
118. Hannouf, Marwa B., Alejandro Padilla-Rivera, Getachew Assefa, and Ian Gates. 2022. Methodological framework to find links between life cycle sustainability assessment categories and the UN Sustainable Development Goals based on literature.
119. van Haaster, Berthe, Andreas Ciroth, João. Fontes, Richard Wood, and Andrea Ramirez. 2017. Development of a methodological framework for social life-cycle assessment of novel technologies. *International Journal of Life Cycle Assessment* 22 (3): 423–440.
120. Norris, Catherine Benoît, Marzia Traverzo, Sabrina Neugebauer, Elisabeth Ekener, Thomas Schaubroeck, and Sara Russo Garrido. 2020. *Guidelines for social life cycle assessment of products and organizations 2020*. United Nations Environment Programme.
121. Kühnen, Michael, and Rüdiger. Hahn. 2017. Indicators in social life cycle assessment: A review of frameworks, theories, and empirical experience. *Journal of Industrial Ecology* 21 (6): 1547–1565.
122. Petti, Luigia, Monica Serreli, and Silvia Di Cesare. 2018. Systematic literature review in social life cycle assessment. *International Journal of Life Cycle Assessment* 23 (3): 422–431.
123. Hardadi, Gilang, and Massimo Pizzol. 2017. Extending the multiregional input-output framework to labor-related impacts: A proof of concept. *Journal of Industrial Ecology* 21 (6): 1536–1546.
124. Zamani, Bahareh, Gustav Sandin, Magdalena Svanström, and Greg M. Peters. 2018. Hotspot identification in the clothing industry using social life cycle assessment-opportunities and challenges of input-output modelling. *International Journal of Life Cycle Assessment* 23 (3): 536–546.
125. Costa, Daniele, Paula Quinteiro, Valdecy Pereira, and Ana Cláudia Dias. 2022. Social life cycle assessment based on input-output analysis of the Portuguese pulp and paper sector. *Journal of Cleaner Production* 330(November 2021).
126. Piñero, Pablo, Martin Bruckner, Hanspeter Wieland, Eva Pongrácz, and Stefan Giljum. 2019. The raw material basis of global value chains: Allocating environmental responsibility based on value generation. *Economic Systems Research* 31 (2): 206–227.

127. Marques, Alexandra, João. Rodrigues, Manfred Lenzen, and Tiago Domingos. 2012. Income-based environmental responsibility. *Ecological Economics* 84: 57–65.
128. Peters, Glen P., Jan C. Minx, Christopher L. Weber, and Ottmar Edenhofer. 2011. Growth in emission transfers via international trade from 1990 to 2008. *Proceedings of the National Academy of Sciences of the United States of America* 108 (21): 8903–8908.
129. Kander, Astrid, Magnus Jiborn, Daniel D. Moran, and Thomas O. Wiedmann. 2015. National greenhouse-gas accounting for effective climate policy on international trade. *Nature Climate Change* 5 (5): 431–435.
130. Oosterhaven, Jan. 2012. Adding supply-driven consumption makes the Ghosh model even more implausible. *Economic Systems Research* 24 (1): 101–111.
131. Oosterhaven, Jan. 1988. On the plausibility of the supply-driven input-output model. *Journal of Regional Science* 28(2).
132. Liang, Sai, Qu. Shen, Zeqi Zhu, Dabo Guan, and Xu. Ming. 2017. Income-based greenhouse gas emissions of nations. *Environmental Science and Technology* 51 (1): 346–355.
133. Lenzen, Manfred, Joy Murray, Fabian Sack, and Thomas Wiedmann. 2007. Shared producer and consumer responsibility - Theory and practice. *Ecological Economics* 61 (1): 27–42.
134. UNEP-SETAC. 2011. *Global Guidance principles for life cycle assessment databases-a basis for greener processes and products*. Paris.
135. Schrijvers, Dieuwertje, Philippe Loubet, and Guido Sonnemann. 2020. Archetypes of goal and scope definitions for consistent allocation in LCA. *Sustainability* 12: 2–3.
136. Schaubroeck, Thomas, Simon Schaubroeck, Reinout Heijungs, Alessandra Zamagni, Miguel Brandão, and Enrico Benetto. 2021. Attributional & consequential life cycle assessment: Definitions, conceptual characteristics and modelling restrictions. *Sustainability (Switzerland)* 13 (13): 1–47.
137. Finnveden, Göran., Michael Z. Hauschild, Tomas Ekvall, Jeroen Guinée, Reinout Heijungs, Stefanie Hellweg, Annette Koehler, David Pennington, and Sangwon Suh. 2009. Recent developments in life cycle assessment. *Journal of Environmental Management* 91 (1): 1–21.
138. Zamagni, Alessandra, Jeroen Guinée, Reinout Heijungs, Paolo Masoni, and Andrea Raggi. 2012. Lights and shadows in consequential LCA. *The International Journal of Life Cycle Assessment* 17 (7): 904–918.
139. Majeau-Bettez, Guillaume, Thomas Dandres, Stefan Pauliuk, Richard Wood, Edgar Hertwich, Réjean Samson, and Anders Hammer Strømman. 2018. Choice of allocations and constructs for attributional or consequential life cycle assessment and input-output analysis. *Journal of Industrial Ecology* 22(4): 656–670.

Chapter 6
Emissions and Mitigation

Abstract The chapter provides an overview of emissions and mitigation strategies for greenhouse gases (GHGs). It starts by discussing the state of global GHG emissions, including trends by gas, source, and country. The chapter then explores mitigation strategies related to electricity generation, covering power technologies and renewable energy sources. The mitigation potential of each technology is evaluated based on life cycle assessment (LCA), including process-based LCA (PLCA) and hybrid LCA (HLCA). The chapter also delves into the use of electric vehicles (EVs) as a means of reducing GHG emissions, comparing them to internal combustion vehicles (ICVs) and exploring their mitigation potential. It also discusses the implications of large-scale deployment of EVs on resource and power requirements. The chapter touches on issues related to intermittence and unreliability of renewable technologies, as well as materials implications of decarbonization. Overall, it provides a comprehensive overview of emissions and mitigation strategies, emphasizing the need for a comprehensive analysis of emissions across the entire supply chain, and highlighting the potential benefits and challenges associated with the deployment of EVs.

6.1 The State of GHG Emissions

6.1.1 The Global GHG Emission by Gases 1970–2020

Figure 6.1 shows global GHG emission distinguished by the major gases from 1970 to 2020. Over those 50 years, global GHG emission increased more than two-fold from 24.0 P (10^{15}) g to 51.7 Pg. Emission declined by around 2 Pg from 2019 to 2020, reflecting the slowing down of global economic activity that resulted from the spread of the COVID-19 pandemic. Of the GHG gases, CO_2 occupies around 76%, followed by CH_4 (around 17%), N_2O (around 5%), and F gases (around 2%). The share of CO_2 increased by around 10%, while the share of CH_4 and N_2O declined by a quarter. While F gases have the smallest amount, they increased at the fastest rate, by a factor of eight.

S. Nakamura, *A Practical Guide to Industrial Ecology by Input-Output Analysis*,
https://doi.org/10.1007/978-3-031-43684-0_6

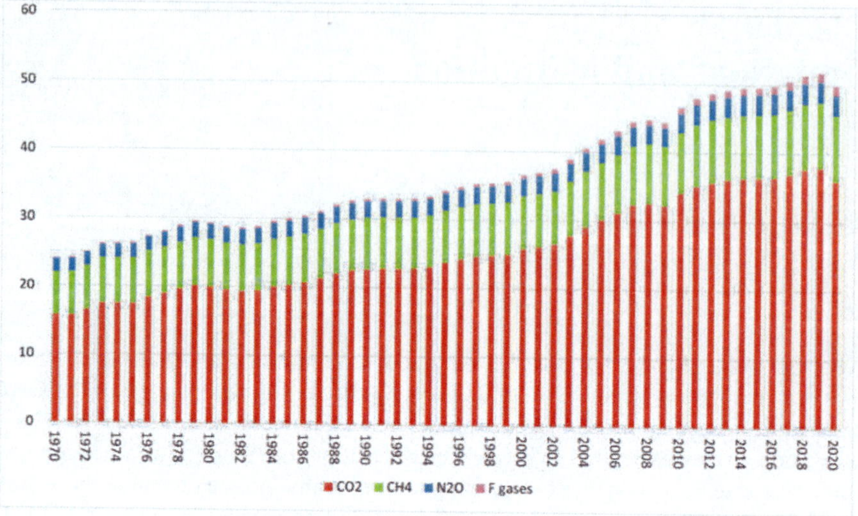

Fig. 6.1 Global GHG by gases Units: Pg-CO_2 equivalents. *Source* Data taken from [1]

6.1.2 *The Global GHG Emission by Sources*

Table 6.1 breaks down the GHG emission for 2019 by sources. The largest source of CO_2 is energy (76%), followed by land-use, land-use change and forestry (LULUCF) (15%), and nonenergy use (7%). The largest source of CH_4 is agriculture (41%), followed by energy (34%), nonenergy use (21%), and LULUCF (5%). The largest source of N_2O is agriculture (58%), followed by energy (23%), LULUCF (14%), and nonenergy use (6%). With all gases combined, energy is the largest source of emission (around 64%), followed by nonenergy use (13%) and LULUCF (12%). CO_2 emission from energy use (fossil-based CO_2) is the largest source of global GHG emission.

Table 6.1 Global GHG emission by gases and sources, 2019

Share in GHG	CO_2	CH_4	N_2O	F gases	All gases combined
	0.76	0.17	0.05	0.02	1.00
	Source share				
Energy	0.76	0.34	0.23	0.00	0.64
Agriculture		0.41	0.58	0.00	0.10
Nonenergy/industrial processes/waste	0.10	0.21	0.06	1.00	0.13
LULUCF	0.15	0.05	0.14	0.00	0.12
Sum	1.00	1.00	1.00	1.00	1.00

1.0 = 100%. *Source* [1, 2]. The combined emission of CH_4 and N_2O from forest and peat fires was allocated to each gas by half. The source of F gases was allocated 100% to industrial processes

6.1.3 *Major Emitting Countries*

Table 6.2 shows for each GHG gas its major emitting countries for 2020. Across the gases, China is the largest emitter accounting for one-third of global emissions for CO_2 and F gas, and 18% to 15% for CH_4 and N_2O, followed by the US and India. These three countries account for 50% of the global emission of CO_2 and F gas and one-third of CH_4 and N_2O. The GHG emission from international water and air transport (InTR) is as large as that of Japan and is larger than that of Germany.

Reflecting the difference in the source of emission among the gases (Table 6.1), the list of major emitting countries for CH_4 and N_2O markedly differs from that of CO_2 and F gases; Brazil, Pakistan, and Argentina are major emitters of CH_4 and N_2O, but not of CO_2 and F gases; Japan and Korea are major emitters of F gases because of their significance of electronics industries, which use those gases. In sharp contrast to the global average, for these two countries F gases have a larger share in the national GHG emission than N_2O. The same applies to Taiwan, as well.

6.1.4 *The GHG Emission by Countries and Sources*

Table 6.3 gives the sectoral breakdown of emissions by the nine major emitting countries. The largest share of emissions is attributed to the power industry. The only exception to this is the US, where transport has a share equal to that of the power industry. Reducing emissions form the power industry is vital to reducing fuel-based CO_2 emissions, the largest component of GHG.

6.2 Mitigating the Emissions Associated with Electricity Generation

We saw above that CO_2 of fuel origin occupies the single largest share of global GHG emissions and that the power sector is the largest fuel user. Mitigating GHG emissions requires a closer look at how electricity is produced.

6.2.1 *Electricity Production and Sources*

Table 6.4 shows the global electricity generation by source and country for the top ten countries, which account for 70% (19 PWh) of global generation (27 PWh). Globally, 62% (17 PWh) of electricity is produced by burning fossil fuels (of which 37% is natural gas and 58% is coal), followed by renewables (28%), and nuclear (10%).

Table 6.2 The top 10 emitting countries, 2020

GHG			CO_2			CH_4			N_2O			F gases		
	Share	Cshare		Share	Cshare		Share	Cshare		Share	Cshare		Share	Cshare
CH	0.29	0.29	CH	0.33	0.33	CH	0.18	0.18	CH	0.15	0.15	CH	0.34	0.34
US	0.11	0.40	US	0.13	0.45	IN	0.08	0.26	US	0.09	0.24	US	0.15	0.48
IN	0.07	0.47	IN	0.07	0.52	US	0.07	0.33	IN	0.09	0.33	IN	0.05	0.53
RU	0.04	0.52	RU	0.05	0.57	BR	0.06	0.39	BR	0.07	0.40	RU	0.04	0.57
BR	0.03	0.54	In	0.03	0.60	RU	0.04	0.43	ME	0.04	0.44	SA	0.04	0.61
JP	0.02	0.56	JP	0.03	0.63	ID	0.04	0.47	ID	0.03	0.47	TH	0.02	0.63
In	0.02	0.59	IR	0.02	0.64	PA	0.02	0.49	RU	0.03	0.49	JP	0.02	0.65
ID	0.02	0.61	DE	0.02	0.66	IR	0.02	0.51	AR	0.02	0.52	ME	0.01	0.66
IR	0.02	0.63	KO	0.02	0.68	NG	0.02	0.54	PA	0.02	0.54	CA	0.01	0.68
SA	0.02	0.64	SA	0.02	0.70	ME	0.02	0.55	AU	0.02	0.56	KO	0.01	0.69
DE	0.02	0.66	ID	0.02	0.71	AR	0.01	0.57	TU	0.02	0.57	UA	0.01	0.71

Source [1]. Cshare: cumulated share. AR: Argentina, AU: Brazil, In: InTR, ME: Mexico, NG: Nigeria, PA: Pakistan, RU: Russia, SA: Saudi Arabia, TH: Thailand, TU: Turkey, UA: UAE. See Table 6.3 for other countries

Table 6.3 Fossil CO_2 by sector and country: Major countries, 2021

	Total	Share of sources				
		Transport	Power industry	Other sectors	OI[a] combustion	Buildings
China (CH)	12466	0.08	0.44	0.16	0.27	0.05
United States (US)	4752	0.35	0.34	0.05	0.14	0.12
India (IN)	2649	0.11	0.48	0.09	0.25	0.07
Japan (JP)	1085	0.17	0.46	0.07	0.20	0.10
Iran (IR)	711	0.19	0.24	0.14	0.22	0.22
Germany (DE)	666	0.22	0.34	0.09	0.17	0.19
South Korea (KO)	627	0.17	0.47	0.10	0.17	0.09
Indonesia (ID)	603	0.22	0.35	0.11	0.27	0.04
Saudi Arabia (SA)	586	0.22	0.41	0.16	0.20	0.01
Sum	24144	0.15	0.42	0.12	0.23	0.08

[a]OI refers to "Other industrial". Units: Tg (Mt)-CO_2. *Source* EDGARv7.0, [3, 4]

The renewables consist mostly of hydroelectricity (54%), followed by wind (23%), solar (13%), and biomass/waste (8%). The power source differs widely among the largest power users. France is the only country where nuclear has the largest share (68%) and fossil fuels have the smallest share (9%). Hydroelectricity is Brazil's and Canada's largest power source, accounting for around 60% of total power production. Japan and Korea are distinguished by their larger share of solar than wind. For China and India, fossil fuels are mostly coal (95%), whereas, for the US, natural gas has a higher share (63%) than coal, and for Japan, their shares are almost equal (51% natural gas and 45 % coal) (36%) [6].

6.2.2 *Outline of Power Technology*

Table 6.5 summarizes the features of 12 major power technologies that are currently available. Of these, ten generate power based on the principle of electromagnetic induction involving the rotation of a turbine, which is further distinguished by the source that provides the rotation energy. Seven of the ten technologies are a thermodynamic process based on vapor power (the Rankine process) or gas turbine (the Brayton process), which uses thermal energy from various sources to rotate the turbine. Four are based on fossil sources, such as coal, oil, and gas, which are nonrenewable, while the remaining three are based on renewable sources. Biomass-fueled power stations obtain thermal energy by combusting biomass. On the other hand, geothermal power stations obtain thermal energy from hydrothermal resources (water and heat). Solar concentrating power stations obtain thermal energy by concentrating solar radiation using mirrors.

Three of the ten power technologies based on electromagnetic induction are not based on a thermodynamic system. They use the kinetic energy of a flow of fluid (water or wind). Solar photovoltaic converts solar radiation into electricity using

Table 6.4 Global electricity generation by source

	World	China	US	India	Russia	Japan	Brazil	Canada	Korea	Germany	France
	TWh/2021										
Generation	27295	8152	4165	1702	1109	955	663	626	587	557	530
	Composition of power source										
Nuclear	0.10	0.05	0.19	0.03	0.20	0.06	0.02	0.14	0.26	0.12	0.68
Fossil fuels	0.62	0.66	0.60	0.77	0.60	0.72	0.22	0.18	0.67	0.47	0.09
Renewables	0.29	0.29	0.21	0.20	0.20	0.22	0.76	0.68	0.08	0.42	0.23
	Composition of renewables										
Hydroelectricity	0.54	0.54	0.29	0.49	0.95	0.38	0.72	0.88	0.06	0.08	0.48
Geothermal	0.01	0.00	0.02	0.00	0.00	0.01	0.00	0.00	0.00	0.00	0.00
Tide and wave	0.00	0.00	0.00	0.00	0.00	0.00	0.00	0.00	0.01	0.00	0.00
Solar	0.13	0.14	0.18	0.20	0.01	0.43	0.03	0.01	0.51	0.21	0.12
Wind	0.23	0.26	0.43	0.23	0.02	0.04	0.14	0.08	0.07	0.49	0.30
Biomass and waste	0.08	0.06	0.08	0.09	0.02	0.13	0.10	0.02	0.35	0.21	0.10

Source [5]

Table 6.5 Characteristics of major power technologies

Power plant type	Electromagnetic induction	Nonrenewable source	Renewable source	Efficiency %	Thermodynamic cycle[a]
Coal-fueled	Yes	Yes		33[c]	Rankine
Oil-fueled	Yes	Yes		33[c]	Rankine
Naturalgas-fueled	Yes	Yes		45[c]	Brayton
Nuclear-fueled	Yes	Yes		33[c]	Rankine
Biomass-fueled	Yes		Yes	20–25[d]	Rankine
Geothermal	Yes		Yes	12[d]	Rankine
Solar-concentrating	Yes		Yes	15–20[e]	Rankine
Hydroelectric	Yes		Yes	90–95[e]	None
Wind	Yes		Yes	59.3 (maximum)	None
Current, tides, and waves	Yes		Yes	80[d]	None
Solar-photovoltaic			Yes	25[d]	None
Fuel cells		Yes[b]	Yes[b]	60[d]	None

[a] A thermodynamic system is a sequence of processes, involving heat and work, that begins and ends at the sate state ([7]). The Rankine cycle consists of a boiler, turbine, condenser, and pump. It is widely used in power plants. The Brayton cycle consists of a compressor, heat exchanger, and turbine. It is used for gas turbine engines and some jet engines. See [7] for further details [b] Depending on how the H_2 is obtained. Currently, H_2 is mostly produced from CH_4, which is nonrenewable. H_2 can be produced without using any nonrenewable resources by water electrolysis with electricity obtained from renewable resources

[c] [8]

[d] [9]

[e] [10] Table 13.1

solid-state devices. A fuel cell is an electrochemical device in which fuel (typically H_2) and oxidizer (normally O_2 from air) undergo a chemical reaction, providing electrical current to an external circuit and producing products (such as H_2O) [7].

6.2.3 *Renewable Energy*

To familiarize readers with the basic features of renewable power technologies, we give some details of them (biomass-fueled ones are excluded because of their large variations [11]). We mostly refer to [10].

6.2.3.1 Hydroelectric Power

Hydroelectric generation utilizes the potential energy difference between elevations. Let h (m) represent the elevation difference, m the mass of water at height h, and g the gravitational acceleration. The potential energy E (J) of water at height h is given by $E = mgh$. For a reservoir with water volume Ω (m^3), water density ρ (kg/m^3), and effective height h (m) (the difference in water level), the potential energy of the reservoir is $E = \rho\Omega gh$. Considering the water flow Q (m^3/s) from the reservoir and the efficiency η (the ratio of output electrical power to kinetic power applied to the turbine), the available power for hydroelectric generation P (kW) is given by ([10], p. 9)

$$P = \eta\rho Qgh \tag{6.1}$$

Hydroelectric power plants typically have high efficiency, ranging from 90% to 95%. For instance, the Hoover Dam in the US, standing 221 m tall, features 17 electric power generators capable of producing a total of 2.2 GW of power, with an efficiency of 90% [10].

Negative and unfavorable aspects of hydropower include spotentially catastrophic dam collapses, displacement of the population (the Three Gorges Dam project in China displaced some 1.3 million people living in more than 1,500 cities, towns, and villages along the Yangtze River), destruction of cultures, and change in ecosystems. An environmentally harmful effect of hydropower dams is also the production of methane from biomass accumulated at the bottom of reservoirs. In the absence of a water reservoir formed by the dam, the biomass decomposes aerobically into CO_2 but when covered by water it decomposes anaerobically, producing CH_4 (see Sect. 2.2.1.2).

6.2.3.2 Tidal Power

Tidal power harnesses the potential energy difference between different elevations, similar to hydro power. However, it utilizes the ebb and flow of tides, influenced by factors such as local winds, water depth, and ocean floor topography (see [10] for more details).

The potential energy stored in a tidal barrage, with an area A and tidal height h, is given by $E = mgh = \rho h^2 Ag/2$ ([10], p. 39), where ρ represents the average density of seawater (1025 kg/m^3) and the division by two accounts for the occurrence of two high tides and two low tides each day. The average power available from the barrage in one day, denoted as P (W), considering the tidal period per day T (s/d) and efficiency η, is given by ([10], p. 39)

$$P = \eta \frac{\rho h^2 Ag}{2T}. \tag{6.2}$$

Turbine efficiencies can exceed 90%, with typical ranges between 70% and 84% [12]. The efficiency of a tidal power plant is approximately 80%.

Tidal power generation using water-filled tidal barrages has a comparable environmental impact to hydroelectric dam power plants. High capital costs are a significant concern for tidal energy [10].

The two largest tidal power stations are La Rance in France and Sihwa Lake in South Korea, with similar net electrical energy outputs of 540 GWh and 553 GWh, respectively. La Rance, built in 1966, has an area A of 22 km^2 and tidal height h of 8.5 m. Sihwa Lake, constructed in 2015, has an area A of 56 km^2 and tidal height h of 5.6 m. For a recent review, please refer to [12].

6.2.3.3 Wind Power

Wind power utilizes the kinetic energy of wind, given by $E = \frac{1}{2}mv^2$ (J = Nm = m^2 kg/s^2), where m (kg) represents the mass of the wind and v (m/s) is the wind speed. The power output (W = J/s = m^2 kg/s^3) can be expressed as $P = \frac{1}{2}v^2\frac{dm}{dt}$ since v is assumed constant. By denoting the air density as ρ (kg/m^3) and the wind volume as V (m^3), we have $m = \rho V$. Considering a wind turbine with rotating blades of length r capturing an area A (m^2), and adjusting with the efficiency factor η, we obtain $\frac{dm}{dt} = \rho A v$, leading to

$$P = \frac{1}{2}\eta\rho A v^3 = \frac{1}{2}\eta\rho\pi r^2 v^3. \tag{6.3}$$

The efficiency of a wind turbine is limited by the power coefficient, which refers to the ratio of the power generated by a turbine to the power of the incoming wind. Betz's law states that the maximum power coefficient is 0.593 ([10], Chap. 3, see [13] for a proof). Consequently, the maximum efficiency of a wind turbine is approximately 0.59.

Equation (6.3) implies that more power can be obtained by increasing r. As of 2022, the world's biggest wind turbine online was with $r = 220$ m and a capacity of 14 MW [14]. While a higher v could generate more power, typically, winds above 25 m/s cannot be used to generate power because of concerns about the damage to the turbine ([10], p. 15). The primary concern with wind power is that it is intermittent and not always available when electricity is needed.

6.2.3.4 Concentrated Solar Power Generation (CSP)

CSP concentrates the solar radiation by using heliostats, mirrors equipped with a two-axis tracking system to track the sun's path, on a solar power tower filled with salt and heats it to over 500 °C; superheated steam from the salt mixed with water drives a turbine generator and produces electricity ([15], p. 32). The efficiency of CSP is around 15–20%, which is the product of the efficiencies of each step of the heat-to-electricity conversion processes, including the collection of sunlight by mirrors and the steam turbine [16].

As of 2022, the world's largest CSP plant is Noor Ouarzazate Solar Complex in Morocco featuring 537,000 parabolic mirrors with an installed capacity of 0.58 GW ([10], Table 13.1). CSP requires water for its operation mainly to convert the exhaust steam into pure water by cooling (CSP is a Rankine system) and to clean the mirrors [15, 17]. The range of water consumption for operation activities is 2.1–3.8 L/kWh, while it is 0.5–4.4 L/kWh for coal power plants [18]. While CSP is most productive in arid areas, lack of water can become an issue for its sustainable operation.

6.2.3.5 Solar Photovoltaics

Solar photovoltaics (PV for short) are solid-state devices that directly convert solar radiation into electricity. They are made of N- and P-type silicon (Si)-based semiconductors joined together. An N-type semiconductor is obtained by adding a material with five valence electrons (electrons in the outermost shell of an atom), like antimony (Sb) or phosphorous (P), to a pure Si crystal. On the other hand, adding material with three valence electrons, like gallium (Ga) or boron (B), to a pure Si crystal gives a P-type semiconductor. This construct (the P-N junction it creates) makes it possible to convert solar radiation into electricity by exploiting the photoelectric effect. The conversion efficiency of PV, the percentage of incident solar energy converted to electricity, is currently around 25%. See [10, 19] for further details of the physics of PV. As of 2022, the world's largest PV plant was Bhadla Solar Park in India, with an installed capacity of 2.25 GW and spanning an area of 57 km^2.

6.2.3.6 Geothermal Power

Geothermal power uses the heat generated in the inner layers of Earth and transferred to the surface. The diffusion of geothermal energy is unevenly distributed on Earth. Regions of high geothermal heat resource potential are located along the plate boundaries, which are connected to form the so-called Ring of Fire, a path along the Pacific Ocean known for frequent earthquakes and active volcanoes [10]. Please see [10, 20] for further details of geothermal energy.

Table 6.6 shows the top ten countries with the largest geothermal resources and their utilization for power generation. US, Indonesia, and Japan have the largest geothermal resources. Despite its large geothermal potential, Japan has the lowest utilization rate and ranks the lowest in installed power capacity. Considering that geothermal is the only abundant energy Japan is endowed with, this fact may appear surprising. The background factor unique to Japan is opposition from the hot spring industry, which fears that their industry could be compromised with the construction of geothermal plants. As of 2022, the largest geothermal power plant is the Geysers geothermal complex in northern California, made up of 18 power plants, with capacity of 900 MW, spreading across 117 km^2.

The efficiency of a geothermal plant is defined as the ratio of net electricity generation to the total mass of fluid (kg/s) multiplied by the average enthalpy (kJ/kg) [22]. Based on 94 geothermal power plants worldwide, [22] obtains an average efficiency of 12%, which is lower than conventional thermal power plants.

Table 6.6 Countries with geothermal resources

Resources (GW)		Power capacity (GW)		Utilization rate %
US	30.0	US	3.70	12.3
Indonesia	27.8	Indonesia	2.30	8.3
Japan	23.5	The Philippines	1.90	31.7
Kenya	7.0	Turkey	1.55	
The Philippines	6.0	Kenya	1.20	17.1
Mexico	6.0	New Zealand	1.06	29.0
Iceland	5.8	Mexico	1.00	16.7
New Zealand	3.7	Italy	0.92	28.0
Italy	3.3	Iceland	0.76	13.1
Peru	3.0	Japan	0.55	2.3

Source [21]

Associated with geothermal power are several environmental impacts: emission of harmful gases brought to the surface, such as CO_2, H_2S, SO_2, and CH_4, noise pollution (mainly during drilling operations), water use and quality, land use (a geothermal facility uses 404 m^2 of land/GWh compared to a coal facility that uses 3,632 m^2/GWh and a wind farm that uses 1,335 m^2/GWh), and the impact on natural phenomena, wildlife, and vegetation ([20], p. 757).

6.2.3.7 Hydrogen Fuel Cells

This section mainly refers to [10]. A fuel cell is a device for the electrochemical conversion of chemical energy into electricity without any intermediate steps, which operate as long as there is a continuous supply of fuel (most commonly gaseous H_2) and oxidant (O_2 from the atmosphere) to electrodes separated by an electrolyte (or a membrane). Fuel reacts on the anode and is oxidized in the reaction while generating electrons

$$2H_2 \rightarrow 4H^+ + 4e^-, \tag{6.4}$$

which then flow to the other electrode through an external circuit and loads, to produce power. The hydrogen ions, H^-, travel through the electrolyte to the cathode where they react with oxygen and electrons to form water

$$O_2 + 4H^+ + 4e^- \rightarrow 2H_2O \tag{6.5}$$

which is the only product of reaction and energy. Currently, platinum (Pt) is the catalyst material used for both anode and cathode, with significant implications for the cost of fuel cells. The efficiency of a fuel cell is around 60% [23].

The biggest challenge of a H_2 fuel cell is that H_2 is not freely available in nature and must first be extracted from the bond state in other substances, such as water or fossil fuels, which requires energy; energy must first be spent to produce hydrogen only to gain less energy back when H_2 is used to generate electricity ([10], p. 90). H_2 is mainly obtained from fossil fuels (mostly CH_4, see (2.7)). H_2 can also be obtained as a by-product of NaOH production, which is based on the reaction

$$2NaCl + 2H_2O \rightarrow 2NaOH + Cl_2 + H_2 \tag{6.6}$$

6.2.4 *LCA of Electricity Generation*

After looking at alternative electricity generation methods, we now turn to their environmental impacts. We first look at major review articles on the LCA of electricity generation to obtain a perspective. Subsequently, we will discuss in detail the application of IO to the LCA of electricity generation for a selection of studies.

6.2.4.1 Literature Reviews on the LCA of Electricity Generation

CSP, PV, Wind Power, Hydropower, and Geothermal Power

Reference [11] reviewed approximately 50 papers related to more than 100 different case studies regarding solar energy (CSP and PV), wind power, hydropower, and geothermal power. Besides GWP in CO_2e, acidity potential (AP) in SO_2e, eutrophication potential (EP) in PO_4^{3-}, photochemical ozone creation potential (POCP) in C_2H_4e, and cumulative energy demand (CED) in MJ were also considered. A harmonization process was applied to reduce data variability and align methodological inconsistencies.

Table 6.7 shows the median values of their harmonized results. Wind power and hydro overall have the least environmental impacts, followed by CSP. While the impacts of PV and geothermal are similar to CSP in terms of GWP, their impacts are very high in terms of EP and AP.

Table 6.7 The environmental impacts of renewable energy technologies

	GWP	EP	AP	CED	POCP
	g CO_2e	mg PO_4^{3-}	mg SO_2e	MJ	mg C_2H_4e
Wind	9.4	4.9	48.9	0.13	4.6
Hydro	11.6	4.8	12.8	0.16	1.5
CSP	30.9	6.8	91.2	0.44	16.4
Geothermal	33.6	46.3	356.6	0.52	22.1
PV	29.2	22.4	293.3	0.61	45.5

Units: Per kWh. *Source* [11] Fig. 6.2, the median values

Compared with conventional power systems based on fossil fuel, renewable energy technologies show significant environmental advantages; a combined cycle natural gas plant has a mean emission of 350–410 g CO_2e/kWh, and a hard coal plant with direct combustion has an emission range of 750–1050 g CO_2e/kWh [24], while all the analyzed renewable technologies are characterized by values lower than 100 g CO_2e/kWh; while an old hard coal plant with direct combustion has an AP range of 2–7 g SO_2e/kWh [24], all the analyzed technologies are characterized by values lower than 1 g SO_2e/kWh.

Biomass and Nuclear

Reference [11] did not consider bio energies because of the great number of existing typologies and technologies that make it impossible to obtain a significant quantity of data for each one of these typologies. Reviewing 33 LCA publications, including 167 case studies of all main electricity generation technologies, including biomass and nuclear, [24] finds the emission factor for biomass power widely varied in the range of 8.5–130 kg-CO_2e per MWh. Pointing to the fact that use of biomass resources for energy purposes will cause competition between energy generation and other uses (e.g., feeding, bedding, plowing back to fields, etc.), [24] observe that neglecting any upstream impacts associated with the biomass (a zero burden assumption) is unlikely to be correct, in particular, for a long-term perspective; the system boundary definition has a large influence on the results because biomass provision accounts for on average 71% of the emissions from biomass systems. Reference [24] also reports for nuclear power emission factor in the range of 3–35 kg-CO_2e per MWh, while the range is 2–20 for hydro, 3–41 for wind power, and 13–190 for PV. As far as GHG emission is concerned, nuclear performs as well as hydro and wind and better than PV.

Contribution of Life Cycle Stages

Another noteworthy result of [24] is the difference among power technologies in their contribution of each life cycle stage to the total life cycle emission; for fossil fuel technologies, fuel combustion represented the majority of total emissions, fuel provision represented the largest contribution for biomass technologies and nuclear power; infrastructures provided the highest impact for renewables.

6.2.4.2 Hybrid LCA of Power Technologies

The above LCA studies are mostly based on PLCA. We now look in detail at some HLCA studies.

Geothermal Power in Japan [25]

While Japan has huge geothermal resources, the country lags behind other countries in terms of utilization for power generation due, among other things, to issues of social acceptance in the local community. With this backdrop, [25] investigated the employment effect over the life cycle of geothermal power generation, 30 years, based on the method of IO-based hybrid analysis (IOH).

The original Japanese IO table with 401 sectors was extended to include five new sectors related specifically to geothermal power generation:

1. Geothermal resource survey,
2. Production well construction,
3. Injection well construction,
4. Steam transport pipe construction,
5. Geothermal power.

These sectors were selected because the specifications of a geothermal power plant are significantly dependent on the natural conditions of the target site, and their investment costs significantly affect the total cost of a geothermal power plant. These five sectors were disaggregated from the "Construction of electric power facilities" and the "Electricity" sectors of the original IO table. The input coefficients of the first four sectors related to the construction of a geothermal power plant were determined based on the 50 MW model plant (assumed in a study commissioned by the Japanese Ministry of Environment), reports from the Japan Geothermal Association, and interviewing business operators, while the coefficients of geothermal power were obtained based on the financial statements of electric power companies, interviews with operators, and reports from the Ministry of the Environment (detailed in [25, 26]).

We now turn to the model. Denote by $\boldsymbol{A}'$ the extended input coefficients matrix of order $(401+5)\times(401+5)$, by $\boldsymbol{\ell}'$ the extended vector of labor input coefficients, and by $\boldsymbol{x}'$ the extended vector of 406 products. Five life cycle stages can be identified for a geothermal power plant, consisting of

1. Resource survey (s),
2. Manufacturing (m),
3. Construction (c),
4. Operation & maintenance (o), and
5. Disposal (d).

Denote by $\boldsymbol{y}_i'$, $i \in \{s, m, c, o, d\}$ an 406×1 vector of inputs at each stage of the 50 MW plant (Table 6.8). The employment impacts associated with the life stages of the 50 MW geothermal power plant, emp', are then given by

$$emp' = \boldsymbol{\ell}'\boldsymbol{L}' \sum_{i\in\{s,m,c,o,d\}} \boldsymbol{y}_i', \tag{6.7}$$

where $\boldsymbol{L}' = (\boldsymbol{I} - \boldsymbol{A}')^{-1}$. The sum of $\boldsymbol{y}_i'$s corresponds to the functional unit. Equation (6.7) refers to the employment induced domestically and abroad. Implicit in (6.7) is the assumption of domestic technology under which the imported goods have the

Table 6.8 Final demand by geothermal plant life cycle stages over 30 years

	y_R Resource survey	y_M Manufactur.	y_C Constr.	y_O Operation and maintenance	y_D Disposal
Turbines	0	1328	0	0	0
Refrigerators and air conditioning apparatus	0	1936	0	0	0
Other general industrial machinery and equipment	0	1133	0	0	0
Rotating electrical equipment	0	965	0	0	0
Transformers and reactors	0	1634	0	0	0
Other industrial electrical equipment	0	1516	0	0	0
Construction of electric power facilities	0	0	3503	0	1458
Geothermal resource survey	1699	0	0	0	0
Production well construction	1529	0	3439	4204	0
Injection well construction	510	0	3057	3312	0
Steam transport pipe construction	0	0	5844	631	0
Geothermal power	0	0	0	137,598	0
Real estate agencies and managers	0	0	1062	0	0
Total	3737	8512	16,906	145,743	1458

Source [25]. Units: Million Japanese yen at 2005 prices

same employment and environmental intensity as goods produced domestically in Japan. The domestic employment impacts, $\boldsymbol{emp}'^d$, were obtained by

$$\boldsymbol{emp}'^d = \boldsymbol{\ell}'(\boldsymbol{I} - \hat{\boldsymbol{\mu}}'^{\mathbf{d}}\boldsymbol{A}') \sum_{i \in \{s,m,c,o,d\}} \boldsymbol{y}_i', \tag{6.8}$$

where $\boldsymbol{\mu}'^d$ is the extended vector of the share of domestic products in total supply (recall from Sect. 4.3.2 that this is a competitive import model with exogenous foreign trade). It was assumed that $\boldsymbol{y}_i$s were 100% domestically produced. The overseas impacts are obtained by $\boldsymbol{emp}' - \boldsymbol{emp}'^d$.

The results indicate that 0.78 to 1.00 person-year per GWh would be required, of which 66% is attributed to the operation phase, 18% to the construction phase, 10% to the manufacturing phase, and 4% to the resource survey phase. The domestic employment accounted for 86% of the total employment. Using the same data and model but replacing $\boldsymbol{\ell}'$ with the vector of extended GHG emission coefficients, $\boldsymbol{f}'$, [26] found the life cycle GHG emission of 31-g CO_2e/kWh, which is close to the value in Table 6.7, of which 59% is attributed to the operation phase, 31% to the construction phase, and 9% to the manufacturing phase.

Renewable Power Technologies in Australia

Reference [27] investigated how effective the large-scale implementation of renewable energy in Australia is likely to be in reducing GHG emissions, including total emissions during the whole life cycle of different technologies.

The following seven renewable technologies were considered

1. Hydro power, run-of-river (diversion): a run-of-river facility channels a portion of a river through a canal and/or a penstock to utilize the natural decline of the river bed elevation to produce energy (US Department of Energy),
2. Hydro power, reservoir (impoundment),
3. Wind power-onshore,
4. Wind power-offshore,
5. Solar photovoltaic (PV),
6. Concentrated solar power (CSP), and
7. Geothermal power.

The Australian IO table was extended column and row-wise to accommodate alternative power technologies analogous to [28, 29], resulting in an IO table with 215 sectors. The process information obtained from external sources (Ecoinvent 3.1 database and data on CSP from an LCA study [30] were then implanted into the extended IO table, with the physical information converted into monetary values based on price data. The whole life cycle of power generation, from raw material mining to decommissioning, was consistently considered, with all inputs expressed as annual averages. This includes transportation and energy requirements during all life cycle stages, construction, operation, and maintenance of plants, and disposal at the end of life. Recycling was not considered.

The carbon footprint (CF) calculation of technologies is based on the standard model identical to the one used above. Denoting by $\boldsymbol{y_j}$ an $n \times 1$ vector of zeros except the jth element, which is y_j, the total carbon footprint of product j, CF_j, is given by

$$CF_j = \sum_i f_i l_{ij} y_j = \boldsymbol{f L y_j} \tag{6.9}$$

where $f_i l_{ij} y_j$ refers to the fraction of CF_j originating from input i.

Concerning how international trade is accounted for, the model of [25] is based on the assumption of domestic technology. Instead of resorting to this strong assumption, a multi-regional input-output (MRIO) framework was used, with the Australian and the Rest-of-the-World (RoW) IO tables linked by tables for imports and exports taken from the Eora database. Denoting by $\boldsymbol{A}^{r,s}$ the matrix of input coefficients from region r to region s and by $\boldsymbol{f}^r$ the intervention vector, with $r, s \in \{\text{Au, RoW}\}$, the impacts taking into account international trade can be given by

$$\begin{pmatrix} \boldsymbol{f}^{\text{AU}} \\ \boldsymbol{f}^{\text{RoW}} \end{pmatrix} \begin{pmatrix} \boldsymbol{I} - \boldsymbol{A}^{\text{AU,AU}} & -\boldsymbol{A}^{\text{AU,RoW}} \\ -\boldsymbol{A}^{\text{RoW,AU}} & \boldsymbol{I} - \boldsymbol{A}^{\text{RoW,RoW}} \end{pmatrix}^{-1} \begin{pmatrix} \boldsymbol{y}_j \\ \boldsymbol{0} \end{pmatrix} \tag{6.10}$$

The MRIO data on the matrices involving RoW was taken from the Eora database, with all the countries excluding Australia aggregated to one RoW region.

The results indicate geothermal electricity has the highest CF intensity (92.2 g CO_2e/kWh), exceeding the results for all other technologies. Run-off-river hydropower (HP) lies on the other end of the range with a CF intensity of 37.2 g $CO2_e$/kWh, while, with 48-g CO_2/kWh, reservoir hydropower performs almost the same as wind power (43–45 g-CO_2e/kWh), followed by PV (73 g-CO_2e/kWh) and CSP (79 g-CO_2e/kWh). Compared to [11], the findings are at the higher end of reported CF intensity ranges and above median values, which can be attributed to Australia's high carbon intensity of production, in particular, electricity production, which is dominated by coal and gas. Generally, the highest contribution to CF is due to electricity inputs ranging from 32% for CSP to 62% for run-off-river and reservoir HP, implying that decarbonization of the overall electricity mix can significantly reduce the CFs of renewable technologies.

The environmental impact potential of imported goods adds up to one-fifth of the total CFs. Geothermal has the lowest relative impact from imports (7.3%), and reservoir HP has the highest (20%). Consistently, imported petroleum, chemical, and nonmetallic mineral products contribute the most to the imports fraction of the CFs (2.8–6.4%).

Integrated LCA of Electricity Supply Scenarios

Generally, renewable technologies require higher initial investments in infrastructure than fossil-based power systems [24]. To assess the trade-offs of increased up-front emissions and reduced operational emissions, [31] conducted the first global, integrated LCA of long-term, wide-scale implementation of electricity generation from renewable sources (PV, CSP, wind, and hydropower) and of carbon dioxide capture and storage (CCS) for fossil power generation based on coal and natural gas.[1] They analyzed the environmental impacts (GHG, PM, ecotoxicity, eutrophication,

[1] CCS is a set of integrated technologies that can capture a large amount of CO_2 from large industrial CO_2 emission sources such as power plants, oil refineries, cement plants, iron & steel plants, and then transport and store the captured CO_2 to the storage sites under pressure for storage [32] LCAs of CCS generally show that while CCS can reduce direct CO_2 emissions from combustion-based electricity generation, the values of nonclimate impact categories increase [32–35].

land occupation) and resource (iron, cement, copper, and aluminum) requirements of the wide-scale global deployment of different low-carbon electricity generation technologies as foreseen in one prominent climate-change mitigation scenario, the International Energy Agency's (IEA) BLUE Map scenario, and compared it with the IEA's Baseline scenario, under which the combined share of solar, wind, and hydropower increases from 16.5% of total electricity generation in 2010 to 39% in 2050 [36].

The inventory data consist of three groups, each corresponding to distinct systems (physical foreground, physical background, and monetary background) and obtained from different sources. The data on the physical foreground system, f, which refers to the energy technology product systems, were obtained based on industry road maps, technology learning curves, and expert opinion. The data on the life cycle inventory of the physical background system (p) was obtained from the Ecoinvent database after adjusting regional differences in electricity mixes. Finally, the MRIO model referring to the background system in monetary units, n, was taken from the EXIOBASE and was used after adjusting to match the nine regions of the IEA Energy Technology Perspectives model. There is no linkage between the physical and economic databases. Please see [37] for further details of the methodology.

The footprint calculation is similar to (6.9), with $\boldsymbol{A}$ consisting of f, p, and n systems, embedded in the MRIO framework. For ease of illustration, we consider a simple case of a two-region model, consisting of region 1 and 2, and denote by $\boldsymbol{A}^{ii}$ the flow within region i, and by $\boldsymbol{A}^{ji}$ the flow from region j into region i. The matrix $\boldsymbol{A}$ involving the three systems and two regions is then given by

$$\boldsymbol{A} = \begin{pmatrix} \boldsymbol{A}^{11} & \boldsymbol{A}^{12} \\ \boldsymbol{A}^{21} & \boldsymbol{A}^{22} \end{pmatrix} = \begin{pmatrix} \begin{pmatrix} \boldsymbol{A}^{11}_{ff} & \boldsymbol{A}^{11}_{fp} & \boldsymbol{A}^{11}_{fn} \\ \boldsymbol{A}^{11}_{pf} & \boldsymbol{A}^{11}_{pp} & \boldsymbol{0} \\ \boldsymbol{A}^{11}_{nf} & \boldsymbol{0} & \boldsymbol{A}^{11}_{nn} \end{pmatrix} & \begin{pmatrix} \boldsymbol{0} & \boldsymbol{0} & \boldsymbol{0} \\ \boldsymbol{0} & \boldsymbol{0} & \boldsymbol{0} \\ \boldsymbol{0} & \boldsymbol{0} & \boldsymbol{A}^{12}_{nn} \end{pmatrix} \\ \begin{pmatrix} \boldsymbol{0} & \boldsymbol{0} & \boldsymbol{0} \\ \boldsymbol{0} & \boldsymbol{0} & \boldsymbol{0} \\ \boldsymbol{0} & \boldsymbol{0} & \boldsymbol{A}^{21}_{nn} \end{pmatrix} & \begin{pmatrix} \boldsymbol{A}^{22}_{ff} & \boldsymbol{A}^{22}_{fp} & \boldsymbol{A}^{22}_{fn} \\ \boldsymbol{A}^{22}_{pf} & \boldsymbol{A}^{22}_{pp} & \boldsymbol{0} \\ \boldsymbol{A}^{22}_{nf} & \boldsymbol{0} & \boldsymbol{A}^{22}_{nn} \end{pmatrix} \end{pmatrix} \quad (6.11)$$

where $\boldsymbol{A}^{11}_{ff}$ refers to the foreground technology matrix in region 1, and $\boldsymbol{A}^{22}_{fn}$ to the input in economic sector in region 2 of the outputs from foreground processes in region 2, and $\boldsymbol{A}^{21}_{nn}$ to the IO matrix of inputs from region 2 to region 1.[2] The intervention matrix corresponding to (6.11) is given by

[2] I owe Thomas Gibon for this exposition.

$$F = \left(F^1 \; F^2\right) \tag{6.12}$$

with

$$F^i = \left(F^i_f \; F^i_p \; F^i_n\right), \quad i = 1, 2.$$

Since the scenarios considered involve the gradual diffusion and replacement of power technologies through capital formation over a span of 50 years, the dynamic IO model with different age capital cohorts ([38], discussed in Sect. 4.4.3.8) was used to assess the evolution of A over time.

The results show that, under the Baseline scenario, emissions of air and water pollutants more than double, whereas the low-carbon technologies introduced in the BLUE Map scenario allow a doubling of electricity supply while stabilizing or reducing pollution. Since renewable technologies tend to have higher material requirements per kWh than fossil-based technologies (per unit generation, PV systems require 11–40 times more copper, and wind power plants require 6–14 times more iron), a substantial increase in material consumption, especially copper, is expected under the BLUE Map scenario. Still, the pollution caused by the higher material requirements of these technologies is found to be small compared with the direct emissions of fossil fuel-fired power plants.

6.2.4.3 Issues of Intermittence and Unreliability

Reliable and Intermittent Power Sources

The LCA studies we discussed above suggest renewable technologies' effectiveness in mitigating GHG and other environmental emissions, particularly wind power. It is important to note that renewable energy sources can be divided into two groups, depending on their reliability to supply power where it is needed at the time of demand [10].

1. The relatively reliable sources whose output can be planned: hydroelectric, biomass, and geothermal energy generation. In Norway, 90% of the electricity demand is covered by hydroelectric power generation [39]. In Iceland, renewable energy provided almost 100% of electricity production, with about 73% coming from hydropower and 27% from geothermal power [40].
2. The intermittent and unreliable sources whose output cannot be planned because of their dependence on weather conditions: solar power generation and wind power (and tidal power to a lesser extent).

In order to provide a stable power supply, the output of each power plant needs to be balanced with ever-changing demand to keep the frequency constant; if this balance is lost, the frequency will fluctuate, affecting the operation of electrical equipment on the consumer side; in the worst case, a large power outage could occur [10]. For power technologies based on intermittent power sources, the actual amount of power to be introduced is limited due to the impact on the power system.

The Effect of Near-100% Intermittent Renewable Scenarios on the Security of Power Supply

Observing that near-100% intermittent renewable scenarios are increasingly proposed as viable power system solutions with underlying assumptions and real-world implications often remaining unexplained and unaccounted for, [41] investigated the effect of such scenarios on the security of power supply. They find that 100% intermittent renewable power systems could be unable to meet basic peak demand and ancillary services requirements, proving inoperable under current system regulations, and they call for modelers and model users to interpret and communicate results carefully to avoid misguiding public opinion and decision-makers. Reference [42] provides the economic evaluation of intermittent renewables and a range of balancing solutions within the western US context. Reference [43] proposes a network of high-voltage direct-current (HVDC) transmission lines as the key enabling technology for the large geographic domains favored for wind and solar power, noting that while electrical storage can also reduce the intermittency of wind and solar, the cost is higher than HVDC transmission lines.

We fully recognize the significance of thoroughly addressing the issue of intermittency when exploring solutions to reduce the environmental consequences of power generation. However, given that this book does not primarily focus on intricate technological discussions, we will refrain from delving into this topic unless it becomes essential.

6.2.4.4 Resource Implications of Renewable Technologies

An energy system powered by renewable technologies generally requires more minerals than its fossil fuel-based counterparts. In their integrated LCA study of electricity supply scenarios involving the ambitious deployment of renewable technologies, [31] addressed this issue by considering resource requirements, besides environmental impacts, and found a significant increase in the demand for copper required for the deployment. The resources they considered were limited to a small number of bulk materials consisting of Al, Cu, Fe, and cement.

Reference [44] investigated the global metal requirements of various nonfossil power technologies, including solar, wind, biomass, hydro, and nuclear, for metals, Al, Cu, Fe, Ni, Zn, Sn, Mo, and Ag. The study revealed that nonfossil power technologies require significantly larger amounts of metals compared to the 2007 mix. When compared to current levels of mine production, the implementation of nonfossil power technologies leads to notable increases in the demand for Al (up to 15%), Ni (up to 250%), Mo (up to 100%), Au (up to 44%), and U (up to 190%).

Mineral resource requirements of renewable technologies are also distinguished by their high dependence on scarce/rare mineral resources. Wind power requires, besides Cu, Zn, Ma, Cr, Ni, and Mo, rare earth elements such as neodymium (Nd), praseodymium (Pr), dysprosium (Dy), and terbium (Tb); PV requires, besides Si, Cd, Ga, and Ag, rare earth elements such as tellurium (Te), indium (In) and selenium

(Se); geothermal requires specialized steel (high in Cr, Mo, Ni, and Ti) to withstand the harsh operating environment characterized by the very high temperatures and potentially corrosive nature of geothermal reservoirs [45].

In their recent study, [46] compared the maximum annual demand for scarce/rare metals in power sector generation infrastructure during the years 2020 to 2050 under 1.5 °C end-of-century warming scenarios with the current annual production rates. The study revealed significant increases in the demand for Dy (310%), Nd (271%), and Te (372%), surpassing the current global production rates by a considerable margin. However, the study also found that the current global reserves of critical materials are expected to be sufficient to meet the future demand from electricity generation infrastructure over the next 30 years, with the potential exception of Te. We will revisit the resource aspects of renewable technologies later in this chapter.

6.3 EV as a Means of GHG Reduction

Fuel combustion in transport is the third largest source of GHG emissions worldwide, next to the power industry and other industrial combustion (Table 6.3). In the US, transport is the largest source of GHG, slightly surpassing the power industry, while in Germany and South Korea, transport is the second largest source of GHG emissions (Table 6.3). Electrification of transport via replacing fossil fuel-based internal combustion vehicles (ICVs) with electric vehicles (EVs) may contribute to substantially reducing GHG emissions when the emission from the power sector is reduced, for instance, by a large-scale deployment of the renewable power technologies we discussed above.

This section is concerned with assessing the potential of EVs as a tool for mitigating GHG emissions, primarily based on LCA. We first look at major studies on the environmental impacts of replacing ICVs with EVs, primarily based on process-based LCA, to obtain a perspective. Subsequently, we will discuss in detail the application of IO to the LCA of EVs for a selection of studies. To provide basic technological prospects, we start with a brief overview of the major differences in the technology between ICVs and EVs, which seem vital in the following discussion.

6.3.1 *Overview of Internal Combustion Vehicles (ICVs) and Electric Vehicles (EVs)*

ICV, EV, and HV

EVs use electric motors for propulsion instead of conventional internal combustion (IC) engines, which work on the combustion principle to get their energy mostly from fossil fuel. While fuel for an ICV is not limited to fossil-based ones but can include biofuel, hydrogen, or ammonia, we henceforth consider the case of fossil-based fuel (petrol), except when otherwise stated. EVs can be classified as Battery Electric Vehicles (BEV) and Hybrid Electric Vehicles (HEV) [47]. Pure EVs have only batteries as their energy source. A vehicle with two or more energy sources and converters is called a hybrid vehicle (HV). An HV with an electrical power train is called a hybrid EV (HEV). The energy sources in HEVs can combine many resources, such as batteries, petrol, biofuels, and fuel cells.

The Glider and Power Train of ICVs and BEVs

Based on key technical functionalities, a vehicle can be divided into two modules, "glider" and "power train"; the power train comprises all the components which are required for generating and transmitting the propulsive energy for the vehicle, while the glider comprises all remaining components of the vehicle which are not strictly related to the propulsion technology: the chassis/body, tires and wheels, seats, belts, windshield and windows, suspension system, etc. [48]. An ICV and a BEV remarkably differ in their powertrain. For a BEV, the power train is composed of the electric motor, the battery, and the power electronics, which process the electric energy in the system and comprise a converter, an inverter, a power distribution unit (PDU), and a charger, whereas for an ICV, the powertrain comprises the engine, gearbox, and tank [48].

Table 6.9 briefly compares the major features of an ICV and a pure BEV. A common feature of all EVs is the capability for Regenerative Braking, a process by which the kinetic energy of the moving vehicle is converted into electricity by reversing the operation of the motor into a generator. Another distinguishable difference between an ICV and a BEV is the remarkably low efficiency compared with electric motors of IC engines, for which around 70% of the energy contained in the fuel is dissipated to the environment as heat, noise (friction), and emissions [49].

Table 6.9 Comparison of ICVs versus BEVs

	ICVs	BEVs
Energy storage	Fuel tank	Battery
Replenish the energy	Petrol pump	Charger
Production of motive force	IC engine	Electric motor
Controls speed & power	Carburetor	Electronic controller
Efficiency	Engine efficiency: ~30%	Motor efficiency ~80%
Braking energy	Not recoverable	Recoverable
Transmission	Complex gear system	One gear

Source [47]

HEVs with Different Hybrid Levels

Falling between these two categories (ICVs and BEVs) are HEVs with different hybrid levels [47]:

1. Series Hybrid (SHEV): the simplest hybrid configuration with the electric motor alone delivering the vehicle traction power as the engine is not connected to the drive train. Example: Nissan e-Power.
2. Parallel Hybrid (ParHEV): IC engine and motor are directly connected to the drive system so that they can individually or jointly propel the vehicle depending on the load. Examples: Honda Civic Hybrid, Mercedes-Benz S400 BlueHYBRID.
3. Series-Parallel Hybrid (SParHEV): a combination of the benefits of both series and parallel architecture. Example: Toyota Prius.
4. Fuel Cell Hybrids (FCHEV, or FCV for short): a series hybrid configuration in which a fuel cell is the energy conversion system, and a battery (or a supercapacitor) is the energy storage system to deliver peak acceleration power. Examples: Toyota Mirai and Hyundai Nexo.
5. Plug-in Hybrid Electric Vehicles (PHEV): Plug-in hybrid EVs are full-hybrids that use a smaller engine, a larger battery, and a larger motor with batteries rechargeable from any external power source. Examples: Honda Clarity PHEV, Toyota RAV4 Prime, Ford Escape PHEV, Chrysler Pacifica PHEV.

6.3.2 Mitigation Potentials of EVs

6.3.2.1 Replacing 100% of US Onroad Vehicles by EVs

We start this section with [50], which assessed the GHG mitigation potential in the US of replacing all the onroad vehicles with EVs (pure BEVs and FCV) under alternative power technologies, consisting of PV, CSP, wind, geothermal, hydroelectric, wave/tidal, nuclear, coal-CCS, and two liquid fuel options (corn-ethanol-E85 and cellulosic-E85). To my knowledge, this is the first systematic study that addressed the mitigating potential of EVs on the entire US economy and deserves attention for its results and methodology.

After considering twelve combinations of energy source-vehicle types, it was found that replacing 100% of US onroad vehicles with BEVs powered by renewable power technologies could reduce the emission from onroad vehicles by around 30%, with wind-BEV showing the best potential of some 33% reduction ([50], Fig. 6.2). The results for FCV were similar, but with slight margins for BEVs.

The Methodology: A Brief Outline

We now briefly discuss the methodology by which this remarkable result was obtained, referring to the Supplementary Information Appendix to [50]. The original derivation was shortened by combining several intermediate steps. The value of items occurring on the right-hand side for the first time was taken from the literature or assumed by the authors (available in the Supplementary Information Appendix to [50]).

In the first step, the amount of net energy to power ICVs in the US (in 2007), $a4$, is obtained, which amounts to 1.050–1.182 (10^{12} kWh or PWh), with variation reflecting the range of ICV efficiency (0.16–0.18). Dividing $a4$ by battery efficiency gives the energy required for batteries in BEVs, $e1$. In the next step, the amount of GHG per kWh of a single wind turbine ($d4$) is obtained based on the electricity required for its construction. Dividing the energy requirement for BEVs, $e1$, by the power output per turbine $d2$ gives the number of turbines needed to propel the BEVs. Finally, multiplying the number of turbines thus obtained, $e2$, with the amount of GHG emissions per turbine, $d4$, gives the emissions of the BEVs, $e3$.

1. Energy required for vehicles:

   a1: Fuel use (L/y) = Vehicle miles traveled (km/y)/Fuel mileage (km/L)
   a2: Energy in fuel (MJ/L) = LHV gasoline (MJ/kg) × fuel density (kg/L)
   a3: Energy needed to power ICVs (MJ/y) = $a1 \times a2$
   a4: Net energy to power ICVs (kWh/y) = $a3 \times$ ICV efficiency/(3600(MJ/kWh))

2. Wind turbine characteristics:

   d1: Single turbine output (kWh/y) = Rated power (kW) × Capacity factor × 8760(h/y) × Efficiency losses
   d2: Energy to manufacture one turbine (kWh/y) =
   Energy to manufacture one turbine (kWh/kW) × Rated power (kW) /Life time of turbine (y)
   d3: Single turbine emissions (kg-CO_2-e/kWh) = d2 × Electricity CO_2 (CO_2-e/kWh)/d1

3. GHG emission of wind-powered BEV

   e1: Energy required for batteries in BEV (kWh/y) = $a4$/Battery efficiency
   e2: Number of turbines required = $e1/d1$
   e3: Wind BEV CO_2-e emissions (kg-CO_2-e/y) = $e2 \times d3$

The emissions associated with other power technologies were calculated in a similar fashion, although details differ among technologies.

A Major Factor Contributing to the Reduction in Carbon Emission

What is remarkable in the results of [50] is that even with BEVs powered by coal-CSS, the upper total emission of which was close to the US grid emission, the US carbon emission could be reduced by 18–26%. In fact, of all the power technologies considered by [50], biofuel was the only case with mixed results (including a net increase in emission). A major factor behind this reduction in carbon emission is the high (tank to wheel) energy efficiency of EVs (75–86%) compared with ICVs, whose efficiency is around 17% ([50], p.170, see [51] for further details of the efficiency of EVs).

The life Cycle Stages Considered

The emission of a wind-powered BEV in [50] corresponds to the emission embodied in the turbines required to produce the power needed to propel the BEV: the emission generated in the production phase of wind turbines is considered. However, the emission generated in the production phase of BEVs, particularly batteries, is not considered. Recall that in LCA, one needs to consider a product's entire life cycle, starting with the resource extraction phase, then the manufacturing and use phases, and ending with the end-of-life (EoL) phase. Regarding the emission from vehicles, [50] considered only the use phase.[3] A life cycle perspective is required to obtain knowledge of the full carbon footprint of EVs.

6.3.2.2 Process-Based LCA of EVs

We now turn to LCA studies of EVs that also consider the impacts generated in life cycle phases other than the use phase, namely the production and EoL phases, starting with studies based on process-based LCA because of their closer proximity with the technical details of EV production. As demonstrated by [50], EVs can greatly contribute to reducing the GHG emissions **in the use phase** when they replace ICVs. A caveat to this claim is that EVs are known to have larger environmental impacts than their ICV counterparts in their production phase, particularly in the production of large volumes of high-performance batteries [52–58]. Table 6.10 gives an example of material compositions for ICV, BEV, and PHEV. Besides its Li-ion batteries, an EV is distinguished from an ICV by the larger use of copper.

The Cumulated GHG Emission of a Vehicle

Corresponding to its three life stages (production, use, and EoL), the total emission of a vehicle over its life, e, can be divided into three components by origin as follows

$$e = e_{production} + e_{use} + e_{EoL} \tag{6.13}$$

Denoting by d^* the lifetime mileage of the vehicle, say 240,000 km [56], by ρ the fuel economy (kWh/km for pure BEVs and L/km for ICVs), and by ϵ the average emission

[3] As we shall see below, the use phase is responsible for the majority of the GWP impact for ICVs. The same will apply to EVs unless electricity is fully decarbonized.

Table 6.10 Material compositions of vehicles with different power trains

	ICV	BEV	PHEV
Steel	793	827	847
Cast iron	139	25	78
Cast aluminum	28	12	66
Wrought aluminum	59	68	23
Copper	24	59	56
Glass	37	44	38
Average plastic	141	151	138
Rubber	29	22	26
Others	23	37	27
Lead-acid battery	15	15	15
Fluid	26	26	26
Tire	36	36	36
Lithium-ion battery	0	286	128
Total	1350	1608	1504

Units: kg. *Source* [58], Table 3

rate (CO_2e/kWh for pure BEVs and CO_2e/L for ICVs) including fuel production, we have

$$e_{use} = \epsilon\rho d^* \tag{6.14}$$

For HVs (see Sect. 6.3.1), ρ and ϵ will be a mixture of several fuel sources and their emissions. Note that $e_{production}$ and e_{EoL} are given for a given vehicle, and are fixed constants, while e_{use} is a variable depending on mileage. Therefore, the cumulated emission of a vehicle with mileage $d \leq d^*$, e_{cum}, can be given by

$$e_{cum} = \alpha + \beta d \tag{6.15}$$

where α and β are constants with $\alpha = e_{production} + e_{EoL}$ and $\beta = \epsilon\rho$. The cumulated emission becomes a linear function of d (km) with the slope β referring to the emission per km.[4] Figure 6.2 gives a schematic illustration of (6.15).

LCA studies on EVs share the finding that the production phase of EVs generally has higher impacts than the comparable ICVs, due mostly to the production of batteries [51, 53, 57–62]. On the other hand, EVs will have a smaller emission in the use phase, owing among other things, to the higher efficiency of electric motors than IC engines. These observations imply the following conditions on the parameters of (6.15)

[4] In reality, ρ may vary depending on the driving mode.; β can be assumed constant for the average value of ρ.

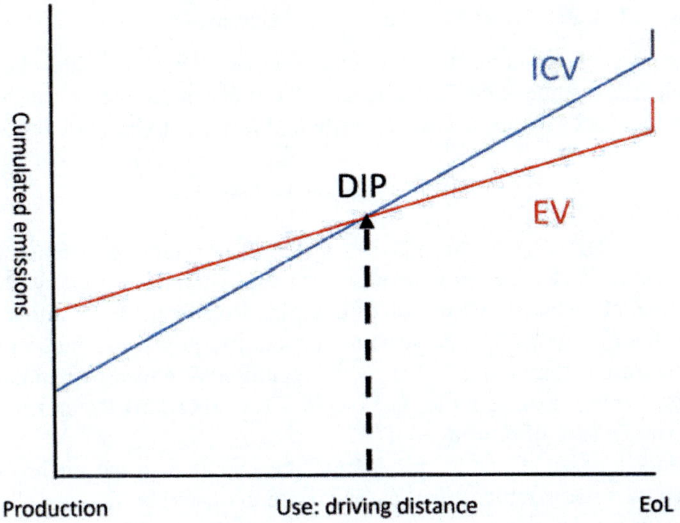

Fig. 6.2 Life cycle impacts of ICVs and EVs: A schematic representation of emissions in a cumulative manner with production, use, and EoL treatment

$$\begin{aligned} \alpha_{EV} &\geq \alpha_{ICV} \\ \beta_{ICV} &\geq \beta_{EV} \end{aligned} \tag{6.16}$$

For instance, ([55], Fig. 6.2) shows, for a medium size car, the following values of α and β

$$\begin{aligned} \alpha_{ICV} &= 5 \times 10^6 \text{g-CO}_2\text{-e}, \quad \alpha_{EV} = 7.1 \times 10^6 \text{g-CO}_2\text{-e} \\ \beta_{ICV} &= 0.14 \times 10^6 \text{g-CO}_2\text{-e/km}, \quad \beta_{EV} = 0.083 \times 10^6 \text{g-CO}_2\text{-e/km}. \end{aligned}$$

Accordingly, the GHG mediation effects of replacing an ICV with an EV take place with the initial GHG emissions "debt" [59] incurred during the vehicle production stage, $\alpha_{EV} - \alpha_{ICV} \geq 0$, being "paid back" with the driving distance, with the distance d° (km) needed to break even given by

$$d^\circ = \frac{\alpha_{EV} - \alpha_{ICV}}{\beta_{ICV} - \beta_{EV}}. \tag{6.17}$$

Depending on the carbon intensity of electricity and the emissions in the production stage, which depends on the vehicle speck, the mileage required to pay back the debt will widely differ.

Extending (6.15): PHEVs and Replacement of Batteries

As an example of extending (6.15) to accommodate HEVs, consider the case of a PHEV, which consumes both gasoline and electricity in the use phase. Accommodating the two fuel sources, the slope coefficient β needs to be extended as

$$\beta = \epsilon_g \rho_g d_g + \epsilon_e \rho_e d_e \tag{6.18}$$

with ϵ_i and ρ_i referring to the emission per unit consumption of fuel i and the consumption of fuel i per km driven, d_i to the distance driven by fuel i, and $i \in$ {gasoline, electricity}. Allocation of the total drive distance, d, into d_e (charge-depleting mode) and d_g (charge-sustaining mode) requires information about the driving/charging patterns of vehicles, which could be provided by a simulator [63] or the Worldwide Harmonized Light Vehicles Test Procedure. Reference [58] considered a simple case of $d_e = d_g = d/2$.

Since the impacts of battery production are included in the intercept term α, the impacts of battery replacement can be accommodated by making the intercept dependent on the replacement. Denoting by d_r the distance for replacing batteries and by α_b the fraction of α referring to battery production, we obtain the following extension of (6.15)

$$e_{cum} = \alpha + \alpha_b D + \beta_{EV} d \tag{6.19}$$

where $D = 1$ for $d^* \geq d \geq d_r$ and $D = 0$, otherwise.

Pay-Back Distances under Alternative Power Technology Scenarios

Using the average European electricity mix (521 g-CO_2 kWh), [55] found that the EV and ICV broke even between 44,000 km and 70,000 km. While the benefit of EVs compared to ICVs increased with distance, compensating for the higher EV production impact requires considerable distances before their emissions break even with the ICVs. The larger EV models were found to require shorter distances to pay back the higher production impacts because of their large β_{ICV} values compared with β_{EV}. On the other hand, regardless of powertrain configuration, the smaller vehicles generally performed better than the larger ones did.

Reference [55] conducted sensitivity analysis involving different carbon intensities of electricity to explore how the carbon intensity of the electricity used for charging influences the life cycle emissions of EVs. When the electricity from the world's average coal power (1029 g CO_2-e/kWh) was used, the benefit of EVs compared to ICVs decreased with distance: replacing ICVs with EVs does not reduce but increases the vehicle GHG emission. On the other hand, when powered by wind-based electricity, compared to ICVs, the life cycle emissions were reduced by 66–70%. In this scenario, the use phase was of low importance because $\beta_{EV} \approx 0$.

International Comparison of Pay-Back Distances with Battery Replacements

Since countries/regions differ in the carbon intensity of electricity for charging EVs, the required distance (the break-even distance) to pay back the EV production stage's initial GHG debt will differ among countries. Reference [61] addressed this issue and estimated the break-even distance or "Distance of Intersection Point" (DIP), where the graphs of cumulated emission for ICVs and BEVs intersect, for the EU, Japan, the US, China, and Australia, using the national/regional grid average GHG intensity for each nation/region. The lifetime driving distance, d^*, was set at 200,000 km across the regions considered. The emission from the production stage was also assumed to be equal across the regions. Besides considering international differences in carbon intensity, [61] also considered the impacts of replacing batteries for EVs after 160,000 km, which were not considered by [55].

No Intersection for Australia?

Their results show a DIP for the EU of 76,500 km, 111,500 km for Japan, 60,800 km for the US, 119.100 km for China, and "no intersection" for Australia. The high carbon intensity of the Australian grid (as already discussed in Sect. 6.2.4.2) was a major reason for this remarkable result. The absence of an intersection (below 200,000 km) occurred because it was necessary to replace batteries before the distance to pay back the impacts of production was reached. Were there no need to replace batteries, the DIP would have been reached somewhere before 200,000 km. Interestingly, their estimate of DIP for the EU exceeds the upper range of [55] by 10%, which can be attributed to the difference in the impacts of the production stage. The fact that the US had a shorter DIP than the EU is attributed to the US's lower fuel economy in ICVs, resulting in a β_{ICV} larger than other countries/regions.

The Pay-Back Distance for Japan Equals the Average Lifetime of a Car

The fact that the DIP is found longer than 100,000 km for Japan may indicate that EVs are not an appropriate option for GHG mitigation, at least under the carbon intensity assumed in their study. The distance almost equals the average lifetime distance of a passenger vehicle in Japan, corresponding to a ten-year lifetime. Further decarbonization of Japan's grid will be imperative for EVs to mitigate GHG emissions.

The Impacts of an EV's Location of Production

The DIP for China is found to be even longer than Japan but is well below the often-assumed lifetime distance for China of 150,000 km [58, 64] or 180,000 km [65]. This result for China is consistent with [58], which find the life cycle CO_2 emission of ICV and BEV to be 33.0–35.5 Mg-CO_2 and 26.8–29.3 Mg-CO_2, and indicate the CO_2 mitigation effects of EVs in China. Observing that different electricity mixes have different impacts on emissions from different types of vehicles and that it may be very difficult to change the national electricity mix in the short term, [58] propose to promote the material production and power battery manufacturing of EVs to be placed in regions with high renewable energy utilization. The impacts of an EV's production location on GHG emissions were also addressed by [62] for facilities located in China, Germany, US, and Italy, and for use in Italy.

Other Power Trains and Other Impacts

So far, our discussion has been focused on comparing the GHG emissions, or the impacts on climate change (CC), of ICVs and BEVs. Variants of power trains exist between ICVs and EVs. CC gets significant attention nowadays, but there are other environmental impacts which deserve due attention.

ICV, HEV, BEV, and FCV in Switzerland

Reference [56] conducted an attributional and process-based LCA, including well-to-wheel (WTW) (energy resource extraction to energy conversion in the vehicle (i.e., driving)) and equipment life cycle (production, manufacturing, maintenance, and end-of-life of vehicle and road infrastructure) to provide a comparative LCA of mid-sized European passenger vehicles operated in Switzerland with different drive train technologies (ICV, HEV, BEV, and FCV) and fuel supply chains representing both modern current and future (reference year: 2030) development status. Fossil fuel for ICVs and HEVs was divided into gasoline, diesel, and compressed natural gas (CNG). The functional unit used for the comparative evaluation was "one kilometer (km) driven" for the lifetime mileage of 240,000 km. The Worldwide harmonized Light vehicles Test Procedure (WLTP) was used as the reference driving cycle for the calculation of vehicle configuration and energy use. Battery lifetime was assumed to be 150,000 km. Besides cumulative GHG emissions, human toxicity potential (HTP), photochemical oxidant formation (POF), terrestrial acidification potential (TAP), and particulate matter formation (PMF) were considered.

Concerning GHG emissions, the results confirm previously published LCA results for BEV in the literature and extend the scope to FCV: BEV and FCV cause substantially less life cycle GHG emissions than conventional fossil-fueled ICV, if and only if they use electricity and hydrogen, respectively, produced from nonfossil energy resources. BEV and FCV can, in the best case, reduce the carbon footprint of current and near-future passenger vehicles by almost 80%. On the other hand, using fossil energy carriers for electricity and hydrogen production could result in an increase in GHG emissions.

Concerning the other environmental burdens analyzed, BEVs and FCVs provide less benefit. In terms of HTP, gasoline ICV was found to always perform better than both BEV and FCV, independently of the fuel generation technology, even with "clean" hydrogen and electricity sources. Responsible for the high burdens for BEV and FCV are the toxic substances mainly released by coal mining and metal mining activities (mainly Ni, Cu, Pt, and Al). Notwithstanding the expected reduction in the carbon intensity of the electricity mix, the HTP caused by the BEV and FCV will still be substantially higher compared to fossil-fueled cars in 2030. That HTP is the major disadvantage of electric powertrains is confirmed by [66–68]. For example, [68] reports that substituting a conventional gasoline vehicle with an electric one in the US, Italy, or France would increase the HTP indicator by one order of magnitude, respectively, by 692%, 534%, and 380%. Also noteworthy with [68] is the finding that EVs have much larger impacts on mineral resource depletion than ICVs.

The BEV charged with electricity from a coal power plant causes the highest burdens in TAP and PMF. However, when charged with electricity from all other

Table 6.11 The environmental impacts of BEV and PHEV compared to ICV (%)

	BEV	PHEV
GHG	−18	−1
VOC	−24	−16
NO_X	−6	−3
$PM_{2.5}$	159	79
SO_2	111	56

Source [58]

sources (both current and near-future), BEVs cause less ATP and less PMF formation than gasoline ICVs. The TAP and PMF burdens of the FCV will remain higher than those of the ICV and HEV: fuel cell (FC) manufacturing and hydrogen production generate relatively high burdens; the majority of the FC-related burdens are due to emissions from mining Pt (used as a catalyst). Compared with gasoline ICVs, BEVs and FCVs cause the lowest burdens in POF.

ICV, PHEV, and BEV in China

Reference [58] compared life cycle impacts of ICVs, PHEVs, and BEVs in terms of GHG, VOCs, NO_X, $PM_{2.5}$, and SO_X for a lifetime drive distance of 150,000 km in 10 years, with the electricity provided by the average Chinese grid. In contrast to [56, 68], who used generic models of power train modules to obtain the inventory for different power trains, [58] chose real vehicle models to represent each power train: Toyota Corolla luxury 2019 (ICV), Nissan Leaf 2019 (BEV) and Toyota Corolla double engine E+pioneer 2019 (PHEV). For PHEV, the mileage ratio consuming electricity (charge-depleting mode) and gasoline (charge-sustaining mode) was set as 50%, respectively.

Table 6.11 summarizes the results. While BEVs and PHEVs contribute to reducing the emissions of GHG, Volatile Organic Compounds (VOX), and NO_X, they increase the emissions of $PM_{2.5}$ and SO_2. Under an ambitious scenario with renewable energy accounting for 85.7% of China's electricity consumption, the CO_2 emissions from BEVs and PHEVs would decline to 107 g/km and 156 g/km from 296 g/km and 228 g/km, indicating a large potential for CO_2 emissions reduction of the use of renewable energy in the electricity mix. Compared to ICVs, the CO_2 emission of BEVs and PHEVs would decrease 53% and 32%, respectively, the VOCs emissions would remain almost unchanged, while the $PM_{2.5}$ emissions would slightly rise. Due to the high biomass power share and emission factor, the SO_2 emissions of BEVs would still remain almost twice the level of ICVs.

6.3.2.3 HLCA-Based Studies on EV

After having reviewed several studies on process-based LCA of EVs, we now turn to LCA studies on EVs based on HLCA.

MRIO-Based HLCA of ICV, HEV, PHEV, and BEV in Australia

Reference [69] assessed the CF of light-duty vehicles (LDV), passenger cars, sport utility vehicles, and light trucks, distinguished by power trains, ICVs (divided into gasoline and diesel), HEVs, PHEVs, and BEVs in Australia, based on the MRIO model they used for assessing the CF of renewable technologies [27] discussed in Sect. 6.2.4.2.

The Australian IO (supply and use table) for 2009 with a sectoral resolution of 440×440 was further disaggregated, the transport sectors into LDVs, trucks, and others, and LDVs into different powertrain sub-sectors by adjusting average input data with technology-specific data. Life cycle inventories (LCI) for ICVs and BEVs were taken from the Ecoinvent 3.1 database, while those for HEVs and PHEVs were modeled based on the gasoline ICV dataset with needed adjustments. The vehicle lifetime was set to 230,000 km [70] and the battery life to 150,000 km [56]. The CF was calculated analogously to (6.9) but augmented with the direct (tailpipe) emission from fuel consumption in the driving stage and the indirect emission from fuel production

$$e_j = \underbrace{\boldsymbol{fLy}_j}_{\text{vehicle production}} + \underbrace{\boldsymbol{fLy}_{\boldsymbol{fuel},j}}_{\text{fuel production}} + \underbrace{\boldsymbol{\epsilon y}_{\boldsymbol{fuel},j}}_{\text{tailpipe}}, \quad j \in \{\text{ICV, HEV, PHEV, BEV}\} \tag{6.20}$$

where $\boldsymbol{y}_j$ is the $n \times 1$ vector referring to the purchase of a vehicle with powertrain j, $\boldsymbol{y}_{\boldsymbol{fuel},j}$ is the $n \times 1$ vector referring to the fuel consumption of j for the lifetime distance (gasoline or diesel for ICVs, gasoline and electricity for PHEVs, and electricity for BEVs), and $\boldsymbol{\epsilon}$ is the $1 \times n$ vector referring to the direct (tailpipe) emission from fuel consumption (zero for BEVs). Recalling that $\boldsymbol{y}_{\boldsymbol{fuel},j}$ is assumed proportional to the driving distance, d, one notices that the first term of (6.20) refers to α and the sum of the remaining terms to βd in (6.15).

Table 6.12 gives the results for the base year 2009.[5] BEVs are characterized by the largest life cycle GHG emissions, followed by ICVs (gasoline-powered) and PHEVs. Remarkably, the direct emission from driving surpasses the indirect emission for BEVs, though to a lesser extent than ICVs. In contrast, PHEVs have a higher share of indirect emissions than direct emissions, while the shares are almost equal for HEVs. The emission from ICVs (diesel-powered) is slightly smaller than PHEVs, whereas HEVs are marked with the lowest emission.

The results indicate that BEVs and PHEVs were not viable for carbon mitigation in Australia in the modeling base year (2009) due to the high-carbon intensity of Australia's electricity grid, a result aligned with [61]. However, the authors also show that BEVs result in notable GHG emissions reductions in scenarios of renewable electricity deployment.

[5] Reference [69] conducted a scenario analysis to explore the CF reduction potential of a large-scale diffusion of electric powertrains up to 2050, given increasing shares of renewable electricity relative to the status quo. We do not discuss the results. The emissions embodied in imports contributed around 8–13%.

Table 6.12 Life cycle GHG emissions of different powertrains based on HLCA

	Direct	Indirect	Total
BEV	241	181	422
PHEV	174	189	363
HEV	130	133	263
ICV-d	222	127	349
ICV-g	233	133	366

Units: g-CO_2/km. *Source* [69], Fig. 6.2b. Direct: emissions from real-world driving, Indirect: emissions from all other. "-d" and "-g" respectively refer to gasoline-driven and diesel-driven. The level of certainties considered in the original result was omitted

Triple Bottom Line Impacts of Alternative Power Trains

Reference [71] examined the macro-level social, economic, and environmental (triple bottom line, TBL) impacts of alternative vehicle technologies (ICV, HEV, PHEV with four charge-depleting ranges, 10, 20, 30, and 40 miles, and BEV) in the US, using the US IO table with 426 sector resolution.

Two Scenarios

In total, 19 macro-level sustainability indicators (two economic indicators, foreign percentage and business surplus, four social indicators, employment, income, tax, and injuries, and 13 environmental indicators, including CC, water withdrawal, total energy, and hazardous waste) were quantified for a scenario in which EVs are charged through the existing US power grid with no additional infrastructure (Scenario 1) and an extreme scenario in which electric vehicles are fully charged with solar charging stations (Scenario 2). The system boundary consisted of vehicle and battery (except ICV) production, vehicle maintenance & repair, electricity generation, gasoline production, charging infrastructure (BEV and PHEV), and vehicle and battery EoL. The lifetimes of vehicles and batteries were assumed to be 150,000 km, and the functional unit was 1 mile (1.61 km) of vehicle travel. Replacement of batteries was not considered.

The Inventory

Vehicles and battery components were calculated separately to distinguish between battery and vehicle manufacturing impacts. The vehicle bodies were assumed to be identical across all the powertrains, with differences stemming from the additional battery and electronics. Battery weights, specific power, and capacity were derived from the GREET 2.7 vehicle cycle model [72]. After the battery weights and specific power requirements were calculated with the model, the costs associated with producing these li-ion batteries were derived from Argonne National Laboratory's cost estimation study for Li-ion batteries. The manufacturing costs of each battery thus obtained were then assigned to the associated NAICS sector ("primary battery manufacturing sector") of y_j. On the other hand, direct impacts such as tailpipe emissions and direct energy consumption while driving, the third term in (6.20), were calculated using process-level data including the GREET model.

The portion driven with electricity (charge-depletion mode) of PHEVs was obtained from the National Household Travel Survey (NHTS). The impacts of the EoL stages for the vehicles and battery were calculated by determining the savings from the recycled materials from each vehicle. For batteries, the materials (steel, aluminum, and copper) were assumed to be 100% recycled. For the vehicle part (except the batteries), the recycling rate was assumed to be 95%, with the materials (steel, aluminum, copper, plastic, rubber, and a small amount of platinum) recycled at 90 to 95%.

The HLCA Model and Results

The calculation method is based on (6.20) with the intervention vector $\boldsymbol{f}$ replaced by the matrix $\boldsymbol{F}$ (and $\boldsymbol{F}_\epsilon$) with 19 rows referring to the 19 sustainability indicators:

$$\boldsymbol{ind}_j = \underbrace{\boldsymbol{FLy}_j}_{\text{vehicle production}} + \underbrace{\boldsymbol{FLy}_{fuel,j}}_{\text{fuel production}} + \underbrace{\boldsymbol{F}_\epsilon \boldsymbol{y}_{fuel,j}}_{\text{tailpipe}}, \quad j \in \{\text{ICV, PHEVs, BEV}\} \tag{6.21}$$

The results for environmental impacts indicate that the vehicle operation phase was the most dominant in all the environmental impact categories, which aligns with the results of [61] for the US. In Scenario 1, the BEV had the second highest GHG emissions after the ICV due to the GHG emission intensity of the electric power generation sector in the US. On the other hand, powering EVs via solar charging stations could reduce their GHG emissions by up to 34%. The least energy-intensive vehicle option was the HEV in Scenario 1, whereas the energy performance of the PHEV-10 was better than the other vehicles in Scenario 2.

Two environmental impact categories favored ICVs against alternative vehicle technologies: the water footprint and PMF. The BEV was the most water-intensive vehicle and had the highest PMF in both scenarios; when powered by solar charging stations, the water footprint of the BEV was significantly reduced. Although the BEV generated the least hazardous waste in Scenario 1, it became the worst alternative in Scenario 2 regarding hazardous waste due to the construction of solar charging stations and the manufacturing of the required materials. The PMF of electric vehicle options was higher than those of ICV and HEV due to air emissions from electric power generation. Although the PMF impact of EVs (PHEVs and BEVs) was reduced under Scenario 2, the reduction fell short of making these options better than the conventional ICV; the construction of solar charging stations and manufacturing the solar panels hindered the PMF savings of EVs. Their results for PMF align with [56, 73].

Sources of Uncertainty

Notwithstanding that the IO table used in this study has one of the world's highest levels of sector resolution, it has only one battery sector, "the primer battery manufacturing sector," which includes various types of batteries besides the Li-ion batteries considered in this study. That the impacts of this sector are averaged impacts of all of the output products of this sector is pointed out by the authors of this study as a possible source of uncertainty in the above results.

I point to another possible source of uncertainty in their results. In [71], the vehicle bodies were assumed to be identical across all the powertrains, with differences stemming from the additional battery and electronics. However, as shown in Table 6.9, the difference between ICVs and EVs in terms of components is not limited to the additional battery and electronics in the latter; a fuel tank, petrol pump, engine, carburetor, and complex gear system do not occur in EVs, while they do in ICVs, manifesting as differences in material compositions among different power trains (Table 6.10). The same assumption was used by [74, 75]. The former considered connected and automated heavy-duty trucks, and the latter the global life cycle material footprints of five types of passenger vehicles (ICV, HEV, PHEV-20, PHEV-40, and BEV) based on ten metals and nine minerals. Besides its extensive study of material footprints, the latter study is noteworthy because of its use of the MRIO database (EXIOBASE). We now look at this study in some detail.

Material Footprint of ICV, HEV, PHEV, and BEV

As we discussed in Sects. 6.2.4.4 and 6.3.2.2, the further deployment of renewable energy and EVs is expected to result in a dramatic increase in the demand for materials that are crucial for their manufacturing. With this backdrop, [74] assessed the US material footprints of vehicles with alternative powertrains (ICV, HEV, PHEV-20, PHEV-40, and BEV) in terms of ten metals ("ores of iron", "bauxite and aluminum", "copper", "lead", "nickel", "tin", "uranium and thorium", "zinc", "precious metal", and "other metals") and nine minerals ("chemical and fertilizer materials", "clays and kaolin", "limestone, gypsum, chalk, dolomite", "salt", "slate", "other industrial minerals", "building stones", "gravel and sand", and "other construction minerals"). Considering that many of these materials are not produced in the US but have to be imported from overseas, an MRIO-based HLCA methodology was used.

The assessment was made for three scenarios: Scenario 1 assumes the current US national average electricity grid mix without an additional infrastructure; Scenario 2 assumes the dominance (60%) of the renewable energy sources; Scenario 3 assumes a large solar charging infrastructure deployment, in which EVs are charged with a solar charging station without depending on the grid.

Life Cycle Inventory Data

Analogous to [71], the manufacturing of ICV was considered as the baseline vehicle, including the vehicle's body, shell, and other miscellaneous parts, with its manufacturing cost calculated based on the average Manufacturer Suggested Retail Price of an average mid-size ICV. Using the data on the price index of automobiles, the manufacturing cost of an average ICV, HEV, PHEV20, PHEV40, and BEV was estimated. The incremental costs of HEV, PHEV, and BEV, over ICV, were allocated to the manufacturing of battery and miscellaneous electrical equipment installed in such vehicles. The production cost of an EV battery was taken from the literature.

The battery energies used in HEV, PHEV-20, PHEV-40, and BEV were obtained based on each type of vehicle that is already commercialized. The inventory thus obtained was assigned to the most closely matching final demand vector in the MRIO model with 163 sectors and 43 countries obtained from the EXIOBASE database. The fuel economy of each power train was obtained from the literature. The EoL stage was considered analogous to [71].

The EoL phase comprised two parts: (1) reprocessing of metals, and (2) landfill and incineration of plastics. The recycling rate of aluminum was assumed to be 90%, whereas it was 95% for steel, iron, copper, and platinum. As for plastics, 95% was assumed to be landfilled, with the rest incinerated.

The MRIO Model

The MRIO-HLCA model is based on (6.22), where $A^{r,s}$ refers to the 163 × 163 input coefficients matrix of inputs from country r into country s, F^i to the 19 × 163 intervention matrix of country i, n_c to the number of countries (=43), and y_j^1 to the final demand vector in country 1 for the vehicle of type j

$$\begin{pmatrix} F^1 \\ F^2 \\ \vdots \\ F^{n_c} \end{pmatrix} \begin{pmatrix} I - A^{1,1} & -A^{1,2} & \dots & -A^{1,n_c} \\ -A^{2,1} & I - A^{2,2} & \dots & -A^{2,n_c} \\ \vdots & \vdots & \ddots & \vdots \\ -A^{n_c,1} & -A^{n_c,2} & \dots & I - A^{n_c,n_c} \end{pmatrix} \begin{pmatrix} y_j^1 \\ 0 \\ \vdots \\ 0 \end{pmatrix} \tag{6.22}$$

The Results

Several remarkable results were obtained. First, under the current techno-economic circumstances, a BEV's life cycle material footprint (LCMF) was found to be 60% higher than that of conventional ICVs. Second, the studied EVs had larger material footprints than a conventional ICV under all circumstances assumed in the analysis. Battery manufacturing was found to be a critical factor in driving the EVs' LCMFs relatively higher, with copper being a major contributor, confirming [52]. Transitioning to the electricity grid mix composed of 60% renewable energy does not bring any significant improvements to the LCMF of BEVs, and transitioning to solar PV-generated electricity worsens it. Shifting entirely to electricity generation through solar PVs increased the LCMF of BEV and PHEVs by about 20%. The study's findings are consistent with those of [68, 76]. Reference [76] reports that PHEVs have 170% higher resource depletion potentials than the corresponding ICVs.

The above results are subject to certain limitations. One limitation arises from the sector resolution of EXIOBASE. The same EXIOBASE sector "Manufacture of electrical machinery and apparatus n.e.c." was used to represent both battery manufacturing and solar modules installation, implying possible uncertainty as to the results. Another source of uncertainty is the assumption that vehicle bodies are the same across all the power trains mentioned above.

An HLCA of Large-Scale Vehicle Electrification in China

The total vehicle stock in China is projected to reach 350–550 million units by 2030, with the EVs stock exceeding 80 million, most of them being passenger EVs (henceforth, EVs refer to passenger EVs except when otherwise stated). With this backdrop, [65] examined whether the initiative of electrifying passenger vehicles enables China to achieve significant CO_2 emission reductions. Three types of power trains were considered, ICVs, PHEVs, and BEVs. BEVs were further divided by two battery types, lithium iron phosphate (LFP) and lithium nickel manganese cobalt oxide (NMC)). Twelve years of lifetime with 15,000 km driving distance per year was assumed.

This study is distinguished by collecting detailed data about almost all EV models sold in 2018, calculating their corresponding life cycle emissions, and obtaining the sales-weighted emissions to reflect the overall performance of each vehicle technology. In contrast, most LCA studies on EVs use subjectively selected representative vehicles or hypothetical ones generated by model calculation to perform comparisons.

The Methodology and Data

The methodology is based on a combined use of process-based LCA and HLCA. The inventory on battery and vehicle assembly processes, use, and end-of-life stages, and the associated CO_2 emissions, as well as tailpipe emissions for ICVs and gasoline-fuelled PHEVs, were calculated based on the GREET model [72]. The columns of input coefficients referring to the assembling of each type of battery and vehicle were then obtained by converting the physical inventory data into money values, by use of price data, and consolidating to the sector classification of a Chinese IO table with 45 sector resolution. The emissions associated with production processes, including the manufacturing of the vehicle (excluding battery), battery, charging infrastructure, and their replacements (if necessary), as well as the upstream production of gasoline and electricity, were calculated based on HLCA, analogously to (6.20), but without the third term. The emissions from the tailpipe and the battery recycling process, calculated by the EverBatt model [77], gives the counterpart to the third term.

The Results

The results indicate that the operation stage dominates the life cycle emissions across all powertrains. For BEVs, CO_2 emissions from the production stage were found to be much higher than those for ICVs, as additional emissions occur from battery-related supply chains, which contribute approximately 40% to BEV production emissions. The results indicate that life cycle CO_2 emissions (g/km) of individual EVs were lower than comparable ICVs across all the scenarios, including the default one of 2018, a further confirmation of previously published LCA results for EVs.

However, the large-scale replacement of ICVs with BEVs in 2018 failed to serve the emission-saving purpose because emissions from manufacturing new EVs, especially from the carbon-intensive battery manufacturing process, were not fully offset by avoided emissions from on-road EVs *within that year*: while on-road EVs saved 2.22 Tg of CO_2 emissions, the production stage resulted in 2.06 Tg of additional emissions. In terms of (6.15), this result can be explained as follows. The increase in emission from the production stage, Δe_{prod}, can be given by[6]

---

[6] Reference [65] uses the αs and βs of individual vehicles weighted by sales volume and also considers the EoL stage, which we neglect for simplicity.

$$\Delta e_{prod} = (\alpha_{EV} - \alpha_{ICV}) \times \text{the number of sales of EVs in year } t \tag{6.23}$$

while the saving in the use phase (of the stock of onroad EVs), Δe_{use}, by

$$\Delta e_{use} = (\beta_{ICV} - \beta_{EV}) \times \text{the onroad stock of EVs in year } t \times d_y. \tag{6.24}$$

where d_y is the mileage per year per vehicle. Using the data on sales volume and stock of EVs for 2018, the result was

$$\Delta e_{prod} - \Delta e_{use} = 2.06 \text{ Tg-CO}_2. \tag{6.25}$$

Under a feasible scenario for 2030, with 45% of electricity produced from renewable sources, manufacturing new EVs would still increase CO_2 emissions by over 10 million tons, but extra emissions could be offset by avoiding emissions from on-road EVs, that is,

$$\Delta e_{prod} - \Delta e_{use} < 0. \tag{6.26}$$

Aligning to [55], this study observes that small-sized EVs, which are important contributors to the current Chinese EV market, correspond to less life cycle emissions than larger high- performance counterparts.

We point out limitations regarding the results. Compared with the above two studies based on IO tables of more than 400 sectors, with 45 sectors, the sector resolution of this study is low. The low resolution hampered the full exploitation of detailed physical information provided by the GREET model. For instance, wrought aluminum, cast aluminum, copper, and magnesium had to be aggregated into "Mining of nonferrous metal ores," and coolant and three types of electrolytes had to be aggregated into "Manufacture of chemical raw materials and chemical products." Since both IO sectors include many other products besides those relevant to EVs, each sector's impacts are averaged impacts of all of the output products of the sector, which may introduce uncertainty to the results obtained.

A MRIO-HLCA of a Li-Ion Battery Pack

As we saw above, LCA studies of EVs identify the production of batteries as a main contributor to the environmental impacts of EVs. Reference [78] conducted a comprehensive life cycle environmental and economic assessment of an advanced battery system (energy storage system (ESS)) for EV applications, following the ISO 14040 and 14044 standards, using the MRIO tables from the world input-output database (WIOD) [79], which allowed the addressing of issues related to fragmentation of production and economic and environmental aspects of the assessed product. This study was the first to provide a global geographical resolution of an ESS's environmental and economic impacts on EVs. The WIOD (2013 version) had a sector resolution of 35 and a regional resolution of 41 countries, including 27 EU countries and 13 major

Table 6.13 Contribution of the top three countries to the total impacts of ESS

	Country percentage contribution					
GWP	Belgium	42	Finland	34	Russia	11
TAP	Belgium	45	Finland	15	RoW	10
POF	Finland	32	RoW	28	Russia	18
PMF	Finland	44	Belgium	18	Russia	9

Source [78]

non EU countries, and one RoW.[7] We henceforth focus on only the environmental impacts.

The Functional Unit and Data

The functional unit (FU) was 1 ESS of a 150,000 km lifetime for EV applications. Final assembly, use stage, and EoL were assumed to take place in Belgium, while some components of the ESS are manufactured around Europe. The main environmental questions to be addressed were:

(a) What are the environmental impacts during the life cycle of the ESS?
(b) Where are those impacts located following the supply chain of the life cycle of the ESS?

The life cycle inventory used to describe the final demand vectors to conduct the MRIO analysis was based on the process-based inventory implemented in [80]. After being converted into monetary units, the process-based inventory in physical mass and energy was assigned to the most closely matching industrial sector of the WIOD, creating the final demand vectors. The use stage was modeled with a driving cycle simulation (NEDC) that estimates the power demand of a vehicle with the assessed battery pack. For the EoL scenario for the ESS, a hydrometallurgical recycling process was assumed. The materials recovered (copper, aluminum, stainless steel, and Li powder) were assumed to substitute primary raw materials in another product system, avoiding the impacts associated with the production of primary raw materials. For the environmental impacts, GWP, TAP, POF, and PMF were considered.

The Results

As for the results of the total impacts, the manufacturing stage was found to be the highest contributor to all the categories assessed in the three life cycle stages, followed by the use stage, aligning with the results of other studies based on PLCA. Around 40% of the total GHG emissions came from the electricity needed for manufacturing, with the production of battery cells having the highest environmental load among the three components of the EES (battery cells, battery control units, and modules). With the impacts on materials processing and raw material extraction reduced by the credit given to recycling, electricity generation was the main contributor in the manufacturing stage. The geographical resolution of the results (Table 6.13) shows

[7] The WIOD we discussed in Sect. 4.3.4 is version 2015 with 56 sectors.

that countries that are not directly involved in the manufacturing stage can have an important role regarding the emissions, i.e., Russia, which was the main supplier of goods and services to Finland. Finland, where the manufacturing of the battery cells is located, had the largest impact concerning POF and PMF; Belgium does not occur in the top three countries concerning POF.

We point out limitations regarding the above results. The low sector resolution of the WIOD, 35 sectors, hampered consideration of the impacts of individual components and materials. For instance, 20 individual components of the battery cells, ranging from the components of the anode to those of the cell container, had to be assigned to only four sectors, "Mining and Quarrying," "Rubber and Plastics," "Chemicals and Chemical Products," and "Basic Metals and Fabricated Metals." Furthermore, it was impossible to distinguish aluminum, copper, and steel from each other.

HLCA of FCV Versus ICV in Japan

Hydrogen energy utilization is expected as an efficient and environmentally benign source of secondary energy. While the renewable energy-derived hydrogen production method is considered environmentally favorable, current hydrogen production is mostly from fossil fuels. With this backdrop, [81] estimated the energy consumption and GHG emissions of the FCV system over the entire life cycle and compared these estimates with those of the existing gasoline-based ICV system to clarify the environmental and energy efficiencies of hydrogen energy derived from fossil fuels. Their estimation was based on HLCA implemented with a Japanese IO table (from 2015) with 390 sectors created by aggregating the original table for this study. Since the resolution of the existing IO table was not high enough for accurately analyzing hydrogen technologies, the table was disaggregated to include hydrogen production, hydrogen shipment (high pressure), lorry transportation, hydrogen station, and gasoline station as new individual sectors (data construction is detailed in [82]).

The FCV and ICV Systems Considered and the Corresponding Functional Units

The FCV and ICV system life cycle stages were as follows

- The FCV system:
  - Production of hydrogen from crude oil steam reformation (naphtha). Representing light naphtha by butane (C_4H_{10}), the chemical reaction for producing hydrogen by steam reforming can be given by ([83])

$$C_4H_{10} + 8H_2O \rightarrow 4CO_2 + 13H_2 \tag{6.27}$$

    For refinery equipment production, only pressure fluctuation adsorption equipment, hydrogen compressors, suction drums, and off-gas compressors required for naphtha reforming were considered.
  - Transport of the hydrogen to eight hydrogen stations as a high-pressure gas.
  - Filling the FCVs with hydrogen gas.

  - The functional unit is 14,200 units of an FCV system with a hydrogen supply of 9.45 Gg-H_2 and a lifetime driving distance of 100,000 km (10,000 km per year)
  - GHG emissions at the FCV utilization stage were set to 0.010 kg-FC/unit/yr, considering only the hydrofluorocarbons emitted by air conditioner use.

- The ICV system:
  - Production of gasoline at a refinery, transportation to 10 gasoline stations (average monthly gasoline sales of 100 kL)
  - The functional unit is 14,200 units of an ICV system with a gasoline supply of 114,300 kL, with a lifetime driving distance of 100,000 km (10,000 km per year)
  - A fuel economy of 12.4 km/L was assumed, which corresponds to the standard ICV model with a weight classification closest to the FCV.

The inputs referring to Maintenance & Construction, Fuel fabrication, Compression (not relevant for the ICV system), Transport, and Station were assigned the vector of final demand $\boldsymbol{f}$ for each vehicle system. The basic structure of an FCV closely resembles an HEV: the engine and gasoline tank are replaced by the FC stack and hydrogen tank. Analogous to [71, 74], the incremental costs of FCV over HEV were allocated to the parts unique to FCVs (the FC stack and hydrogen tank). The automobile costs thus obtained were assigned in $\boldsymbol{y}$.

The Model and Results

The HLCA model is given by (6.20), with $j \in \{ICV, FCV\}$. Life cycle energy consumption per km was estimated to be 4.3 MJ/km and 6.0 MJ/km for the FCV- and ICV systems, of which 64% originated from vehicle manufacturing for FCVs and 33% for ICVs. Life cycle GHG emissions per km were estimated to be 0.34 kg/CO_2e for the FCV system and 0.48 kg/CO_2e for the GV system, with the share of the manufacturing stage almost identical to that of energy consumption. FCVs exhibited lower energy consumption and GHG emissions than GVs of similar vehicle weights.

A sectoral breakdown of energy consumption revealed that for the FCV system, 60% was attributed to electric power consumption, of which 70% was attributed to vehicle manufacturing, 17% to hydrogen station boosting and cooling equipment, and 10% to transportation boosting and light oil. The second largest sector next to electric power consumption was FCV traveling (hydrogen production). For the ICV system, the largest fraction of energy consumption went to the driving phase, followed by the power sector. While vehicle manufacturing and usage accounted for the largest share of power consumption, around 45% of electric power was used to refine oil from crude oil and to operate gas stations.

The study concludes that fossil fuel-derived hydrogen production, which provides a stable and economical energy supply, could contribute to improving environmental and energy efficiencies, making it a viable low-carbon power source option for the near future.

We point out possible limitations regarding the above results. The use of a high-resolution IO table expanded to include processes specific to the FCV system enabled the authors to consider details of the FCV system that otherwise would have been possible only with a detailed PLCA, but at the cost of introducing ambiguity in the system boundary. Still, the authors pointed to possible uncertainty in the results due to the use of price-based data instead of physical data, particularly for the power sector. Another concern is the focus of this study on only GWP. As indicated by the studies we discussed above, the use of precious metals in an FC, such as Pt needed for the catalyst, could cause significant impacts other than GWP. Closely related to this point is the neglect in this study of the impacts of imports or the limited consideration of global impacts. Given the high dependence of the FCV system on imports, the global impacts of the extraction and processing of these imports may be nonnegligible, providing another source of uncertainty.

6.3.2.4 A Synthesis of HLCA Based on WIO

The HLCA of ICVs and EVs discussed above is based on (6.21), with the first term referring to the impacts from the production stage, and the second and third terms to those from the use phase. Recalling that the life cycle of a product consists of the production, use, and EoL stages, we notice that the EoL stage is missing in (6.21). Of the HLCA-based studies on EVs we discussed above, [74, 78, 84] considered the EoL stage, where most metals in EoL vehicles and batteries are recovered and recycled. However, they explain how the EoL stage was considered using only descriptions, without presenting rigid mathematical formulas. The absence of exact mathematical formulas makes it difficult to reproduce the previous LCA results or conduct new LCA studies on similar topics: an LCA calculation needs exact mathematical formulas.

In Sect. 5.4, we showed that by extending the standard IO model to include the EoL stage, besides the production and use stages, the WIO made it possible to represent the whole life cycle of a product. We now show how the HLCA model (6.21) can be extended, within the framework of the WIO, to include the EoL stage explicitly. Recall from Sect. 5.4.2.2 that the WIO model in its standard form is given by (see Sect. 5.4.2.2 for the notations)

$$\begin{aligned} \boldsymbol{x}^{\mathrm{I}} &= \boldsymbol{A}^{\mathrm{I}}\boldsymbol{x}^{\mathrm{I}} + \boldsymbol{A}^{\mathrm{II}}\boldsymbol{x}^{\mathrm{II}} + \boldsymbol{y}^{\mathrm{I}} \\ \boldsymbol{w} &= \boldsymbol{G}^{\mathrm{I}}\boldsymbol{x}^{\mathrm{I}} + \boldsymbol{G}^{\mathrm{II}}\boldsymbol{x}^{\mathrm{II}} + \boldsymbol{w}^{y} \end{aligned} \tag{6.28}$$

and that the impacts originating from the final demand for goods & services, $\boldsymbol{y}^{\mathrm{I}}$, and from the discard of waste and EoL products from the final demand, $\boldsymbol{w}^{y}$, are given by

$$\begin{aligned} \boldsymbol{e} &= \begin{pmatrix} \boldsymbol{F}^{\mathrm{I}} & \boldsymbol{F}^{\mathrm{II}} \end{pmatrix} \begin{pmatrix} \boldsymbol{I} - \boldsymbol{A}^{\mathrm{I}} & -\boldsymbol{A}^{\mathrm{II}} \\ -\boldsymbol{S}\boldsymbol{G}^{\mathrm{I}} & \boldsymbol{I} - \boldsymbol{S}\boldsymbol{G}^{\mathrm{II}} \end{pmatrix}^{-1} \begin{pmatrix} \boldsymbol{y}^{\mathrm{I}} \\ \boldsymbol{S}\boldsymbol{w}^{y} \end{pmatrix} + \boldsymbol{F}_{\epsilon}\boldsymbol{y}_{fuel} \\ &= \begin{pmatrix} \boldsymbol{F}^{\mathrm{I}} & \boldsymbol{F}^{\mathrm{II}} \end{pmatrix} \begin{pmatrix} \boldsymbol{L}^{\mathrm{I,I}} & \boldsymbol{L}^{\mathrm{I,II}} \\ \boldsymbol{L}^{\mathrm{II,I}} & \boldsymbol{L}^{\mathrm{II,II}} \end{pmatrix} \begin{pmatrix} \boldsymbol{y}^{\mathrm{I}} \\ \boldsymbol{S}\boldsymbol{w}^{y} \end{pmatrix} + \boldsymbol{F}_{\epsilon}\boldsymbol{y}_{fuel} \end{aligned} \tag{6.29}$$

We now turn to implement (6.29) into (6.21).

The WIO Representation of (6.21)

Denote by $\boldsymbol{y}_j$ an $n_{\mathrm{I}} \times 1$ vector referring to the purchase of a vehicle of type j, by $\boldsymbol{y}_{fuel,j}$ an $n_{\mathrm{I}} \times 1$ vector referring to the use of fuel to power vehicle j, by $\boldsymbol{y}_{waste,j}$ an $n_w \times 1$ vector referring to the discard of waste in the use stage of vehicle j, and by $\boldsymbol{y}_{EoL,j}$ an $n_w \times 1$ vector referring to the generation of EoL vehicle j at the end of its life. Examples of $\boldsymbol{y}_{waste,j}$ include waste oil, waste acid (from lead batteries), and EoL batteries. Extended to include the EoL stage, we obtain the following WIO representation of (6.21)

$$\boldsymbol{e} = \begin{pmatrix} \boldsymbol{F}^{\mathrm{I}} & \boldsymbol{F}^{\mathrm{II}} \end{pmatrix} \begin{pmatrix} \boldsymbol{L}^{\mathrm{I,I}} & \boldsymbol{L}^{\mathrm{I,II}} \\ \boldsymbol{L}^{\mathrm{II,I}} & \boldsymbol{L}^{\mathrm{II,II}} \end{pmatrix} \left(\begin{pmatrix} \boldsymbol{y}_j \\ \boldsymbol{0} \end{pmatrix} + \begin{pmatrix} \boldsymbol{y}_{fuel.j} \\ \boldsymbol{S}\boldsymbol{y}_{waste,j} \end{pmatrix} + \begin{pmatrix} \boldsymbol{0} \\ \boldsymbol{S}\boldsymbol{y}_{EoL,j} \end{pmatrix} \right) + \boldsymbol{F}_{\epsilon} \boldsymbol{y}_{fuel,j} \tag{6.30}$$

Besides the occurrence of waste/EoL products, $\boldsymbol{y}_{waste,j}$ and $\boldsymbol{y}_{EoL,j}$, (6.30) is distinguished from (6.21) in several respects. First, it explicitly considers the generation and recycling of process (primary) and secondary waste via $\boldsymbol{G}^{\mathrm{I}}$ and $\boldsymbol{G}^{\mathrm{II}}$. The positive elements of $\boldsymbol{G}^{\mathrm{I}}$ refer to the generation of primary waste, while the negative elements refer to recycling. On the other hand, $\boldsymbol{G}^{\mathrm{II}}$ refers mostly to the generation of secondary waste, such as shredded waste, generated upon submitting primary waste to respective treatment processes. Treatment processes are represented by relevant columns in $\boldsymbol{A}^{\mathrm{II}}$, with the allocation of primary wastes to them given by the relevant rows of allocation matrix $\boldsymbol{S}$.

It is usually the case that primary wastes are recycled not in their original form but after being transformed by waste treatment processes into secondary wastes. For instance, the WIO table developed by the Japanese Ministry of the Environment [85] identifies ten treatment processes, 48 items of primary waste, and 51 items of secondary waste, with the former mostly recycled after having been transformed into the latter (Sect. 5.4.2.1). The generation of primary wastes is represented as positive elements in $\boldsymbol{G}^{\mathrm{I}}$. Waste treatment processes transform them into secondary wastes and residues, represented as positive elements in $\boldsymbol{G}^{\mathrm{II}}$. The recycling of secondary wastes is represented by negative elements of $\boldsymbol{G}^{\mathrm{I}}$.

Secondly, wastes in the use phase, such as waste oil, waste acid (from lead batteries), waste tires, and end-of-life batteries, can be represented as positive elements in $\boldsymbol{y}_{waste,j}$. For instance, the EoL stage of Li-Ion batteries considered by [78] can be represented by the EoL battery pack occurring as a positive element in $\boldsymbol{y}_{waste,j}$, submitted to the column in $\boldsymbol{A}^{\mathrm{II}}$ referring to the hydrometallurgical recycling process, with recovered materials and treatment residues occurring as positive elements in $\boldsymbol{G}^{\mathrm{II}}$ and recycling thereof as negative elements of $\boldsymbol{G}^{\mathrm{I}}$.

Thirdly, the EoL vehicle (ELV) occurs as a positive element in $\boldsymbol{y}_{EoL,j}$, with the treatment (for example, disassembling and shredding) to which it is subjected given by the corresponding row elements of $\boldsymbol{S}$ and the treatment processes occurring as column elements in $\boldsymbol{A}^{\mathrm{II}}$. As for the recovery and recycling of parts, components, and materials, the same as that of EoL batteries described above applies.

Treatment Residues

Any waste treatment will generate residues (represented as positive elements in $\boldsymbol{G}^{\mathrm{II}}$), such as vehicle shredding residues, which need further treatment. The treatment processes to which residues are submitted are given by the corresponding row elements of $\boldsymbol{S}$. If the process to which residues are submitted happens to be the final disposal, that is, landfill, there will be no further treatment. If submitted to an intermediate process, such as incineration, the residues will undergo a further transformation, say into ash, before they end up in a landfill (ash could still be recycled in cement production).

Reuse, Re-manufacturing, and Refurbishment

The modeling represented by (6.30) is by no means limited to recycling materials recovered from an EoL product, with the original functionality of its parts and components being mostly lost. Reuse, re-manufacturing, and refurbishment of EoL products/components, which aim at preserving their original functionality [86, 87], can also easily be considered by appropriate modeling of the relevant processes and extending the row elements of $\boldsymbol{G}$s to accommodate the flow (generation and use) of these items. It is worthwhile to note that reuse, re-manufacturing, and refurbishment of EoL products/components are a vital ingredient of Circular Economy (CE) [88] (Sect. 5.4.4.5).

6.3.2.5 Electrification, Not a Silver Bullet!?

As we have seen above, LCA studies broadly agree with the effectiveness of EVs in mitigating CO_2 emissions from fossil fuel consumption when renewable technologies generate electricity. As we briefly pointed out in Sect. 6.2.4.3, the intermittent nature of renewable technologies such as wind power, PV, and CSP, needs to be properly taken into account in assessing their ability to meet the security of the power supply. None of the LCA studies we discussed above considered this aspect of renewable power technologies. Most LCA studies on EVs use the impacts per vehicle or km driven as the functional unit and assume a stable supply of electricity even under scenarios with 100% reliance on intermittent power sources. These LCA studies provide perspectives at a vehicle level, but they fall short of examining the large-scale implications of vehicle electrification [89]. The impacts of large-scale deployment of EVs are seldom considered. Reference [50] is a noticeable exception, addressing also the issue of intermittence to some extent.[8] While [65] considered the impacts of large-scale vehicle electrification in China, no consideration was given to its implications for the power supply.

---

[8] Reference [50] proposes methods of reducing intermittency or its effects, including interconnecting geographically dispersed intermittent energy sources, using smart meters to charge EVs optionally, and storing the electric power for later use. See [90] for a further update along the line of this study and [91] for a debate about the methodology used.

With this backdrop, [89] analyzed the energy and resource implications for the US LDV fleet remaining within sectoral CO_2 emission budgets consistent with a 2 °C target with a clear focus on electrification. A backcasting procedure was used to quantify the timing and volumes of EVs required to remain within suitable CO_2 emission budgets, the resulting electricity use and potential impacts on the electricity system, and the material flows of the electric batteries. Backcasting (a concept opposite to forecasting) analysis is a normative method that involves working backwards from a particular desirable future end-point to the present to determine the physical feasibility of that future and what policy measures would be required to reach that point [92, 93].

In this study, the desirable future was represented by a 2 °C global temperature change target at the end of the century compared to pre-industrial levels. An optimization model was used to build the backcasting procedure, with the objective function of minimizing the stock of EVs in the US LDV fleet from 2020 to 2050, in terms of the EV deployment level, under the constraints on the levels of CO_2 emissions from LDVs to remain within the CO_2 emission budget allowed. The size of the US LDV fleet, the associated CO_2 emissions and material flows, and the CO_2 emission budget pertaining to the fleet were obtained from the Global Change Assessment Model (GCAM) [94].

The following were the major findings of this extensive study:

Required EVs: Under a business-as-usual LDV fleet and current policies, up to 350 million EVs would need to be on the road in the US in 2050, or up to 90% of the on-road LDV fleet. A 100% market share of EVs by 2050, or possibly as early as 2035, would be necessary to attain this order of EV penetration.

Required electricity to power the EV fleet: Powering the fleet of 350 million on-road EVs in the US would imply an annual electricity demand of up to 1,730 TW hours, equivalent to 41% of the 2018 annual national electricity generation. Furthermore, the shape of peak residential demand could also be dramatically affected, calling for coordination between the EVs and their driving/charging behaviors to avoid technical instability in power systems.

Required critical materials to produce the EV fleet: Up to 5.0, 7.2, and 7.8 Tg (Mt), respectively of Li, Co, and Mn would need to be extracted between 2019 and 2050 for the US LDV fleet alone, or 8% and 29% of the identified world terrestrial resources of Li and Co in 2019.

Based on these results, [89] conclude that betting solely on EVs to bridge the US LDV fleet's CO_2 mitigation gaps is not realistic, that is, "*electrification is not a silver bullet*," and calls for the need to develop a wide range of policies combined with a willingness to drive less with lighter, more efficient vehicles, including transit-oriented land-use policies, deployment of new public transport options, innovative taxes on fuel, parking, congestion, and road use, and subsidies for public transportation.[9]

[9] The sales-weighted average mass of any new US LDV has been increasing in the last decades, due, among others, to the increasing share of light-trucks among new vehicles, and better performance (more horsepower) [89].

6.3.2.6 Materials Implications of Decarbonization

We complement the above results for the metals requirement of EVs with a recent study by [95] on possible bottlenecks of future demand (for metals) versus geological availability.

Assessing the supply risks for 31 materials under a business-as-usual scenario about the world development of wind power, solar photovoltaic, solar thermal power, and passenger EVs for the 2016–2050 time period, [95] found high supply risks for 13 elements: Cd, Cr, Co, Cu, Ga, In, Li, Mn, Ni, Ag, Te, Sb, and Zn, with Te (mostly demanded for the manufacture of solar photovoltaic cells) presenting the highest risk. Improving recycling rates was suggested as a means to overcome these constraints.

Reference [96] provides global estimates of end-of-life recycling rates for 60 metals and metalloids, circa 2008:

- (<1%) Li, Be, B, Sc, V, Ga, Ge, As, Se, Sr, Y, Zr, In, Te, Ba, Hf, Ta, Os, Tl, Bi, and the lanthanides,
- (1−10%) Sb and Hg,
- (10−25%) Ru, Cd, and W
- (25−50%) Mg, Mo, and IR,
- (>50%) Al, Ti, Cr, Mn, Fe, Co, Ni, Cu, Zn, nb, Rh, Pd, Ag, Sn, Re, Pt, Au, and Pb.

Notably, of the 13 metals identified as critical, the In, Li, and Te recycling rate was below 1%.

6.4 Summary and Remarks

Many LCA studies confirm the effectiveness of replacing fossil-based power technologies with those based on renewable sources, such as wind, PV, CSP, geothermal, hydro, and tidal. While other mitigation options include nuclear and CCS, the former still has problems with the final disposal of nuclear waste. CCS effectively reduces CO_2 emissions, but at the cost of increasing other emissions. Since CCS requires electricity for its operation, the electricity required to meet a given demand for electricity has to increase.

Analogous to replacing fossil-based power technologies with those based on renewables, electrification of the transport sector is regarded as an effective means to mitigate GHG emissions. The above LCA studies show that the extent to which EVs contribute to the mitigation depends on the driving distance, the emission embodied in the vehicle, and how the electricity to power the vehicle is produced; mere replacement of ICVs with EVs, neglecting these factors, may be counterproductive.

A caveat to renewables is that some are powered by intermittent sources, which applies to wind and solar, and to a lesser extent to geothermal, hydro, and tidal. None of the LCA studies we discussed has addressed the issue of intermittence. Instead, it was assumed that whatever the energy source, the electricity is steadily supplied without intermittence.[10] LCA studies are mostly concerned with the impacts per EV, km, or kWh but seldom consider the impacts of large-scale deployment of these technologies. Consideration of intermittence, an issue beyond the scope of this book, will be an important area of future research in IE.

It is important to notice that GHG emissions are by no means the only environmental concern to be addressed. We saw above that renewable technologies and EVs require more materials than conventional fossil-based ones, particularly scarce and rare metals, in the production stage. The backcasting analysis by [89] indicates that the global scale replacement of ICVs with EVs could have enormous implications for the global availability of minerals such as Li, Co, and Mn. Extending the scope to renewable power technologies, [95] identified high supply risks for 13 metals.

With skyrocketing growth in the demand for these minerals amid limited availability and uneven geographical distribution, it has become increasingly important to look for these metals in urban mines, that is, in EoL products, besides natural mines. For some critical metals, including the one with the highest supply risk, Te, the recycling rate remains below 1% [96], leaving much room for exploitation.

Effective exploitation of metals in urban mines calls for good knowledge about the flow of metals in the economy. For instance, we need to know in which products the metals of concern initially occur and how the product location of the metals evolves over the life cycles of products involving recycling into applications different from the initial one. Equally important is the information about the combination of metals in a product; metals seldom occur isolated in a product but simultaneously as alloys or mechanically/chemically bound together. These are the subject of Material Flow Analysis (MFA), which is the topic of the next chapter.

References

1. Olivier, Jos G.J. 2021. Trends in global CO_2 and total greenhouse gas emissions. *Summary Report*, 2022. PBL Netherlands Environmental Assessment Agency, The Hague: Technical report.
2. Japan Ministry of the Environment. 2022. National Greenhouse Gas Inventory Report of JAPAN. Technical report, Ministry of the Environment, Japan Greenhouse Gas Inventory Office of Japan (GIO), CGER, NIES.
3. Crippa, M., D. Guizzardi, E. Solazzo, M. Muntean, E. Schaaf, F. Monforti-Ferrario, M. Banja, J.G.J. Olivier, G. Grassi, S. Rossi and E. Vignati. 2021. GHG emissions of all world countries - 2021 Report, EUR 30831 EN. Technical report, Publications Office of the European Unio, Luxembourg.
4. Crippa, M., D. Guizzardi, M. Banja, E. Solazzo, M. Muntean, E. Schaaf, F. Pagani, F. Monforti-Ferrario, J. Olivier, R. Quadrelli, A. Risquez-Martin, P. Taghavi-Moharamli, G. Grassi, S. Rossi, D. Jacome-Felix-Oom, A. Branco and E. San-Miguel-Ayanz. 2022. CO_2 emissions of all world countries - 2022 Report, EUR 31182 EN. Technical report, Publications Office of the European Union.

[10] In their voluminous book on LCA, one with more than 1000 pages, [97], the term "intermittence" occurs on only a single page, p. 652.

5. US Energy Information Administration. 2023. https://www.eia.gov/international/data/world. Accessed on 2023-07-01.
6. IEA. 2023. https://is4ie.org/about/what-is-industrial-ecology. Accessed on 2023-05-31.
7. Moran, Michael J., Howard N. Shapiro, Daisie D. Boettner, and Margaret B. Bailey. 2010. *Fundamentals of engineering thermodynamics*. Wiley.
8. US Energy Information Administration. 2023. https://www.eia.gov. Accessed on 2023-07-01.
9. US Department of Energy. 2023. https://www.nrc.gov/docs/ML1002/ML100280723.pdf. Accessed on 2023-07-01.
10. Hossain, Eklas, and Slobodan Petrovic. 2021. *Renewable energy crash course*. Springer.
11. Asdrubali, Francesco, Giorgio Baldinelli, Francesco D'Alessandro, and Flavio Scrucca. 2015. Life cycle assessment of electricity production from renewable energies: Review and results harmonization. *Renewable and Sustainable Energy Reviews* 42: 1113–1122.
12. Wang, Z.J., and Z.W. Wang. 2019. A review on tidal power utilization and operation optimization. *IOP Conference Series: Earth and Environmental Science* 240(5).
13. Ragheb, Magdi, and Adam M. Ragheb. 2011. Wind turbines theory-the betz equation and optimal rotor tip speed ratio. *Fundamental and Advanced Topics in Wind Power* 1(1): 19–38.
14. The world's biggest wind turbines. 2022. https://www.evwind.es/2022/06/04/the-worlds-biggest-wind-turbines/86393. Accessed on 2023-05-31.
15. Richter, Christoph, Daniel Lincot, and Christian Gueymard (eds.) *Solar energy*. Springer.
16. Munz, Benjamin, and James Hays. 2009. Concentrating solar trough modeling: Calculating efficiency. *A burns & McDonnell.*
17. Guillén-Lambea, Silvia, and Mónica Carvalho. 2021. A critical review of the greenhouse gas emissions associated with parabolic trough concentrating solar power plants. *Journal of Cleaner Production* 289.
18. Fthenakis, Vasilis, and Hyung Chul Kim. 2010. Life-cycle uses of water in U.S. electricity generation. *Renewable and Sustainable Energy Reviews* 14(7): 2039–2048.
19. Kaushika, N.D., Anuradha Mishra, and Anil K. Rai. 2018. *Solar photovoltaics technology, system design: Reliability and viability*. Springer.
20. Kaltschmitt, Martin, Nickolas J. Themelis, Lucien Y. Bronicki, Lennart Söder, and Luis A. Vega. 2013. *Renewable energy systems*. New York, NY, USA: Springer.
21. JOGMEC. 2023. https://geothermal.jogmec.go.jp/. Accessed on 2023-07-01.
22. Moon, Hyungsul, and Sadiq J. Zarrouk. 2014. Efficiency of geothermal power plants: A worldwide review. *Geothermics* 51(November 2012): 142–153.
23. U.S. Department of Energy. 2023. Fuel cell technologies office. https://www.energy.gov/. Accessed on 2023-05-31.
24. Turconi, Roberto, Alessio Boldrin, and Thomas Astrup. 2013. Life cycle assessment (LCA) of electricity generation technologies: Overview, comparability and limitations. *Renewable and Sustainable Energy Reviews* 28: 555–565.
25. Hienuki, Shunichi, Yuki Kudoh, and Hiroki Hondo. 2015. Life cycle employment effect of geothermal power generation using an extended input-output model: The case of Japan. *Journal of Cleaner Production* 93: 203–212.
26. Hienuki, Shunichi. 2015. *Environmental and socio socio-economic impacts analysis of the introduction of renewable energy technologies using input-output tables*. Ph.D. thesis, Yokohama National University.
27. Wolfram, Paul, Thomas Wiedmann, and Mark Diesendorf. 2016. Carbon footprint scenarios for renewable electricity in Australia. *Journal of Cleaner Production* 124: 236–245.
28. Malik, Arunima, Manfred Lenzen, RÔmulo. Neves Ely, Erik Dietzenbacher, The case of biorefining in Australia. 2014. Simulating the impact of new industries on the economy. *Ecological Economics* 107: 84–93.
29. Wiedmann, Thomas O., Sangwon Suh, Kuishuang Feng, Manfred Lenzen, Adolf Acquaye, Kate Scott, and John R. Barrett. 2011. Application of hybrid life cycle approaches to emerging energy technologies-the case of wind power in the UK. *Environmental Science & Technology* 45(13): 5900–5907.

30. Burkhardt, John J., Gravin A. Heath, and Craig S. Turchi. 2011. Life cycle assessment of a parabolic trough concentrating solar power plant and the impacts of key design alternatives. *Environmental Science and Technology* 45 (11): 2457–2464.
31. Hertwich, Edgar G., Thomas Gibon, Evert A. Bouman, Anders Arvesen, Sangwon Suh, Garvin A. Heath, Joseph D. Bergesen, Andrea Ramirez, Mabel I Vega, and Lei Shi. 2015. Integrated life-cycle assessment of electricity-supply scenarios confirms global environmental benefit of low-carbon technologies. *Proceedings of the National Academy of Sciences* 112(20): 6277–6282.
32. Wang, Yan, Zhen Pan, Wenxiang Zhang, Tohid N. Borhani, Rui Li, and Zhien Zhang. 2022. Life cycle assessment of combustion-based electricity generation technologies integrated with carbon capture and storage: A review. *Environmental Research* 207: 112219.
33. Marx, J., A. Schreiber, P. Zapp, M. Haines, J-Fr Hake, and J. Gale. 2011. Environmental evaluation of CCS using life cycle assessment-a synthesis report. *Energy Procedia* 4: 2448–2456.
34. Singh, Bhawna, Anders H. Strømman, and Edgar G. Hertwich. 2011. Comparative life cycle environmental assessment of CCS technologies. *International Journal of Greenhouse Gas Control* 5 (4): 911–921.
35. Schreiber, Andrea, Petra Zapp, and Josefine Marx. 2012. Meta-analysis of life cycle assessment studies on electricity generation with carbon capture and storage. *Journal of Industrial Ecology*, 16(SUPPL.1).
36. International Energy Agency (IEA). 2010. Energy technology perspectives: Scenarios and strategies 2050.
37. Gibon, Thomas, Richard Wood, Anders Arvesen, Joseph D. Bergesen, Sangwon Suh, and Edgar G. Hertwich. 2015. A methodology for integrated, multiregional life cycle assessment scenarios under large-scale technological change. *Environmental Science and Technology* 49 (18): 11218–11226.
38. Pauliuk, Stefan, Richard Wood, and Edgar G. Hertwich. 2015. Dynamic models of fixed capital stocks and their application in industrial ecology. *Journal of Industrial Ecology* 19 (1): 104–116.
39. Norwegian Ministry of Petroleum and Energy. 2023. Norway's energy supply system. https://energifaktanorge.no/en/norsk-energiforsyning. Accessed on 2023-05-31.
40. Ministry of the Environment, Energy, and Climate, Governement of Iceland. 2023. https://www.government.is/topics/business-and-industry/energy. Accessed on 2023-05-31.
41. Clara Franziska Heuberger and Niall Mac Dowell. 2018. Real-world challenges with a rapid transition to 100% renewable power systems. *Joule* 2 (3): 367–370.
42. Mileva, Ana, Josiah Johnston, James H. Nelson, and Daniel M. Kammen. 2016. Power system balancing for deep decarbonization of the electricity sector. *Applied Energy* 162: 1001–1009.
43. MacDonald, Alexander E., Christopher T.M. Clack, Anneliese Alexander, Adam Dunbar, James Wilczak, and Yuanfu Xie. 2016. Future cost-competitive electricity systems and their impact on US CO2 emissions. *Nature Climate Change* 6(5): 526–531.
44. Kleijn, René, Ester Van der Voet, Gert Jan Kramer, Lauran Van Oers, and Coen Van der Giesen. Metal requirements of low-carbon power generation. *Energy* 36(9): 5640–5648.
45. International Energy Agency. 2021. *The Role of Critical Minerals in Clean Energy Transitions*. International Energy Agency: Technical report.
46. Wang, Seaver, Zeke Hausfather, Steven Davis, Juzel Lloyd, Erik B. Olson, Lauren Liebermann, Guido D. Núñez-Mujica, and Jameson McBride. 2023. Future demand for electricity generation materials under different climate mitigation scenarios. *Joule*, 7(2): 309–332.
47. Vidyanandan, K. 2018. Overview of electric and hybrid vehicles. *Energy Scan* 3: 7–14.
48. Del Duce, Andrea, Marcel Gauch, and Hans Jörg Althaus. 2016. Electric passenger car transport and passenger car life cycle inventories in ecoinvent version 3. *International Journal of Life Cycle Assessment* 21(9): 1314–1326.
49. U.S. EPA. All-Electric Vehicles. 2023. https://www.fueleconomy.gov/feg/evtech.shtml. Accessed on 2023-06-11.

50. Jacobson, Mark Z. 2009. Review of solutions to global warming, air pollution, and energy security. *Energy & Environmental Science* 2(2): 148–173.
51. Faria, Ricardo, Pedro Moura, Joaquim Delgado, and Anibal T. De Almeida. 2012. A sustainability assessment of electric vehicles as a personal mobility system. *Energy Conversion and Management* 61: 19–30.
52. Notter, Dominic A., Marcel Gauch, Rolf Widmer, Patrick Wäger, Anna Stamp, Rainer Zah, and Hans Jörg Althaus. 2010. Contribution of Li-Ion batteries to the environmental impact of electric vehicles. *Environmental Science and Technology* 44(19): 6550–6556.
53. Hawkins, Troy R., Bhawna Singh, Guillaume Majeau-Bettez, and Anders Hammer Strømman. 2013. Comparative environmental life cycle assessment of conventional and electric vehicles. *Journal of Industrial Ecology* 17(1): 53–64.
54. Dunn, J.B., L. Gaines, J.C. Kelly, C. James, and K.G. Gallagher. 2015. The significance of Li-ion batteries in electric vehicle life-cycle energy and emissions and recycling's role in its reduction. *Energy and Environmental Science* 8 (1): 158–168.
55. Ellingsen, Linda Ager Wick. 2016. The size and range effect: Life-cycle greenhouse gas emissions of electric vehicles. *Environmental Research Letters* 11: 1–8.
56. Bauer, Christian, Johannes Hofer, Hans Jörg Althaus, Andrea Del Duce, and Andrew Simons. 2015. The environmental performance of current and future passenger vehicles: Life Cycle Assessment based on a novel scenario analysis framework. *Applied Energy* 157: 871–883.
57. Lombardi, Lidia, Laura Tribioli, Raffaello Cozzolino, and Gino Bella. 2017. Comparative environmental assessment of conventional, electric, hybrid, and fuel cell powertrains based on LCA. *International Journal of Life Cycle Assessment* 22 (12): 1989–2006.
58. Yang, Lai, Biying Yu, Bo Yang, Hao Chen, Gabriel Malima, and Yi Ming Wei. 2021. Life cycle environmental assessment of electric and internal combustion engine vehicles in China. *Journal of Cleaner Production* 285: 124899.
59. Dunn, Jennifer B., Linda Gaines, John Sullivan, and Michael Q. Wang. 2012. Impact of recycling on cradle-to-gate energy consumption and greenhouse gas emissions of automotive lithium-ion batteries. *Environmental Science and Technology* 46 (22): 12704–12710.
60. Ellingsen, Linda Ager Wick , Guillaume Majeau-Bettez, Bhawna Singh, Akhilesh Kumar Srivastava, Lars Ole Valøen, and Anders Hammer Strømman. 2014. Life cycle assessment of a lithium-ion battery vehicle pack. *Journal of Industrial Ecology* 18(1): 113–124.
61. Kawamoto, Ryuji, Hideo Mochizuki, Yoshihisa Moriguchi, Takahiro Nakano, Masayuki Motohashi, Yuji Sakai, and Atsushi Inaba. 2019. Estimation of CO_2 emissions of internal combustion engine vehicle and battery electric vehicle using LCA. *Sustainability (Switzerland)* 11(9).
62. Franzò, Simone, and Alessio Nasca. 2020. The environmental impact of electric vehicles: A novel life cycle-based evaluation framework and its applications to multi-country scenarios. *Journal of Cleaner Production* 315: 2021.
63. Tribioli, Laura, Michele Barbieri, Roberto Capata, Enrico Sciubba, Elio Jannelli, and Gino Bella. 2014. A real time energy management strategy for plug-in hybrid electric vehicles based on optimal control theory. *Energy Procedia* 45: 949–958.
64. Hao, Han, Xiang Cheng, Zongwei Liu, and Fuquan Zhao. 2017. Electric vehicles for greenhouse gas reduction in China: A cost-effectiveness analysis. *Transportation Research Part D: Transport and Environment* 56(August 2017): 68–84.
65. Xiong, Siqin, Yunshi Wang, Bo Bai, and Xiaoming Ma. 2020. A hybrid life cycle assessment of the large-scale application of electric vehicles. *Energy* (xxxx): 119314.
66. Hawkins, Troy R., Ola Moa Gausen, and Anders Hammer Strømman. 2012. Environmental impacts of hybrid and electric vehicles: A review. *The International Journal of Life Cycle Assessment* 17(8): 997–1014.
67. Nordelöf, Anders, Maarten Messagie, Anne-Marie Tillman, Maria Ljunggren Söderman, and Joeri Van Mierlo. 2014. Environmental impacts of hybrid, plug-in hybrid, and battery electric vehicles-what can we learn from life cycle assessment? *The International Journal of Life Cycle Assessment* 19(11): 1866–1890.
68. Lombardi, Simone, Laura Tribioli, Giulio Guandalini, and Paolo Iora. 2020. Energy performance and well-to-wheel analysis of different powertrain solutions for freight transportation. *International Journal of Hydrogen Energy* 45 (22): 12535–12554.

69. Wolfram, Paul, and Thomas Wiedmann. 2017. Electrifying Australian transport: Hybrid life cycle analysis of a transition to electric light-duty vehicles and renewable electricity. *Applied Energy* 206 (August): 531–540.
70. Weymar, Elisabeth, and Matthias Finkbeiner. 2016. Statistical analysis of empirical lifetime mileage data for automotive LCA. *International Journal of Life Cycle Assessment* 21 (2): 215–223.
71. Cihat Onat, Nuri, Murat Kucukvar, and Omer Tatari. 2014. Towards life cycle sustainability assessment of alternative passenger vehicles. *Sustainability (Switzerland)* 6(12): 9305–9342.
72. Argonne National Laboratory. 2018. *The greenhouse gases, regulated emissions, and energy use in transportation model (GREET)*. US Department of Energy: Technical report.
73. Yang, Yi., and Reinout Heijungs. 2018. On the use of different models for consequential life cycle assessment assessment. *The International Journal of Life Cycle Assessment* 23 (4): 751–758.
74. Sen, Burak, Nuri C. Onat, Murat Kucukvar, and Omer Tatari. 2019. Material footprint of electric vehicles: A multiregional life cycle assessment. *Journal of Cleaner Production* 209: 1033–1043.
75. Sen, Burak, Murat Kucukvar, Nuri C. Onat, and Omer Tatari. 2019. Life cycle sustainability assessment of autonomous heavy-duty trucks. *Journal of Industrial Ecology*.
76. Henßler, Martin, Vanessa Bach, Markus Berger, Matthias Finkbeiner, and Klaus Ruhland. 2016. Resource efficiency assessment-comparing a plug-in hybrid with a conventional combustion engine. *Resources* 5(1).
77. Dai, Qiang, Jeffrey Spangenberger, Shabbir Ahmed, Linda Gaines, Jarod C. Kelly, and Michael Wang. 2019. EverBatt: A Closed-loop battery recycling cost and environmental impacts model. *Argonne National Laboratory* (pp. 1–88).
78. Sanfélix, Javier, Cristina de la Rúa, Jannick Hoejrup Schmidt, Maarten Messagie, and Joeri Van Mierlo. 2016. Environmental and economic performance of an li-ion battery pack: A multiregional input-output approach. *Energies* 9(8).
79. Dietzenbacher, Erik, Bart Los, Robert Stehrer, Marcel Timmer, and Gaaitzen de Vries. 2013. The construction of world input-output tables in the WIOD project. *Economic Systems Research* 25 (1): 71–98.
80. Sanfélix, Javier, Maarten Messagie, Noshin Omar, Joeri Van Mierlo, and Volker Hennige. 2015. Environmental performance of advanced hybrid energy storage systems for electric vehicle applications. *Applied Energy* 137: 925–930.
81. Hienuki, Shunichi, Haruka Mitoma, Mari Ogata, and Izumi Uchida. 2021. Environmental and energy life cycle analyses of passenger vehicle systems using fossil fuel-derived hydrogen. *International Journal of Hydrogen Energy* (xxxx).
82. Hienuki, Shunichi. 2017. Environmental and socio-economic analysis of naphtha reforming hydrogen energy using input-output tables: A case study from Japan. *Sustainability (Switzerland)* 9(8): 1376.
83. Niitsuma, Takuya. 2008. Hydrogen production and utilization in refineries. *Hydrogen Energy System* 33 (2): 26–29.
84. Cihat Onat, Nuri, Murat Kucukvar, and Omer Tatari. 2015. Conventional, hybrid, plug-in hybrid or electric vehicles? State-based comparative carbon and energy footprint analysis in the United States. *Applied Energy* 150: 36–49.
85. Nakamura, Shinichiro. 2020. Tracking the product origins of waste for treatment using the WIO data developed by the japanese ministry of the environment. *Environmental Science and Technology* 54 (23): 14862–14867.
86. Gharfalkar, Mangesh, Zulfiqur Ali, and Graham Hillier. 2016. Clarifying the disagreements on various reuse options: Repair, recondition, refurbish and remanufacture. *Waste Management and Research* 34 (10): 995–1005.
87. Erfan Shafiee Roudbari, S.M.T., Fatemi Ghomi, and Mohsen S. Sajadieh. 2021. Reverse logistics network design for product reuse, remanufacturing, recycling and refurbishing under uncertainty. *Journal of Manufacturing Systems* 60(November 2020): 473–486.

88. Commission, European. 2015. *Circular Economy Action Plan*. European Union: Technical report.
89. Milovanoff, Alexandre, I. Daniel Posen, and Heather L. MacLean. 2020. Electrification of light-duty vehicle fleet alone will not meet mitigation targets. *Nature Climate Change* 10(12): 1102–1107.
90. Jacobson, Mark Z., Mark A. Delucchi, Mary A. Cameron, and Bethany A. Frew. 2015. Low-cost solution to the grid reliability problem with 100% penetration of intermittent wind, water, and solar for all purposes. *Proceedings of the National Academy of Sciences of the United States of America* 112 (49): 15060–15065.
91. Clack, Christopher T.M., Staffan A. Qvist, Jay Apt, Morgan Bazilian, Adam R. Brandt, Ken Caldeira, Steven J. Davis, Victor Diakov, Mark A. Handschy, Paul D.H. Hines, and Others. 2017. Evaluation of a proposal for reliable low-cost grid power with 100% wind, water, and solar. *Proceedings of the National Academy of Sciences* 114(26): 20161038.
92. Robinson, John B. 1990. Futures under glass. A recipe for people who hate to predict. *Futures* 22(8): 820–842.
93. Karl, H. 1996. Dreborg. Essence of backcasting. *Futures* 28 (9): 813–828.
94. Calvin, Katherine, Pralit Patel, Leon Clarke, Ghassem Asrar, Ben Bond-Lamberty, Ryna Yiyun Cui, Alan Di Vittorio, Kalyn Dorheim, Jae Edmonds, Corinne Hartin, Mohamad Hejazi, Russell Horowitz, Gokul Iyer, Page Kyle, Sonny Kim, Robert Link, Haewon Mcjeon, Steven J. Smith, Abigail Snyder, Stephanie Waldhoff, and Marshall Wise. GCAM v5.1: Representing the linkages between energy, water, land, climate, and economic systems. *Geoscientific Model Development* 12(2): 677–698.
95. Valero, Alicia, Antonio Valero, Guiomar Calvo, and Abel Ortego. 2018. Material bottlenecks in the future development of green technologies. *Renewable and Sustainable Energy Reviews* 93 (May): 178–200.
96. Graedel, Thomas E., Julian Allwood, Jean-Pierre Birat, Matthias Buchert, Christian Hagelüken, Barbara K. Reck, Scott F. Sibley, and Guido Sonnemann. 2011. What do we know about metal recycling rates? *Journal of Industrial Ecology* 15: 355–366.
97. Hauschild, Michael Z., Ralph K. Rosenbaum, and Stig Irvin Olsen (eds.) *Life cycle assessment theory and practice*. Springer.

Chapter 7
Material Flow Analysis

Abstract The chapter provides an in-depth exploration of Material Flow Analysis (MFA) models, covering both static and dynamic approaches. Starting with static MFA models, the discussion centers around the transfer matrix, a crucial element that governs the movement of materials between different production and use stages. The application of the Markov chain is then introduced as a way to analyze these static MFA models. Furthermore, the chapter delves into the IO-based static MFA, with a specific focus on the WIO-MFA model. This model facilitates the conversion of monetary input-output tables into physical material flows. The extensions of the WIO-MFA model are explored, highlighting its ability to accommodate diverse pathways such as packaging and containers. Several case studies are provided to showcase the practical implementation of the WIO-MFA model, including materials like PVC, bisphenol A, iron and steel, ferroalloys in end-of-life vehicles (ELV), copper, the remanufacturing of automobile engines, and plastic packaging and containers. Additionally, the chapter introduces the UPIOM method, which offers an MFA perspective on a specific material induced by a particular product. By rearranging the sequence of sectors based on the degree of fabrication, the UPIOM method can generate a triangulated IO flow matrix. Shifting focus to dynamic MFA models, the chapter provides a brief overview of the standard model based on stock growth and lifetime distributions. Moreover, the MaTrace models are presented as comprehensive tracking mechanisms that trace materials over time and products within open-loop recycling systems. These models explicitly consider losses and the quality of scrap. The chapter concludes by discussing the extensions of the MaTrace models to incorporate global multiregional interdependency, alloys, and multi-material scenarios. To enhance comprehension and practical application, case studies centered around metals are presented throughout the chapter. Summary and remarks conclude the chapter.

S. Nakamura, *A Practical Guide to Industrial Ecology by Input-Output Analysis*,
https://doi.org/10.1007/978-3-031-43684-0_7

7.1 The Static MFA: Methods and Applications

We will start our discussion with a static MFA, taking metal production as an example.

7.1.1 The MFA Model Based on Transfer Coefficients

The MFA Model

MFA is concerned with tracing how the input of raw materials is ultimately transformed into final products and accumulated as stock after undergoing a series of intermediate processes where the outputs of the preceding stage become the inputs to the next production stage, with a fraction of the inputs ending up as losses in each production stage. A fraction of accumulated stock is eventually discarded, which undergoes end-of-life (EoL) treatment to become secondary material to enter the material cycle anew or disposed to a landfill. The law of mass conservation implies that the initial input of raw materials equals the sum of the final products and the losses. Figure 7.1 gives a simplified representation of the flow of aluminum from alumina to the final products, such as cars, industrial machines, and buildings, with process waste generated in each production stage recycled. Four processes are identified: electrolysis, recasting, rolling and forming, and fabrication. The inputs entering each process are transformed into outputs, consisting of products to become inputs to the next process and losses. Mass conservation implies that the mass of alumina used as the raw material equals the sum of aluminum occurring in the final products and all the losses except those recycled.

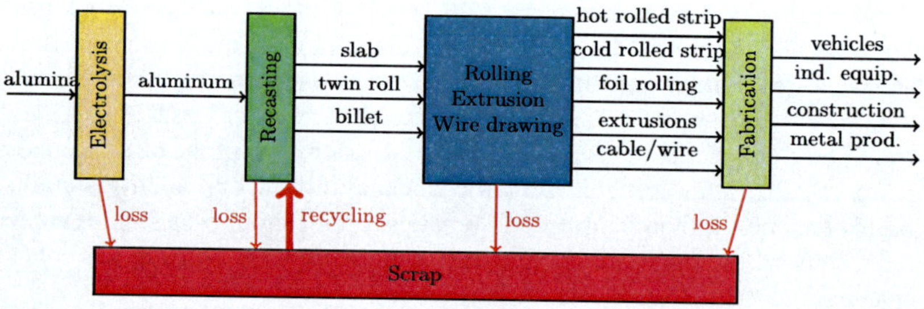

Fig. 7.1 A schematic representation of the transfer of raw materials to final products: aluminum. A schematic representation of the flow of aluminum from alumina to final products. This figure is a grossly simplified representation of Fig. 1 from [1], and it is intended solely for expositional purposes

MFA and Material Footprint

MFA differs from the material footprint (MF), which is the material counterpart of carbon footprint (CF). The MF refers to the "global allocation of used raw material extraction to the final demand of an economy, enumerating the link between the beginning of a production chain (where raw materials are extracted from the natural environment) and its end (where a product or service is consumed), but does not record the actual physical movement of materials within and among countries, which is the object of an MFA [2]." The integrated global LCA study on the wide-scale global deployment of renewables by [3] we discussed in Sect. 6.2.4.2 is concerned with MF but not with MFA; their results indicate a substantial increase in material consumption, especially copper, under IEA's BLUE Map scenario.

7.1.1.1 The Transfer of Inputs to Outputs at a Process Level

Transfer Coefficients

Since MFA is a quantitative discipline, like LCA, we need a mathematical formulation to flesh out the qualitative construct (or skeleton) as in Fig. 7.1 with actual data. In Fig. 7.1, inputs entering each process are "transferred" to outputs leaving the process. In a standard MFA, this transfer (or transformation) of inputs, $i = 1, \ldots, m_k$, in a process, say process k, into outputs, $j = 1, \ldots, n_k$, is represented by *transfer coefficients*, $t_{k;i,j}$ [4, 5]. Denoting by $x_{k;I,i}$ the input i entering process k, and by $x_{k;O,j}$ the output j leaving the process, we have

$$x_{k;O,j} = \sum_{i=1}^{m_k} x_{k;I,i} t_{k;i,j} \tag{7.1}$$

or for all inputs and outputs of the process simultaneously considered

$$\boldsymbol{x}_{k;O} = \boldsymbol{x}_{k;I} \boldsymbol{T}_k \tag{7.2}$$

where $\boldsymbol{x}_{k;J} = (x_{k;1,J}, x_{k;2,J}, \ldots)^\top$, $J \in \{I, O\}$, and $\boldsymbol{T}_k$ is a matrix of $m_k \times n_k$ with $\boldsymbol{T}_k = [t_{k;i,j}]$. With its elements being transfer coefficients, the matrix $\boldsymbol{T}_k$ can be termed the *transfer matrix*. Mass balances imply

$$\sum_j t_{k;i,j} = 1, \quad \text{or} \qquad \boldsymbol{T}_k \boldsymbol{\iota}_O = \boldsymbol{\iota}_I \tag{7.3}$$

Considering Losses in Transfer

The vector of outputs, $\boldsymbol{x}_{k;O}$, in (7.2) includes losses. We now modify (7.2) to exclude losses. Denote by $\gamma_{k;ij} \in [0, 1)$ the yield of input i in its transfer to output j in process k, that is, the fraction of input i that becomes a component of output j. Since a perfect process could not exist, we expect $\gamma_{k;ij} \in [0, 1)$. Applying $\gamma_{k;ij}$ to (7.1), the output net of losses, say $\bar{x}_{k;O,j}$, can be given by

$$\bar{x}_{k;O,j} = \sum_{i=1}^{m_k} x_{k;I,i}\gamma_{k;i,j}t_{k;i,j} \tag{7.4}$$

Denoting by $\boldsymbol{\Gamma}_k = [\gamma_{k;ij}]$ a matrix of equal size as $\boldsymbol{T}_k$, we have from (7.2)

$$\bar{\boldsymbol{x}}_{k;O} = \boldsymbol{x}_{k;I}(\boldsymbol{\Gamma}_k \odot \boldsymbol{T}_k) \tag{7.5}$$

where $\odot$ refers to the Hadamard product, that is, $\boldsymbol{\Gamma}_k \odot \boldsymbol{T}_k = [\gamma_{k;ij}t_{k;ij}]$. Recalling that $1 - \gamma_{k;ij}$ gives the fraction of i that leaves process k as loss, the loss vector, $\bar{\boldsymbol{x}}_{k;L}$, can be given by

$$\bar{\boldsymbol{x}}_{k;L} = \boldsymbol{x}_{k;I}\left((\boldsymbol{J}_k - \boldsymbol{\Gamma}_k) \odot \boldsymbol{T}_k\right) \tag{7.6}$$

where $\boldsymbol{J}_k$ is a matrix of ones with equal size as $\boldsymbol{\Gamma}_k$ (and $\boldsymbol{T}_k$). Note that (7.5) and (7.6) correspond to Eqs. (1) and (2) in [1], except for the use of different notations.

7.1.1.2 The Transfer of Inputs to Outputs at a System Level

The above formulations (7.1)–(7.6) refer to each process in the system. To obtain a system-wide picture of the transfer of inputs into outputs across all the processes, we need information about how the processes are interconnected: how $\boldsymbol{x}_{k;O}$, $\boldsymbol{T}_k$, and $\boldsymbol{x}_{k;I}$ are interrelated for different ks. Taking the system given by Fig. 7.1 as an example, we now consider how $\boldsymbol{T}_k$s are connected.

The Case of No Losses

Neglecting process waste for the sake of simplicity and denoting by $\boldsymbol{x}_0$ the initial inputs to the system and by $\boldsymbol{x}_k$ the output of process k, we obtain the following simple relationship among the four processes

$$\begin{aligned}
\boldsymbol{x}_1 &= \boldsymbol{x}_0\boldsymbol{T}_1 \\
\boldsymbol{x}_2 &= \boldsymbol{x}_1\boldsymbol{T}_2 = \boldsymbol{x}_0\boldsymbol{T}_1\boldsymbol{T}_2 \\
\boldsymbol{x}_3 &= \boldsymbol{x}_2\boldsymbol{T}_3 = \boldsymbol{x}_0\boldsymbol{T}_1\boldsymbol{T}_2\boldsymbol{T}_3 \\
\boldsymbol{x}_4 &= \boldsymbol{x}_3\boldsymbol{T}_4 = \boldsymbol{x}_0\boldsymbol{T}_1\boldsymbol{T}_2\boldsymbol{T}_3\boldsymbol{T}_4
\end{aligned} \tag{7.7}$$

The (i, j) element of matrix $\boldsymbol{T}_1\boldsymbol{T}_2\boldsymbol{T}_3$ gives the fraction of initial input i that becomes j after three transitions, while the ith row of $\boldsymbol{T}_4$ gives the final state of initial input i.

Alternatively, denoting by $\boldsymbol{T}$ the following matrix

$$\boldsymbol{T} = \begin{pmatrix} \mathbf{0} & \boldsymbol{T}_1 & \mathbf{0} & \mathbf{0} & \mathbf{0} \\ \mathbf{0} & \mathbf{0} & \boldsymbol{T}_2 & \mathbf{0} & \mathbf{0} \\ \mathbf{0} & \mathbf{0} & \mathbf{0} & \boldsymbol{T}_3 & \mathbf{0} \\ \mathbf{0} & \mathbf{0} & \mathbf{0} & \mathbf{0} & \boldsymbol{T}_4 \\ \mathbf{0} & \mathbf{0} & \mathbf{0} & \mathbf{0} & \boldsymbol{I} \end{pmatrix} \tag{7.8}$$

and writing $\check{\boldsymbol{x}}_0 = (\boldsymbol{x}_0 \quad \mathbf{0} \quad \mathbf{0} \quad \mathbf{0} \quad \mathbf{0})$, we have

$$\begin{aligned} (\mathbf{0} \quad \boldsymbol{x}_1 \quad \mathbf{0} \quad \mathbf{0} \quad \mathbf{0}) &= \check{\boldsymbol{x}}_0\boldsymbol{T} \\ (\mathbf{0} \quad \mathbf{0} \quad \boldsymbol{x}_2 \quad \mathbf{0} \quad \mathbf{0}) &= (\mathbf{0} \quad \boldsymbol{x}_1 \quad \mathbf{0} \quad \mathbf{0} \quad \mathbf{0})\boldsymbol{T} = \check{\boldsymbol{x}}_0\boldsymbol{T}^2 \\ (\mathbf{0} \quad \mathbf{0} \quad \mathbf{0} \quad \boldsymbol{x}_3 \quad \mathbf{0}) &= (\mathbf{0} \quad \mathbf{0} \quad \boldsymbol{x}_2 \quad \mathbf{0} \quad \mathbf{0})\boldsymbol{T} = \check{\boldsymbol{x}}_0\boldsymbol{T}^3 \\ (\mathbf{0} \quad \mathbf{0} \quad \mathbf{0} \quad \mathbf{0} \quad \boldsymbol{x}_4) &= (\mathbf{0} \quad \mathbf{0} \quad \mathbf{0} \quad \boldsymbol{x}_3 \quad \mathbf{0})\boldsymbol{T} = \check{\boldsymbol{x}}_0\boldsymbol{T}^4 \end{aligned} \tag{7.9}$$

Note that

$$(\mathbf{0} \quad \mathbf{0} \quad \mathbf{0} \quad \mathbf{0} \quad \boldsymbol{x}_4)\boldsymbol{T} = (\mathbf{0} \quad \mathbf{0} \quad \mathbf{0} \quad \mathbf{0} \quad \boldsymbol{x}_4) \tag{7.10}$$

which shows that the final products $\boldsymbol{x}_4$ are the final destination of $\boldsymbol{x}_0$: having once reached the final state, state 4, $\boldsymbol{x}_0$ stays in that state with no further transfer.

Recycling of Process Waste

The presence of losses or the generation of process waste can be considered by replacing $\boldsymbol{T}_k$ with $\boldsymbol{\Gamma}_K \odot \boldsymbol{T}_k$ and $\boldsymbol{x}_k$ with $\bar{\boldsymbol{x}}_k$ in the above formulation. It is usually the case that process wastes are mostly recycled, with a fraction ending up in a landfill. We now extend the above formulation by including the generation of process wastes and their recycling and final disposal to a landfill. Denote by $\boldsymbol{T}_{k,w}$, an $m_k \times n_w$ matrix, the generation of waste in process k, by $\boldsymbol{T}_{w,k}$ the use of waste (recycling) in process k, and by $\boldsymbol{t}_{w,L}$, an $n_w \times 1$ vector, the transfer vector of waste to a landfill. We then have the following expression of $\boldsymbol{T}$

$$\boldsymbol{T} = \begin{pmatrix} \mathbf{0} & \boldsymbol{T}_{1,w} & \boldsymbol{T}_1 & \mathbf{0} & \mathbf{0} & \mathbf{0} & \mathbf{0} \\ \mathbf{0} & \boldsymbol{T}_{w,1} & \boldsymbol{T}_{w,2} & \boldsymbol{T}_{w,3} & \boldsymbol{T}_{w,4} & \mathbf{0} & \boldsymbol{t}_{w,L} \\ \mathbf{0} & \boldsymbol{T}_{2,w} & \mathbf{0} & \boldsymbol{T}_2 & \mathbf{0} & \mathbf{0} & \mathbf{0} \\ \mathbf{0} & \boldsymbol{T}_{3,w} & \mathbf{0} & \mathbf{0} & \boldsymbol{T}_3 & \mathbf{0} & \mathbf{0} \\ \mathbf{0} & \boldsymbol{T}_{4,w} & \mathbf{0} & \mathbf{0} & \mathbf{0} & \boldsymbol{T}_4 & \mathbf{0} \\ \mathbf{0} & \mathbf{0} & \mathbf{0} & \mathbf{0} & \mathbf{0} & \boldsymbol{I} & \mathbf{0} \\ \mathbf{0} & \mathbf{0} & \mathbf{0} & \mathbf{0} & \mathbf{0} & \mathbf{0} & 1 \end{pmatrix} \tag{7.11}$$

where T_ks, $k = 1, \ldots, 4$, are understood to exclude losses. The second column refers to waste generation, while the second row refers to the transfer thereof among production processes (recycling) and a landfill. The last row of T refers to the state of a landfill; once process waste moves to a landfill, it is transferred to no other state.

The MFA of Global Aluminum and Steel from Liquid Metal to Final Products

References [1, 6] respectively traced the global flows of aluminum and steel from liquid metal to final products, revealing for the first time a complete map of the global system for these metals. Their methodology is based on the transfer matrix as given by (7.11) but different from it in two respects; first, besides ore, EoL scrap was considered as another source of metal resources.; secondly, a landfill was not considered an absorbing state. Their results for aluminum indicate the poor material efficiency of the industry, where approximately 50% of all cast aluminum is discarded as scrap during manufacturing without reaching a final product, and the significant "dilution" and "cascade" flows from high-grade aluminum to lower purity alloys to make up for the shortfall in scrap supply and to obtain the desired alloy mix. Possible solutions for addressing the problems of dilution and down-cycling were proposed, including better solutions for segregating and sorting of EoL scrap, and exploring reuse options.

Like aluminum, five processes were identified concerning steel: Reduction (blast furnaces and electric arc furnaces (EAFs)), Steel making, Casting, Rolling/Forming, and Fabrication (into four groups of final products), with iron resources coming from ore and EoL scrap. Their findings include that one-quarter of all liquid steel produced never makes it into end-use goods; instead, it is discarded as scrap during the casting, forming, and fabrication processes, implying wide scope for a possible increase in material efficiency.

7.1.1.3 *The Transfer Matrix and Markov Chain

The Transfer Model Involving Different States

The transfer model can, in general, be represented as

$$x_O = x_I T \tag{7.12}$$

with the rows of T representing sources and the columns representing destinations. The transfer coefficient t_{ij} represents the transfer of material from state i to state j; the material moves from the state of i, say aluminum slab, to the state of j, say aluminum rolled strip. Each time the inputs are multiplied by the transfer coefficients, they are transferred to the next state. Therefore, it is legitimate to regard (7.12) as a process that transforms the state of matter to the next stage, say from state i to state $i + 1$. Denoting by "(s)" the state of material, an alternative representation of (7.12) in terms of states can be given by

$$\boldsymbol{x}(s+1) = \boldsymbol{x}(s)\boldsymbol{T} \tag{7.13}$$

In the literature on stochastic processes, (7.13) is known as the Markov chain, one of the most widely used models of stochastic processes [7–9].[1] This section discusses features of the Markov chain that are highly relevant to MFA.

Transition Probability Matrix and the Markov Chain

In the Markov chain, t_{ij} refers to the probability of material in state i moving to state j in the next step. Denoting by $s = 0, 1, \ldots$ the state, with $s = 0$ the initial state, we have from (7.13)

$$\boldsymbol{x}(s) = \boldsymbol{x}(s-1)\boldsymbol{T} = \boldsymbol{x}(0)\boldsymbol{T}^s \tag{7.14}$$

As we noted above, the (i, j)-element of $\boldsymbol{T}^s$ gives the the probability, $P_{ij}(s)$, of a material in initial state i coming to be in state j through s transitions ([7], Theorem 2.3.2)

$$P_{ij}(s) = [\boldsymbol{T}^s]_{i,j} \tag{7.15}$$

Absorbing Markov Chains

An important class of the Markov chain highly relevant to MFA is "absorbing" chains. The states of a Markov chain are classified into two groups: a transient state and an absorbing state.[2] The former, once left, are never entered again; while the latter, once entered, are never again left [7]. For example, in the system given by Fig. 7.1, intermediate products such as slab and billet in the production of aluminum products are in a transient state; once they get further fabricated and reach the next state, such as hot rolled strips or extrusions, there is no way for them to return to the previous state (except as process waste, which, however, represents another state). On the other hand, final products, such as vehicles, industrial machinery, and construction, refer to an absorbing state.

If i is an absorbing state, $t_{ii} = 1$ and $t_{ij} = 0$ for all $j \neq i$. A Markov chain, all of whose nontransient states are absorbing, is called *an absorbing Markov chain*. A static MFA model, where all materials eventually end up in final products, is an absorbing Markov chain; having once become final products, the materials are "locked in" there with no further transfer to other states.[3] The model we considered

[1] A Markov chain is a special form of a Markov process, whose transition probabilities, p_{ij}, do not depend on the number of steps. See [7] for more details.

[2] To be exact, an absorbing state is a special case of a nontransient (or ergodic state); if a state is the only element of a nontransient set, it is called an absorbing state. See [7] for details.

[3] In a dynamic MFA, where the EoL processes of final products are explicitly considered, not final products but landfill will be the absorbing state.

above in Sect. 7.1.1.2 is an absorbing Markov chain: the identity matrix at the south-east corner of $\boldsymbol{T}$ in (7.8) concerns the absorbing state.

The Canonical Form of $\boldsymbol{T}$ of an Absorbing Markov Chain

For a given set of materials, let there be n_a absorbing states, final products or final disposals, such as landfill, and n_t transient states, including resources, intermediate products, and process waste. Let $n := n_a + n_t$. The $n \times n$ matrix $\boldsymbol{T}$ can then be put into canonical form as follows, with the rows of $\boldsymbol{T}$ representing sources, while the columns represent destinations

$$\boldsymbol{T} = \begin{pmatrix} \boldsymbol{Q} & \boldsymbol{R} \\ \boldsymbol{0} & \boldsymbol{I} \end{pmatrix} \tag{7.16}$$

The $n_t \times n_t$ matrix $\boldsymbol{Q}$ concerns the process as long as it stays in transient states, the $n_t \times n_a$ matrix $\boldsymbol{R}$ concerns the transition from transient to absorbing states, and the identity matrix $\boldsymbol{I}$ of order n_a refer to the absorbing state.

The Fundamental Matrix

Recall from (7.15) that the probability of i being in transient state j after s transitions, $P_{ij}(s)$, $i, j \in$ transient state, is given by

$$P_{ij}(s) = [\boldsymbol{Q}^s]_{ij} \tag{7.17}$$

By definition of a transient state, we have

$$\lim_{s\to\infty} [\boldsymbol{Q}^s]_{ij} = 0, \quad \forall i, j \tag{7.18}$$

that is,

$$\lim_{s\to\infty} \boldsymbol{Q}^s = \boldsymbol{0} \tag{7.19}$$

From which follows the important result ([7], Theorem 3.2.1) that for any absorbing Markov chain $\boldsymbol{I} - \boldsymbol{Q}$ has an inverse, and

$$(\boldsymbol{I} - \boldsymbol{Q})^{-1} = \sum_{k=0}^{\infty} \boldsymbol{Q}^k \tag{7.20}$$

The inverse is defined as the *fundamental matrix* and denoted by $\boldsymbol{N}$

$$\boldsymbol{N} = (\boldsymbol{I} - \boldsymbol{Q})^{-1} \tag{7.21}$$

The matrix N has an important probabilistic interpretation, which states that

> *The mean of the total number of times the process is in a given transient state is finite with the means given by* N ([7], Theorem 3.2.4)

The (i, j) element of N is the average number of times transient state j is encountered in the transition from transient state i to an absorbing state [9].

The Number of Transitions from the Virgin Materials to Landfill

Reference [10] applied the above modeling based on the Markov chain to assess the average residence time in use (years) of the element of iron in Japan. In their analysis, not final products but landfill corresponds to the absorbing state: a fraction of EoL final products is transferred to a landfill (the absorbing state). There were 13 transient states: blast furnace (BF) crude iron, BF in-house scrap, BF steel products, Use in construction, Use in machines, Use in automobiles, Use in containers, Use in other products, Obsolete scrap, Electric furnace (EF) crude iron, EF in-house scrap, EF steel products, and Industrial scrap. Their analysis uses the fact that the total average number of times a material (iron) is used in products from the initial state s until ultimately being landfilled through an unlimited number of transitions, τ_s, is

$$\tau_s = \sum_{i \in P} N_{si} \tag{7.22}$$

where P refers to product categories, construction, machines, cars, containers, and other products [8]. Their main concern was estimating the average residence time of the elements of iron in society. Based on information about the average lifetimes of products τ_i in each state $i \in P$ (30 years for construction, 12 years for machines, 13 years for automobiles, one year for containers, and 12 years for others), and the recovery ratio of steel scrap (ranging from 50% for construction to 92% for automobiles), they calculated the average residence time of elements of iron in society from initial state $s = 1$ (blast furnace crude iron) until ultimately being landfilled after an unlimited number of transitions as

$$\theta_1 = \sum_{i \in P} T_i [N]_{1i} \tag{7.23}$$

It was found that the average number of times used and average residence time in society were, respectively, around 2.7 times and 63 years under the steel consumption structure in Japan in 2000 and that the cyclical use of steel was greatly affected by the construction steel recovery rate; raising the recovery rate of construction steel from 50 to 60% resulted in values of 3.2 for the average number of times used and around 76 years for the average residence time in society.[4]

---

[4] An increase in the recovery rate of "construction scrap" is represented by an increase in the transfer of "construction" to "obsolete scrap" and the associated reduction in that transferred to "landfill."

The Matrix $\boldsymbol{B} = \boldsymbol{NR}$

Another quantity of high relevance to MFA, which is derived from the fundamental matrix $\boldsymbol{N}$, is the following $n_t \times n_a$ matrix

$$\boldsymbol{B} = \boldsymbol{NR} \tag{7.24}$$

the (i, j) element of which gives the probability that the process starting in transient state i ends up in absorbing state j ([7], Theorem 2.3.7). Since in an absorbing Markov chain any transient state eventually ends up in an absorbing state ([7], Theorem 3.1.1), we have

$$\boldsymbol{B}\boldsymbol{\iota}^{\top} = \boldsymbol{\iota} \tag{7.25}$$

that is, the row sums of $\boldsymbol{B}$ are equal to 1.

Nonstochastic Interpretation of the Markov Chain and its Application to MFA

Originally developed as a model of stochastic processes, the Markov chain is intrinsically stochastic, with t_{ij} referring to transition probabilities. However, one can also interpret transition probabilities in a nonstochastic manner and apply a Markov chain to a nonstochastic model; in [7], Sect. 7.7, the IO model is interpreted as a Markov chain, with the columns (processes) referring to the states and the matrix of input coefficients to the transition matrix.

Ghosh Input-Output Absorbing Markov Chains

An important IO study based on the nonstochastic interpretation of the Markov chain is [9]. They developed an IO-based absorbing Markov chain for describing the network of paths through an industrial system taken by an embodied resource (say, ore) from extraction through intermediate products and, consumer (final) products. In their Markov chain, states represent resources, intermediate products, and consumption goods. Let there be n_R resources, such as ore, and n_P products, including both intermediate and consumption (final) products. Following our notations for the IO model, denote the flow of resources to products by $\boldsymbol{Z_{RP}}$ (processing of ore to metals) and that of products to products by $\boldsymbol{Z_{PP}}$. In terms of the canonical form (7.16), the matrix $\boldsymbol{Q}$ of order $(n_R + n_P) \times (n_R + n_P)$ is given by

$$\boldsymbol{Q} = \begin{pmatrix} \boldsymbol{0} & \bar{\boldsymbol{A}}_{\boldsymbol{RP}} \\ \boldsymbol{0} & \bar{\boldsymbol{A}}_{\boldsymbol{PP}} \end{pmatrix} \tag{7.26}$$

where $\bar{\boldsymbol{A}}_{\boldsymbol{RP}} = \hat{\boldsymbol{x}}_{\boldsymbol{P}}^{-1}\boldsymbol{Z_{RP}}$ and $\hat{\boldsymbol{A}}_{\boldsymbol{PP}} = \hat{\boldsymbol{x}}_{\boldsymbol{P}}^{-1}\boldsymbol{Z_{PP}}$. The transfer coefficients from resources to resources, the $n_R \times n_R$ matrix in the upper left, and from products to resources, the $n_P \times n_R$ matrix in the lower left, are zeros because "neither resources nor intermediate products flow to resources [9]." The transfer matrices $\bar{A}_{RP}$ and $\bar{A}_{PP}$ correspond to the Ghosh matrix (see Sect. 5.5.3.1 for the Ghosh matrix). Because of this feature, their model is termed "Ghosh input-output absorbing Markov chains" by [11]. The $(n_R + n_P) \times n_P$ matrix $\boldsymbol{R}$ takes the form

$$\boldsymbol{R} = \begin{pmatrix} \boldsymbol{0} \\ \boldsymbol{R}_P \end{pmatrix} \tag{7.27}$$

where $\boldsymbol{R}_P$ is a diagonal matrix of order n_P with its (i, i) element referring to the share of product i destined to consumption product i (the final use). From (7.24), it then follows

$$\boldsymbol{B} = \begin{pmatrix} \boldsymbol{B}_R \\ \boldsymbol{B}_P \end{pmatrix} = \begin{pmatrix} \bar{\boldsymbol{A}}_{RP}(\boldsymbol{I} - \bar{\boldsymbol{A}}_{PP})^{-1}\boldsymbol{R}_P \\ (\boldsymbol{I} - \bar{\boldsymbol{A}}_{PP})^{-1}\boldsymbol{R}_P \end{pmatrix} \tag{7.28}$$

Of particular relevance for MFA is the matrix $\boldsymbol{B}_R$, the (r, j) element of which gives the resource r embodied in product j. However, the above formulation does not remove waste flows and may also include services, both of which do not physically enter the final products.[5] To the extent that these points apply, their model would translate to a consumption-based footprint perspective [11]. Reference [9] implemented the framework to a numerical example of a world trade model involving three regions.

7.1.1.4 The Data Situation in Standard MFAs

Like an LCA study, an MFA study is highly data-intensive; however theoretically sophisticated, an MFA without being fleshed out with actual data is of little use for IE. As a major drawback of the available material flow data, [11] points to the fact that they either refer to specific products or report total economy-wide material consumption without distinguishing any products or end-uses [12–14]. They urge the need to improve the resolution and coverage of end-use products in MFA.

We saw above that the standard MFA (*à la* [4]), is based on the transfer matrix $\boldsymbol{T}$, which is closely related to an absorbing Markov chain. As demonstrated by [9], the matrix $\boldsymbol{T}$ referring to transient states corresponds to the Ghosh (output coefficients) matrix. Conceptual affinity and the public availability of IO tables make combining MFA and IO a promising means to improve the resolution and coverage of end-use products in MFA. With this backdrop, we now turn to the IO-based modeling of MFA.

[5] Waste flows can easily be excluded by use of the yield filter as in (7.5).

7.1.2 The IO-Based Modeling of MFA

It is interesting that, despite the close affinity between MFA and IO, one finds little mention of it in mainstream textbooks of MFA and IO; for instance, [15] makes no mention of MFA at all, and while [4] briefly mentions IO, no MFA-related details are shown. Even when one acknowledges the close affinity between the transfer matrix and the Ghosh matrix, actual application to the MFA of an IO table can be hampered by the fact that an IO table is expressed in monetary units and, depending on the level of sector resolution, its sector is an aggregate of many materials/products, while MFA is concerned with the physical flows of well-defined materials.[6] This situation resembles the early days of hybrid LCA when the LCA community started adopting the IO but with some reservation because, among others, IO tables are in monetary units. The development of the method of WIO-MFA [16, 17] opened a new way to combine the MFA with IO by providing an easy way to convert the monetary flow in an IO table into a physical flow distinguished by materials.

7.1.2.1 The WIO-MFA

Since the method of WIO-MFA is detailed elsewhere [11, 16–18], our exposition of it will be brief. The very naming of WIO-MFA indicates the way the method was conceived. As we saw in Sect. 5.4, the WIO observes the mass balances of waste in two respects. First, the row-wise balance, under which the amount of waste net of recycling/reuse is equal to that of waste for treatment

$$\sum_{j=1}^{n}(w_{ij}^{+} - w_{ij}^{-}) = w_i \tag{7.29}$$

where n is the number of all sectors including final demand. Second, the column-wise balance of waste in a waste treatment sector, say k

$$\underbrace{\sum_{i \in \text{waste}} s_{ki}\, w_i}_{\text{waste allocated to treatment sector } k} = \underbrace{\sum_{j \in \text{waste}} (w_{jk}^{+} - w_{jk}^{-})}_{\text{waste generated by treatment sector } k} \tag{7.30}$$

where s_{ki} refers to the fraction of waste i allocated to treatment k, with $\sum_k s_{ki} = 1$. On the other hand, the WIO does not satisfy the "column-wise mass balance" in the conventional goods and service-producing sectors,

[6] The latter does not apply to bulk MFAs, where one is concerned not with the mass of each material, but the mass of all materials entering and leaving the economy.

$$\sum_i \text{the mass of input } i \text{ entering } j = \sum_k \text{the mass of output } k \text{ (excluding waste) from } j + \sum_l \text{the mass of waste } l \text{ from } k \quad (7.31)$$

because it relies on the conventional monetary IO data for the inter-industry flow of goods and services among the conventional goods- and service-producing sectors. The WIO-MFA was developed to make WIO satisfy Eq. (7.31).

Deriving the Standard WIO-MFA

Products, Materials, and Resources

Let there be n endogenous sectors, with N denoting the set of them. We partition the set n into mutually exclusive and exhaustive sets of *products* (P), *materials* (M), and *resources* (R). The objects of interest in the MFA are classified as *materials* and the classification of other sectors depends on which sectors are classified as *materials*. The sectors, whose degrees of fabrication are lower than those of *materials* (i.e., sectors whose outputs cannot be made of *materials*) are classified as *resources*, and the sectors whose degrees of fabrication are higher than those of *materials* are called *products*. *Products* are made of *products* and *materials*, *materials* are made of *resources*, while *resources* are not produced within the system but are given from outside the system.

It is important that *materials* be defined such that only *resources* but no *materials* are used to produce *materials*, which implies that all the materials must be of the same order of fabrication. Otherwise, double counting would occur, which violates mass balances (see [17] for a proof).

The term *resources* in WIO-MFA is a general one that refers to inputs whose fabrication stage immediately precedes that of *materials*, and hence can diverge considerably from its original meaning: Depending on the choice of materials, ores as well as alloys, such as stainless steel, can be resources.

Mass and Yield Filters

Not all inputs can become physical components of a product with mass. Inputs with no mass, such as services and energy, are typical examples of such inputs. Some inputs with mass do not physically enter a product; ancillary inputs, such as abrasives, fuels, and lubricants, are typical examples. Packaging is another important example, which we will discuss in detail in Sect. 7.1.2.2. Let ϕ_{ij} be an index which takes 1 when the flow (i, j) can become a physical component of j, and 0 other wise; $\phi_{ij} = 0, \forall j$, for service i, and $\phi_{ij} = 0$ when i is used as ancillary input to produce j, and $\mathbf{\Phi} = [\phi_{ij}]$. Since resources and materials have a mass, $\phi_{ij} = 1, \forall i, j \in \{R, M\}$. For this reason, $\Phi_{IJ}, I, J \in \{R, M\}$, do not occur below.

Fractions of physical inputs that are not used as ancillary inputs end up as process waste without becoming components of a product (we consider packaging in

Sect. 7.1.2.2). Denote by $\boldsymbol{\Gamma} = [\gamma_{ij}]$ the yield matrix, with $\gamma_{ij} \in (0, 1]$ referring to the yield of input i in the production of j (we already encountered this matrix on Sect. 7.1).

Denote by $\boldsymbol{A}$ the $n \times n$ matrix of input coefficients, and by $\boldsymbol{A}_{IJ}, I, J \in \{P, M, R\}$ its partitioned matrices based on the above partition of N. Using $\boldsymbol{\Phi}$ and $\boldsymbol{\Gamma}$, also partitioned in terms of $\{P, M, R\}$, we can then the obtain the matrix, $\tilde{\boldsymbol{A}}$ referring to the flows that become components of P

$$\begin{aligned}\tilde{\boldsymbol{A}} &= \boldsymbol{\Gamma} \odot \boldsymbol{\Phi} \odot \boldsymbol{A} \\ &= \begin{pmatrix} \boldsymbol{\Gamma}_{PP} \odot \boldsymbol{\Phi}_{PP} \odot \boldsymbol{A}_{PP} & \boldsymbol{0} & \boldsymbol{0} \\ \boldsymbol{\Gamma}_{MP} \odot \boldsymbol{A}_{MP} & \boldsymbol{0} & \boldsymbol{0} \\ \boldsymbol{0} & \boldsymbol{\Gamma}_{RM} \odot \boldsymbol{A}_{RM} & \boldsymbol{0} \end{pmatrix} \\ &= \begin{pmatrix} \tilde{\boldsymbol{A}}_{PP} & \boldsymbol{0} & \boldsymbol{0} \\ \tilde{\boldsymbol{A}}_{MP} & \boldsymbol{0} & \boldsymbol{0} \\ \boldsymbol{0} & \tilde{\boldsymbol{A}}_{RM} & \boldsymbol{0} \end{pmatrix}\end{aligned} \tag{7.32}$$

The Material Composition of Products

We are interested in converting the $\boldsymbol{A}$ in monetary units into a physical one, based on the material composition of products in physical units. For this aim, it is necessary to convert the matrix $\boldsymbol{A}_{MP}$ in monetary units into one in physical units, $\boldsymbol{A}^*_{MP}$. One option is to use the unit price of materials, $\boldsymbol{p}_M = [p_i, i \in \boldsymbol{M}]$, to convert the monetary value into physical quantity[7]

$$\boldsymbol{A}^*_{MP} = (\hat{\boldsymbol{P}}_M)^{-1}\boldsymbol{A}_{MP} \tag{7.33}$$

Alternatively, $\boldsymbol{A}^*_{MP}$ can be directly obtained from various sources, including life cycle inventory databases and industry statistics.

With $\boldsymbol{A}_{MP}$ replaced by $\boldsymbol{A}^*_{MP}$, we obtain from (7.32) the following expression for the material composition of products, $\boldsymbol{C}_{MP}$

$$\boldsymbol{C}_{MP} = \tilde{\boldsymbol{A}}^*_{MP}(\boldsymbol{I} - \tilde{\boldsymbol{A}}_{PP})^{-1} \tag{7.34}$$

The (i, j)-element of this matrix, say $c_{MP,ij}$, gives the amount of material i that is contained in a unit of product j. If all the materials are measured in a common mass unit, say kg, then $\sum_i c_{MP,ij}$ gives the weight of a unit of product j in kg (that is, the sum of all the materials that are contained in the product), where the unit of j can be physical or monetary.

With $\boldsymbol{C}_{MP}$ thus obtained, we can convert $\boldsymbol{A}_{PP}$ into physical units in material i by

$$\boldsymbol{A}^*_{PP(i)} = \textbf{diag}(\boldsymbol{C}_{MP,i.})\boldsymbol{A}_{PP} \tag{7.35}$$

[7] This is the formulation used by [19].

where $\boldsymbol{C}_{\boldsymbol{MP},i.}$ refers to the ith row elements of the matrix $\boldsymbol{C}_{\boldsymbol{MP}}$. Analogously, the physical counterparts to $\boldsymbol{Z}_{\boldsymbol{PP}}$ and $\boldsymbol{Z}_{\boldsymbol{Py}}$ in terms of material i can be respectively given by

$$
\begin{aligned}
\boldsymbol{Z}^{*}_{\boldsymbol{PP}(i)} &= \mathbf{diag}(\boldsymbol{C}_{\boldsymbol{MP},i.})\boldsymbol{Z}_{\boldsymbol{PP}} \\
\boldsymbol{Z}^{*}_{\boldsymbol{Py}(i)} &= \mathbf{diag}(\boldsymbol{C}_{\boldsymbol{MP},i.})\boldsymbol{Z}_{\boldsymbol{Py}}
\end{aligned}
\tag{7.36}
$$

For illustrative purposes, a simple numerical example is presented in Appendix C.

Waste Materials

The above formula of WIO-MFA can easily be extended to accommodate the use of waste materials either as *materials* or *resources*. Denoting by W the partition of M referring to waste and by $\tilde{\boldsymbol{A}}_{\boldsymbol{WP}}$ the matrix of waste inputs, (7.32) is extended to include waste as follows

$$
\tilde{\boldsymbol{A}} = \begin{pmatrix} \tilde{\boldsymbol{A}}_{\boldsymbol{PP}} & \mathbf{0} & \mathbf{0} \\ \tilde{\boldsymbol{A}}_{\boldsymbol{MP}} & \mathbf{0} & \mathbf{0} \\ \tilde{\boldsymbol{A}}_{\boldsymbol{WP}} & \mathbf{0} & \mathbf{0} \\ \mathbf{0} & \tilde{\boldsymbol{A}}_{\boldsymbol{RM}} & \mathbf{0} \end{pmatrix}
\tag{7.37}
$$

with the material composition matrix including waste, $\boldsymbol{C}_{(\boldsymbol{MW})\boldsymbol{P}}$, given by

$$
\boldsymbol{C}_{(\boldsymbol{MW})\boldsymbol{P}} = \begin{pmatrix} \tilde{\boldsymbol{A}}^{*}_{\boldsymbol{MP}} \\ \tilde{\boldsymbol{A}}_{\boldsymbol{WP}} \end{pmatrix} (\boldsymbol{I} - \tilde{\boldsymbol{A}}_{\boldsymbol{PP}})^{-1}
\tag{7.38}
$$

where it is understood that $\tilde{\boldsymbol{A}}_{\boldsymbol{WP}}$ is measured in physical units. The case of waste used as *resources* can easily be considered by adding the matrix $\tilde{\boldsymbol{A}}_{\boldsymbol{WM}}$ below $\tilde{\boldsymbol{A}}_{\boldsymbol{RM}}$.

Case Studies

PVC Flows in Japan

Polyvinyl chloride (PVC), the first industrially produced general-purpose plastic, is one of the most widely used thermoplastic polymers, ranging from civil and construction materials to consumer products. However, the occurrence of chlorine and additives, such as plasticizers, stabilizers, flame retardants, and fillers, calls for the proper life cycle management of PVC products, in particular in the end-of-life (EoL) phase. Mixing EoL PVC products with different additives spoils their resource value and can limit the use of recycled material to lower value uses. With this backdrop, [20] conducted an extensive MFA of PVC in the Japanese economy based on WIO-MFA fed with the IO data and a set of detailed physical data on the flows of PVC provided by industry associations.

Table 7.1 and Fig. 7.2 show major results. Table 7.1 shows the PVC composition of final products, which are major users of PVC, in terms of PVC products. PVC resins

Table 7.1 The composition (%) of PVC products in final products

Final products	PVC products						
	WP	FS	PPB	WS	OFC	OTH	STU
Public construction (rivers, drainage, and so on)	0	1	97	1	0	1	14
Other civil engineering and construction	0	1	96	2	0	1	9
Residential construction (wooden)	15	23	55	4	0	3	7
Nonresidential construction (nonwooden)	3	50	36	8	1	3	7
Residential construction (nonwooden)	9	30	53	5	1	3	6
Agricultural public construction	0	1	96	2	0	1	3
Public construction (roads)	0	5	85	8	0	2	3
Passenger motor cars	1	19	1	31	0	48	3
Telecommunication facilities construction	0	0	54	30	15	1	2
PVC films and sheets	0	100	0	0	0	0	2
Electric wires and cables	0	0	0	100	0	0	2
House rent (imputed rent)	9	28	55	4	0	4	1
Flowers and plants	0	100	0	0	0	0	1
Preserved agricultural foodstuffs	0	99	0	0	0	0	1

Source [20], Table 3.
WP wallpaper, *FS* films and sheets, *PPB* plates, pipe, and bars, *WS* wire sheathing, *OFC* optical fiber cables, *OTH* others, *STU* shares in total use

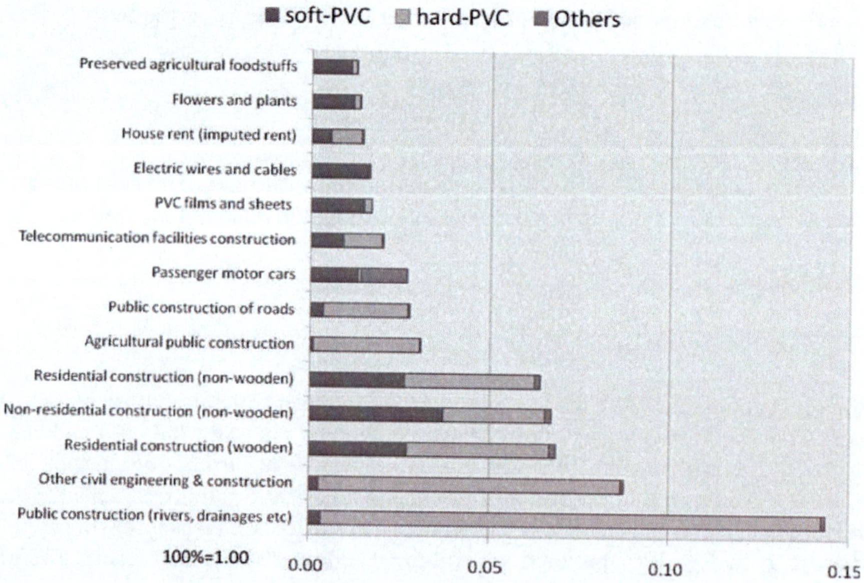

Fig. 7.2 Final destination of polyvinyl chloride (PVC) by PVC products (top 60%). *Source* [20], Fig. 4

enter public construction and civil engineering, the largest users of PVC products, in the form of plates, pipes, and bars, concrete examples of which are rainwater gutters and pipes. In construction (both residential and nonresidential), the second largest user of PVC, PVC resins also occur, as well as in plates, pipes, and bars, in the form of films and sheets for flooring and other construction applications and in the form of wallpaper as well. In passenger motor cars and telecommunication construction, wire sheeting represents a significant portion of the PVC being used.

Figure 7.2 shows the final destinations of PVC, distinguished by three categories, Soft-PVC, Hard-PVC, and Others (PVC scrap, PVC resins, and PVC paints). The PVC products entering public construction and civil engineering, the largest users of PVC products, are almost exclusively made of hard PVC. In construction (residential, nonresidential, and telecommunication facilities), the second largest user of PVC, both hard- and soft-PVC products, can be found. The results show that about 50% of the domestic accumulation occurs in the form of PVC plates, pipes, and rods (hard-PVC); about 30% occurs in the form of PVC films and sheets (soft-PVC); and about 8% occurs in the form of wire sheathing, suggesting the importance of recycling PVC products used for construction and civil engineering applications to manage EoL PVC products.

**UPOIM*

The main concern of MFA studies is quantifying the actual flow of materials (including substances) in an economy.[8] On the other hand, an LCA is usually concerned with identifying impacts associated with a well-defined functional unit, say, a product system. Quantifying the flow of materials/substances associated with the product system can thus provide useful information for LCA and contribute to extending the scope of complementarity between LCA and MFA, the two major tools of IE. With this backdrop, [21] proposed a new methodology based on IO for identifying the physical IO flow of individual materials associated with the production of a given product, the Unit Physical Input-Output by Materials (UPIOM).

From the basic IO calculation, it follows that the inter-industry flow matrix $\boldsymbol{Z}_{xx}$ is given by

$$\boldsymbol{Z}_{xx} = \boldsymbol{A}\,\mathbf{diag}(\boldsymbol{x}) = \boldsymbol{A}\,\mathbf{diag}((\boldsymbol{I} - \boldsymbol{A})^{-1}\boldsymbol{y}) = \boldsymbol{A}\,\mathbf{diag}(\boldsymbol{L}\boldsymbol{y}) \tag{7.39}$$

which implies that $\boldsymbol{Z}_{xx}$ refers to the inter-industry flow that is needed to meet $\boldsymbol{y}$. Which inter-industry flow would be necessary to meet one unit of final demand for product j? Replacing $\boldsymbol{y}$ in (7.39) by an $n \times 1$ unit vector with its j element equal to one and the remaining elements equal to zero, we obtain

$$\boldsymbol{A}\,\mathbf{diag}(\boldsymbol{L}_{\cdot j}) \tag{7.40}$$

---

[8] While the term "substance flow analysis" (SFA) is often used to refer to the MFA of substances in the literature, we use the term MFA to include SFA.

where $L_{.j}$ refers to the jth column elements of L. By use of the material composition matrix C_{MP}, this quantity can be converted to an $(n_P + 1) \times n_P$ matrix of physical flow of material m, say, $U(m, j)$, as

$$U(m, j) = \begin{pmatrix} \text{diag}(C_{MP,m.})\tilde{A}_{PP} \\ (\tilde{A}^*_{MP})_{m.} \end{pmatrix} \tag{7.41}$$

where $(\tilde{A}^*_{MP})_{m.}$ refers to the direct input of m in physical units. The (i, k)-element of $U(m, j)$, $(u(m, j))_{ik}$, gives the mass of material m (e.g., iron) that is embedded in product j (e.g., a car) as product k (e.g., ball bearings) in the form of input i (e.g., hot rolled steel). Therefore, we have (see [21] for a proof))

$$\sum_m \sum_i (u(m, j))_{ij} = \text{the mass of product } j \tag{7.42}$$

Figure 7.3 gives a $U(m, j)$ with m = pig iron and j = passenger motor car, which was triangulated based on the degree of fabrication by use of the algorithm due to [22] (see [21] for details). Figure 7.3 shows how pig iron is embedded in a passenger car after being transformed into parts and components. The column in the far left referring to the car manufacturing sector indicates that the largest portion of pig iron is embedded in a passenger car in the form of "Motor vehicle bodies", followed

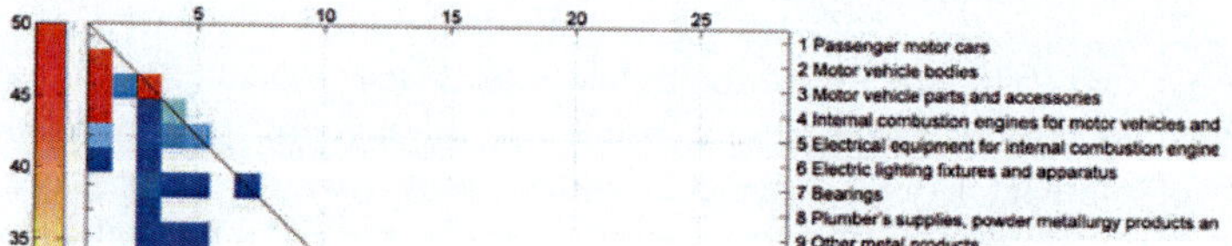

Fig. 7.3 UPIOM: the flow of iron associated with car production. The flow of pig iron associated with the production of a passenger car (one million Japanese yen value of at 2005 prices with a total iron content of 460 kg) triangulated by the Simpson-Tsukui algorithm [22]. Each square refers to an element of the UPIOM, with different colors indicating its relative magnitude based on the color scale on the left. Elements larger than 50 kg are displayed as equal to 50 kg. Pig iron occurs in the row at the bottom. The row at the bottom represents the input of pig iron, which is considered a material. Consequently, there is no corresponding column for pig iron. *Units* kg. *Source* [21], Fig. 3

by "Parts and accessories," and "Engine and its parts." Tracing the second column referring to the manufacturing of car bodies shows that pig iron becomes car bodies mostly in the form of "Coated steel" (row 17). Reading down the column referring to the manufacturing of "Coated steel" (column 17), one finds that "Coated steel" is made of "BOF ordinary steel." Compared to car bodies, the manufacturing of parts and accessories given by column 3 is characterized by using a significantly larger number of inputs, more than 20 out of the 28 inputs, reflecting the more complicated nature of these products than car bodies. Of these 21 ferritic inputs, "Iron and steel shearing and slitting" occupy, with around 40 kg, the largest share (except the internal inputs among parts and accessories). Reading the column of "Iron and steel shearing and slitting" downward shows that it is mostly made of "Cold-finished steel" and "Ordinary steel strip."

Iron and steel scrap constitute an alternative source of ferrous materials. Figure 7.4 gives the UPIOM of iron and steel scrap associated with producing a unit of a passenger car, where the sectors were ordered using the same sequence as in Fig. 7.3. Comparison with Fig. 7.3 reveals that the flow of iron and steel scrap tends to be smaller in magnitude and scope, reflecting the small share (around 15%) of secondary material in the total ferrous material used to produce a passenger car. The row at the bottom represents a significant difference from Fig. 7.3: iron and steel scrap are mostly used to produce "EAF special steel" and "Cast, and forged materials". Due to its higher quality requirements, the production of EAF special steel requires iron and steel scrap of higher quality than EAF ordinary steel, the major ferrous material of construction steel. The above result indicates that the iron and steel scrap used in car production is of higher quality than that used in construction.

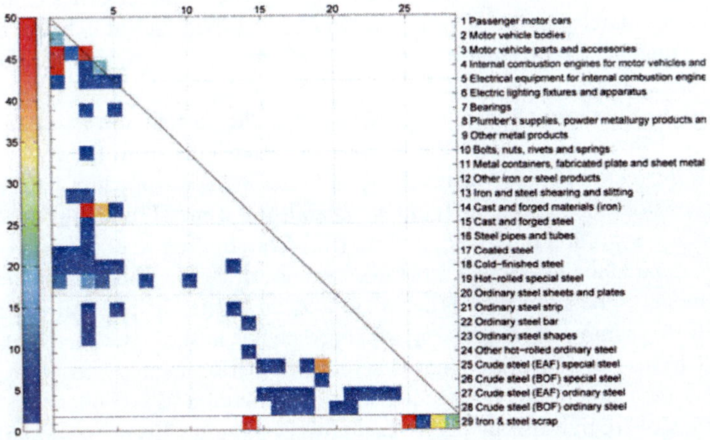

Fig. 7.4 UPOIM: the flow of iron and steel scrap associated with car production. See Fig. 7.3 for explanation. *Unit* kg. *Source* [21], Fig. 4

Bisphenol A (BPA) Flows in China

Bisphenol A (BPA), a widely used petrochemical compound, has become an emerging global environmental management challenge because its leakage is associated with potential environmental and human health impacts. Information about its supply chains is vital to understanding the drivers and the fate of BPA flows and designing effective policies to regulate the use and leakage of BPA. While information about the flows related to direct use, such as BPA used in the production of paint or wire coating, is available, little is known about indirect flows. To fill this gap in information, [23] delineated direct, and indirect BPA flows for the 2012 Chinese economy. A Chinese IO Table (139 sectors) was used, which was converted to material flows using WIO-MFA with BPA as *material*.

The results show that construction, production of educational and recreational products, and automobile manufacturing are the most BPA-intensive sectors in total BPA flows, indicating the dominance of direct effects. However, it was found that the public management and health sectors incur significant indirect BPA flows (as embedded and inter-sectoral BPA placed into use), even though direct BPA use by these sectors is limited; currently overlooked indirect BPA flows were revealed.

Metal Recycling from EoL Vehicles

In end-of-life vehicles (ELV) treatment, the steel components of a car are usually shredded and pressed to mix them as scrap, which is subsequently remelted in EAF to produce secondary steel [24]. Unintentional mixing of different metal species in scrap is inevitable in this process for two reasons. First, separating steel components perfectly from other metal components of a car, such as copper (Cu) and tin (Sn) in electronics, is difficult. Secondly, as well as iron (Fe), steel contains carbon, and other alloying metals, such as chromium (Cr), nickel (Ni), and molybdenum (Mo). Therefore, metal scrap originating from an ELV will be a mixture of various metal species, mixed unintentionally [25].

Because of the thermodynamic distributions of the various metals in the mixed scrap, it is difficult to separate the various metal species from the molten scrap under current economic and technical conditions ([26], see Appendix 1 for thermodynamic consideration about the refining capabilities of metal remelting processes). For instance, separation of Cu, Ni, and tin (Sn) from molten steel scrap is difficult, whereas Cr and zinc (Zn) can be separated because of the thermodynamic tendency of Cr to move to the solid phase (i.e., slag) and Zn to the gas phase (i.e., captured in a dust filter). Unintentional mixing of metal species in scrap destined for an EAF can result in loss of function values when high-function metals end up in carbon steel where their function is not needed and contamination of secondary steel when the presence of foreign metal species compromises the quality of secondary steel. A well-known example of the latter is the contamination of steel by Cu and Sn [27, 28].

With this backdrop, [29] analyzed the composition of ELV-derived scrap to estimate the possibility of unintentionally inputting substances and alloying elements

Table 7.2 Compositions and contents of alloying elements in untreated and treated passenger car

Before EoL treatment

	Automobile parts		Direct inputs		Automobile body		Combustion engine		Total
Total	316.1	32	221.3	22	291.5	29	162.6	16	991.4
Cr	2.09	31	2.18	33	0.86	13	1.51	23	6.64
Ni	0.54	36	0.45	29	0.20	13	0.32	21	1.51
Mo	0.06	30	0.07	33	0.03	13	0.05	24	0.21

After EoL treatment

	ELV-derived scrap		Parts removed for reuse		Parts removed for nonferrous metal recycling		Combustion engine		Total
Total	594.8	60	156.5	16	77.5	8	162.6	16	991.4
Cr	3.97	60	1.03	16	0.12	2	1.51	23	6.64
Ni	0.87	57	0.27	18	0.05	3	0.32	21	1.51
Mo	0.13	60	0.03	15	0.00	1	0.05	24	0.21

Source [29], Table 4. For each item, the numbers in the first column refer to kg and in the second column to %, with the row sum equal to 100%

via ELV-derived scrap into EAF-based steel recycling. They developed a WIO-MFA table with a 619 × 519 sector resolution, including 58 types of crude-steel and 29 steel materials, the highest resolution IO table ever constructed for an IO-based MFA of iron and steel products (the hot-rolled special steel sector was further disaggregated into 19 types).

Table 7.2 shows the composition and contents of the alloying elements in each component of a passenger car, estimated by WIO-MFA. The amounts of Cr, Ni, and Mo in a passenger car were estimated as 6.64, 1.51, and 0.21 kg, respectively. The higher Cr content is attributed to the use of special steels, such as stainless steel, structural alloy steel, and heat-resistant steel, all of which contain relatively high quantities of Cr. It was estimated that around 0.50–0.69 kg of Cr in the ELV-derived scrap from one unit of an ELV was lost to the slag, which amounts to 1.824–2.487 Gg per year. No losses to the slag phase occur for Ni and Mo. However, it was found that a significant portion of them ends up in carbon steel, where their functions are not needed.

In a subsequent study, [30] extended the coverage of metals to include manganese (Mn) and Mo and increased the resolution of ELV-derived scrap by disaggregating the parts and components of an ELV. Figure 7.5 shows the estimated results for the car-part compositions of a middle-class passenger car obtained by WIO-MFA. The concentration of alloying elements differs among car parts, indicating that mixing will result in diluting the higher concentrations of some parts with the lower concentrations of other parts. Exhausts are characterized by high concentrations of Cr, Ni, and Mo because of the use of stainless steel. The Mn concentration in the suspension is higher than the average concentration in ELV-derived steel scrap. On the

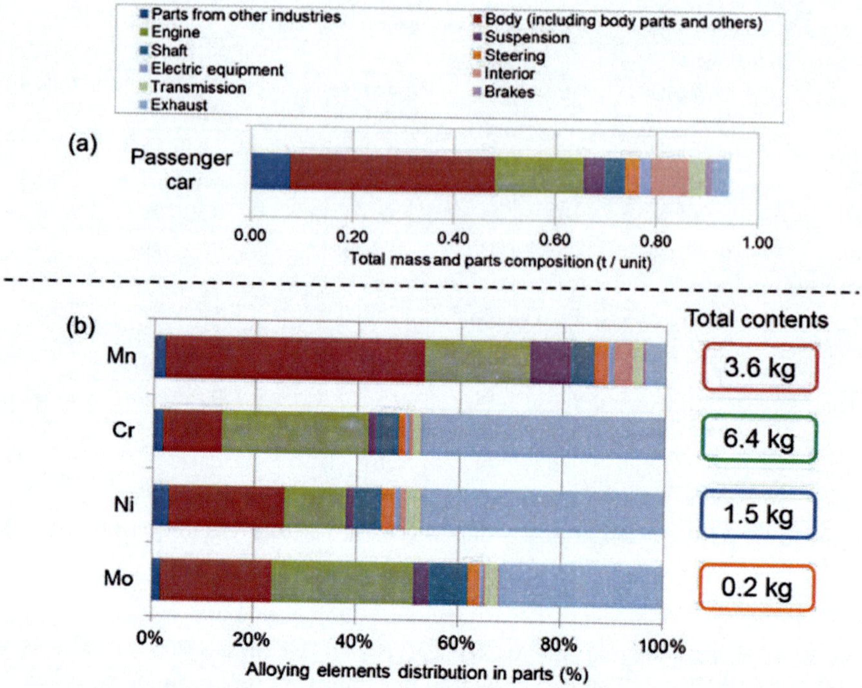

Fig. 7.5 Car-part composition percentages in a passenger car unit, with the associated concentrations of alloying elements as obtained by WIO-MFA. **a** details the composition of the parts and components, while **b** gives the percentages of alloying elements in the corresponding parts. The parts from other industries category includes all the intermediate products (i.e., those not produced by automobile-related sectors) used in the production of a passenger car. *Source* [30], Fig. 3

other hand, the parts, the body, and the brakes have low concentrations of alloying elements. Consequently, mixing different parts can cause a loss in the quality of alloying elements and steel. It was estimated that sorting ELV-derive steel scrap by parts can result in a significant recovery of alloying elements; more specifically, a 10-fold saving of them was achieved by sorting exhaust parts. The recoverable mass of alloying elements from sorted ELV-derived steel scrap was found to correspond to 8.2% of the annual consumption of alloying elements in Japan as virgin resources in EAF steel making.

The Copper Flows from Different Sources Across the US Economy

Copper demand has more than doubled in the past 40 years, a trend likely to continue in the coming decades; renewable power technologies require more copper than the current technologies based on fossil fuels (see Sect. 6.2.4.4). For the sustainable management of copper resources, it is crucial to understand the economy-wide flows of copper from different sources, including scraps of different qualities and the associated energy consumption.

With this backdrop, [31] mapped the copper flows from different sources across the US economy and quantified the energy-saving potentials of recycling various copper scrap qualities by applying WIO-MFA to the 2011 US-IO with 411 endogenous sectors. Their IO-based MFA is distinguished by exploring the detailed manufacturing process in terms of first-stage manufacturers (e.g., of metal rods, wire, and castings), component manufacturers (e.g., of insulated cables, motors, and valves), and final-stage manufacturers (e.g., of buildings and automobiles). Six types of copper sources were considered: primary production, new scrap, and four EoL scrap types including no. 1 scrap, no. 2 scrap, low-grade copper-bearing scrap, and copper alloy scrap. Writing $\boldsymbol{y_P}$ for the vector representing the US total final demand of products sectors, the calculation of copper requirement was based on

$$\boldsymbol{C_{total}} = \boldsymbol{A}^*_{\boldsymbol{MP}}(\boldsymbol{I} - \boldsymbol{A_{PP}})^{-1}\,\mathbf{diag}(\boldsymbol{y_P}) \tag{7.43}$$

with M referring to the six copper types, and that of energy requirement on

$$\text{energy} = \mathbf{diag}(\boldsymbol{f})\boldsymbol{C_{total}} \tag{7.44}$$

where $\boldsymbol{f}$ is the 1×6 vector representing life cycle cumulative primary energy demand per unit of copper material from the six sources.

The total copper requirement in the 2012 US economy, $c_{total} = \boldsymbol{\iota}_{\boldsymbol{M}}^{\top}\boldsymbol{C_{total}}\boldsymbol{\iota_P}$, was estimated to be 2.347 Mt (2.347 Tg) with a recycling input rate (the proportion of metal that is produced from both new (production waste) and old (EoL) scrap) of 33%. It was estimated that about 38% of the total copper used in the 2012 US economy did not reach the final demand but was either lost during the manufacturing process or was used in an intermediate use that does not have physical copper output to final demand. Scenario analysis involving higher collection rates and better sorting of copper scrap (implemented via changes in $\boldsymbol{A}^*_{\boldsymbol{MP}}$ and $\boldsymbol{A_{PP}}$) showed that if all potentially recyclable copper scrap were recycled, energy consumption associated with copper production would decrease by 15%, with alloy scrap as the largest contributor. However, further energy benefits from increased recycling are limited by the lower quality of the scrap yet to be recycled.

7.1.2.2 *Extending the WIO-MFA to Accommodate Diverse Pathways

The WIO-MFA model presented above focuses on the material composition of a product, such as alloys, parts, and components, and its destinations to final products such as automobiles. It lumps together inputs that do not constitute products as waste; for a physical nonancillary input, there are only two destinations: to enter the mass of a product or not to enter and discard as process waste. However, in reality, the destinations (pathways) of inputs are more diverse. For example, packaging and containers do not physically enter a product but accompany it to its next destination, a pathway not considered by the WIO-MFA.

The Methodology

To cope with this problem, [19] extended the WIO-MFA model to accommodate diverse pathways of inputs. They considered the following three additional pathways

(S) Input i accumulates in sector j This represents the case in which input i does not constitute the mass of output j but enters the stock at sector j. Actual examples are the inputs of machine tools and communication equipment (mobile phones), with transaction values not large enough to be counted as fixed capital formation.

(A) Input i accompanies output j This represents the case in which input i accompanies output j and they flow into another sector (k) together. A typical example is a case in which input i is used as containers or packaging of output j (e.g., an input from the "plastic products" sector to the "beverages and foods" sector used for plastic bottles).

(D) Input i dissipates at sector j This represents the case in which input i is dissipated at sector j in such a form that recovery for recycling is almost impossible. Typical examples are pesticides.

It is important to note that S and A do not exclude the flow of *products* to *materials* and or *resources*. The composition matrix $\boldsymbol{C_{MP}}$ is not affected by this extension of the model because it refers to the inputs that become physical components of a product; the above flows in S and A do not belong to M and do not become components of $\boldsymbol{C_{PM}}$.

In this section, we henceforth consider only the flow of physical and nonancillary inputs, that is, the flow (i, j) with $\phi_{ij} = 1$. Denote by $\boldsymbol{\Gamma_C} = [\gamma_{C,ij}]$ the fraction of (physical and nonancillary) input i that either enters product j or is discarded as process waste, and by $\boldsymbol{\Gamma_K} = [\gamma_{K,ij}]$, $K \in \{S, A, D\}$ the fractions corresponding to the three pathways introduced above. $\boldsymbol{\Gamma_C}$ gives the destinations of i that are covered by the standard WIO-MFA, that is, $\boldsymbol{\Gamma}$. Since the four pathways are exclusive and mutually exhaustive, we have

$$\sum_{K \in \{C,S,A,D\}} \gamma_{K,ij} = 1 \tag{7.45}$$

where C refers to the pathways considered by the standard WIO-MFA, accumulated in products or discarded as process waste. We then obtain the following expression for the accumulation of material i in endogenous sectors $j \in N$

$$\mathbf{diag}(\boldsymbol{C_{MP,i.}})\,(\boldsymbol{\Gamma_{S;PN}} \odot \boldsymbol{Z_{PN}}) \tag{7.46}$$

and for the accumulation of material i as accompaniment in endogenous and final demand sectors

$$\mathbf{diag}(C_{MP,i.})\left(\boldsymbol{\Gamma}_{A;PN} \odot \mathbf{Z}_{PN}\right)\boldsymbol{B} \tag{7.47}$$

where $\boldsymbol{B}$ is an $n \times (n + n_y)$ matrix, with $b_{ij} = z_{ij}/x_i, i \in N, j \in N \cup N^y$ (N^y refers to the set of final demand sectors and n_y to the number of its elements).

Reference [19] applied the above methodology to copper flows in Japan, with copper as *materials*. Results include that accumulation of copper in endogenous sectors was not small and negligible (it accounted for 12.2% of the overall flow) and that accumulation of copper as accompaniments to products, such as containers and packaging, was very small.

Application to Plastic Flows in Japan

Amid the increasing global awareness of the environmental impacts of plastic waste [32–34], the Japanese government laid out ambitious targets including a reduction of 25% for single-use plastic waste and the reuse/recycling of 60% for plastic containers and packaging by 2030. However, little was known about the current usage situation of single-use plastics including containers and packaging, which should be the basis of such policies. Reference [35] analyzed the current usage situation of single-use plastics including containers and packaging in Japan, identifying the nationwide material flow of plastics based on the extended WIO-MFA of [19].

The amount of plastic consumption associated with domestic demands and exports, including losses, in Japan was estimated to be 15.5 Tg (Mt) in 2015. With exports excluded, the inflow of plastics into Japanese households and industries (that is, plastics that are considered to be used and disposed of domestically) was estimated to be 8.4 Tg, of which 1.6 and 2.5 Tg were estimated for containers and packaging comprising household and industry inflows, respectively, through the purchase/procurement of products, services, and raw materials. The amount of plastic container and packaging inflow into industries was substantial enough to merit attention: the containers and packaging accompanying the procurement of raw materials and products flowed into industry sectors at each stage of the supply chain. Food containers and packaging that flowed into the food-processing and food service sectors accounted for 15% of the inflow of containers and packaging into industries. Furthermore, industries were found to have markedly lower recycling rates of plastic containers and packaging, 12–15%, than households, 60%. The results indicate the importance of increasing the collection of plastic food packaging from the food industry to meet the ambitious recycling/reuse target.

7.2 The Dynamic MFA: Methods and Applications

While a static MFA model is time invariant, a dynamic MFA model is a function of time. A dynamic MFA (dMFA) is usually represented by difference equations, due to the discrete nature of most data available for MFA studies.

7.2.1 The Standard Model of dMFA

A classical example of dMFA is [36], which serves as the basic model [37, 38].

7.2.1.1 The Basic Model of Stock Growth

Denoting by $S(t)$ the stock of a material at the end of t, by $I(t)$ the inflow (gross addition) in t, and by $O(t)$ the outflow (discarded/decommissioned) in t, the stock of the material at the end of t is given by

$$S(t) = S(t-1) + I(t) - O(t) \tag{7.48}$$

Alternatively, denoting by $S(0)$ the amount of in-use stock at time 0 when the system started, we have

$$S(t) = \sum_{r=1}^{t} (I(r) - O(r)) + S(0) \tag{7.49}$$

The amount of discard $O(t)$ depends on past inflows and the rate at which they are discarded. Denoting by $\psi(t-t') \in [0, 1], t \geq t'$ the fraction of $I(t')$ that is discarded in $t \geq t'$, with $\psi(0) = 0$ and $\sum_{r\geq 0}^{\infty} \psi(r) = 1$, we have

$$O(t) = \sum_{t' \leq t} \psi(t-t') I(t') \tag{7.50}$$

Since the stock is the sum of past inflows net of outflows, we have

$$S(t) = \sum_{t' \leq t} (1 - \psi(t-t')) I(t') \tag{7.51}$$

Reviewing around 60 studies on dMFA, [37] found that they feature similar basic modeling principles but very diverse extrapolation methods. Basic principles include the calculation of outflows of the in-use stock based on inflow like (7.50) and using a lifetime distribution function to obtain $\psi()$. Most dMFA studies are *prospective*: they are concerned with the future stocks and flows not with past stocks and flows (*retrospective*). Suppose one is at time t_1 and interested in the evolution of stocks and flows up to time $t_2 > t_1$. From (7.49) and (7.50), we have

$$S(t_2) = \sum_{t_1 \leq s \leq t_2} (1 - \psi(t_2 - s)) I(s) + S(t_1) \tag{7.52}$$

Diverse methods are used to extrapolate the future flow of $I(s)$, including constant, linear, exponential, and logistic models or approaches based on socioeconomic variables, such as regression models or the intensity-of-use hypothesis. The most commonly used exogenous variables in extrapolations are population and life-style (the amount of material stock per population).

7.2.1.2 Modeling the Lifetime Distribution

The lifetime of a product is mostly represented by parametric probabilistic distributions, such as normal [36, 39], log-normal [40], Weibull [41–43], or generalized gamma distributions [44, 45]. The Weibull distribution is one of the most widely used ones in dMFA.

The Weibull distribution, different from the normal distribution, has a closed form expression for its (cumulative) distribution function F (Appendix D gives details of the Weibull distribution). Using the Weibull distribution, the probability that the life of a product will not exceed t is

$$\text{Prob[product life} < t] = F(t) = 1 - \exp\left(-\left(\frac{t}{\eta}\right)^{\beta}\right) \tag{7.53}$$

Since $1 - F(t)$ represents the fraction of the stock remaining at time t, the fraction that was discarded during year t, $q(t)$, can be formulated as

$$q(t) = F(t) - F(t-1) \tag{7.54}$$

The total amount of items that were discarded during t, $O(t)$, can then be given by

$$O(t) = \sum_{r \le t} q(r) I(t - r) \tag{7.55}$$

The above formulation was derived from the (cumulative) distribution function F. Alternatively, q can be directly obtained from the density function f.

A product's life can be altered by adopting different product design, materials, and maintenance- and use patterns, which can be accommodated in the Weibull distribution by changing the values of β and η. Reference [46] investigated the environmental and economic impacts of extending a passenger car's average life by altering the value of η. A similar procedure was used by [44, 45] but for the general gamma distribution function, which includes the Weibull distribution as a special case.

7.2.1.3 Case Studies

Retrospective MFA of the 20th Century

Reference [13] analyzed the global stocks, inflows, and outflows of all materials (steel, Al, Cu, an aggregate of other metals and industrial minerals, concrete, asphalt, bricks, primary and down-cycled aggregates, paper, solid-wood products, and plastics) and their relationship to economic growth, energy use, and CO_2 emissions from 1900 to 2010. Over this period, global material stocks increased 23-fold, reaching 792 Pg in 2010. A considerable proportion of all primary materials used globally has accumulated in growing material stocks in the built environment of cities and of rural areas: the twentieth century was a century of massive stockpiling. Despite efforts to improve recycling rates, it was found that continuous stock growth precludes closing material loops, with recycling only contributing 12% of inflows to stocks.

Prospective MFA of Global Metals to Meet a Climate Goal up to 2100

Reference [47] developed global targets for the flows, stocks, and use intensity for six major metals (Fe, Al, Cu, Zn, Pb, and Ni) in the global economy out to 2100, which are consistent with emissions pathways to achieve a 2 °C climate goal. Their methodology is based on a combined use of global dMFA and linear programming (LP), where LP is used to determine the maximum production available under the annual carbon budget, while minimizing the divergence between the projected supply and baseline demand. Results indicate that despite advances in low-carbon metal production, a transformative system change to meet society's needs with less metal is required to remain within a 2 °C pathway. Their results indicate that, globally, the demand for goods and services over the 21st century needs to be met with approximately 7 Mg (t)/capita of metal stock, roughly half the current level in high-income countries.

An MFA of Automotive Aluminum under Various EoL Management Scenarios

Currently, CV engines are the largest user of cast aluminum, the primary destination of aluminum scrap. Since an EV has no engine, the expected large-scale replacement of ICVs with EVs could result in a significant change in the demand for aluminum scrap. The risk of a future mixed scrap surplus has been exposed by [48, 49]. With this backdrop, [39] explored various interventions in EoL management and recycling of automotive aluminum, using a global dMFA of automotive aluminum and its alloying elements with resolution on component (14 component groups) and alloy (seven alloy types) level, combined with an LP to quantify the scrap surplus and recycling paths under maximum scrap utilization. This MFA is a prospective one covering the period from 2010 to 2050. For each year, the model determines (1) the stock of vehicles in use from the population and vehicle ownership; (2) vehicles leaving use by lifetime distribution (a normal distribution) and production in previous years; (3) demand for new vehicles in five segments (mini/small cars, small family cars, large family cars, executive cars, and luxury cars) from a balance equation of stock change and outflow; (4) aluminum metal in new components, and the alloys needed for these; (5) availability of scrap of different compositions by past alloy use and ELV management criteria; (6) an LP-based optimal blending of scrap and primary metals to produce the alloys needed, which minimizes the use of primary aluminum and alloying elements. Since scrap streams contain a variety of alloys, a cascading of recycling occurs where some alloys (such as high-Cu cast alloys) absorb most of the scrap, and others act as sources of scrap. Scenario analysis involving different EoL management showed that increased component dismantling before vehicle shredding could be an effective intervention in the medium term, especially if combined with the development of safety-relevant components such as wheels from secondary material. Reference [50] updated this study by considering the recent surge in the penetration of EVs and the different aluminum needs thereof from ICVs (EVs require more wrought aluminum).

7.2.2 *MaTrace Models*

MFA studies with high metal/alloy resolution, including those we discussed above, [24, 25, 29–31, 39], indicate the possibility of qualitative degradation of metals/alloys in the recycling/EoL phase due to the unintentional (uncontrolled) mixing of metal species, resulting in the loss of functions or contamination of secondary materials. The former loss occurs when metals with a specific function, such as Ni in stainless steel, end up in carbon steel, where the function is unnecessary and is wasted. The latter happens when tramp elements, such as Cu or Sn, get intruded into secondary steel, which can compromise its performance in processing [27, 28]. When this contamination happens, dilution by primary metal may be needed, resulting in additional environmental impacts and costs [51].

Degradation of scrap quality in the EoL phase due to unintended mixing makes open-loop (or cascaded) recycling more common than closed-loop recycling. Open-loop recycling is subject to loss of functionality of original materials, dissipation in forms that are difficult to recover, and recovered metals might need dilution with primary metals to meet quality requirements. Sustainable management of metal resources calls for the minimization of these losses. Imperative to this is quantitative tracking of the fate of materials over successive rounds of open-loop recycling involving multiple products, processes, and losses.

Reference [52] developed the MaTrace model, an IO-based dynamic model of MFA, that can trace the fate of materials over time and across products in open-loop recycling, with explicit consideration of losses and quality of scrap. Figure 7.6 depicts the architecture of the MaTrace model. MaTrace is concerned with tracking the fate of materials that initially occur in a final product, say product A, over its subsequent life stages, involving EoL processing, consisting of collection, disassembling/demolition, and sorting/separation into scraps, metallurgical processes, such as remelting and or smelting, where scraps are converted to secondary materials,

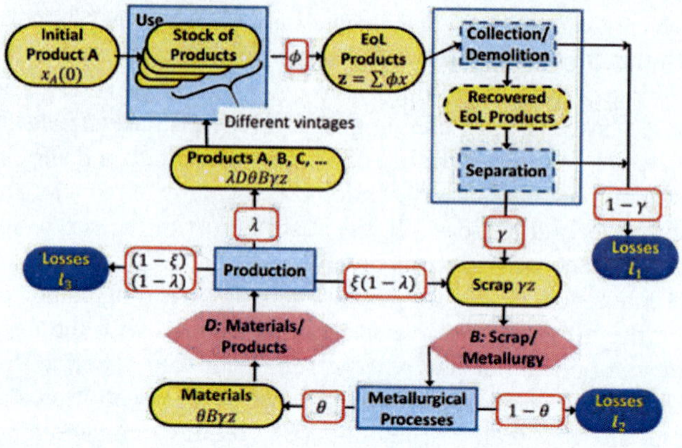

Fig. 7.6 The architecture of a MaTrace model. The ovals denote the flow of inputs and outputs, rectangles denote processes where inputs are transformed into outputs, and hexagons denote allocation processes. Collection- and separation processes are combined into a single compartment due to data availability and simplicity. Greek symbols indicate parameters referring to relevant processes: γ, θ, λ, and η are process yield ratios, $\boldsymbol{B}$ is allocation of scraps to refinery processes, and $\boldsymbol{D}$ is allocation of refined materials to products. *Source* [52], Fig. 1

fabrication into products, including those other than A, and their accumulation as stocks, with losses occurring at each of the transformation and use phases.

7.2.2.1 The Mathematical Structure of MaTrace

This section gives a brief outline of the derivation of MaTrace. For further details, please see [52].

The Amount of Material in EoL Products

Denoting by $x_j(r)$ the mass of the material under consideration in final product $j \in N_y$ produced in time r, the amount of material that occurs in EoL product j generated in t, $z_j(t)$, is given, as (7.50), by

$$z_j(t) = \sum_{r=0}^{t} \psi_j(t-r)x_j(r) \tag{7.56}$$

where the initial value $x_j(0)$ is given.

Scraps Recovered from EoL Products

Denoting by $\boldsymbol{\Gamma} = [\gamma_{sj}]$ an $n_s \times n_y$ matrix with γ_{sj} the fraction of $z_j(t)$ that is recovered as scrap s, we have

$$\text{Scraps recovered from EoL products in } t = \boldsymbol{\Gamma z}(t) \tag{7.57}$$

with $\boldsymbol{\iota}^{\top} - \boldsymbol{\iota}^{\top}\boldsymbol{\Gamma} \geq \mathbf{0}$ giving the recovery losses.

Refining of Scrap into Secondary Materials

Recovered scrap thus obtained is then allocated to n_{R} refining (recycling) processes, each of which produces a single refined material. Denoting by $\boldsymbol{B}$ an $n_{\mathrm{R}} \times n_{\mathrm{s}}$ matrix with $b_{rs} \in [0, 1]$ referring to the share of recovered scrap s allocated to the refining process r, and by $\theta_r \in [0, 1)$ the yield of this process, the amount of refined materials obtained from EoL products is given by the $n_{\mathrm{R}} \times 1$ vector

$$\hat{\boldsymbol{\theta}}\boldsymbol{B\Gamma z}(t) \tag{7.58}$$

with $(\boldsymbol{I} - \hat{\boldsymbol{\theta}})\boldsymbol{B\Gamma z}(t)$ giving refinery losses, such as metals ending up in slag.

Destination of Secondary Materials

After further processing, the secondary materials become final products, except the fractions discarded as production losses. Denoting by $\boldsymbol{D}$ *the end-use share matrix*, with d_{ij} representing the output share of material i (e.g., EAF steel) as product j (e.g., a house) to final demand, we obtain the final products produced in t as [9]

$$\boldsymbol{x}(t) = \boldsymbol{D}\hat{\boldsymbol{\theta}}\boldsymbol{B\Gamma z}(t) \tag{7.59}$$

or, for its kth element, final product k

$$x_k(t) = \underbrace{\sum_{m=1}^{n_{\mathrm{R}}} d_{km} \underbrace{\theta_m \sum_{s=1}^{n_{\mathrm{S}}} b_{ms} \underbrace{\sum_{j=1}^{n_y} \gamma_{sj} \underbrace{\sum_{r=0}^{t} \psi_j(t-r)\, x_j(r)}_{\text{EoL products}}}_{\text{Scrap recovered}}}_{\text{Refined materials}}}_{\text{Refined materials allocated to final products}} \tag{7.60}$$

Note that the matrix $\boldsymbol{D}$ corresponds to the transfer matrix $\boldsymbol{T}$ we discussed above in Sect. 7.1.1. Matrix $\boldsymbol{D}$ was obtained based on the method of WIO-MFA we discussed above in Sect. 7.1.2.1 as

[9] This expression refers to when there are no losses in producing final products.

$$\boldsymbol{D} = \mathbf{diag}(\boldsymbol{y_P})\boldsymbol{C}\,\mathbf{diag}(\boldsymbol{y}_P^\top \boldsymbol{C})^{-1} \tag{7.61}$$

where $\boldsymbol{y_P}$ is the $n_y \times 1$ vector of final demand, and $\boldsymbol{C}$ is an $n_y \times n_r$ matrix with c_{ir} the amount of refined material r destined to final product i: $\boldsymbol{C}$ is the transpose of $\boldsymbol{C_{MP}}$ used in WIO-MFA. Because of the way matrix $\boldsymbol{D}$ is obtained, MaTrace can be called an IO-based dMFA.

MaTrace: The Final form with All Losses Considered

Considering losses in the production of final products, with $\lambda_i \in [0, 1]$ referring to the yield ratio of product i, the overall equation of the evolution of materials, including the recycling of both process- and EoL wastes, becomes (see [52], the supporting information (SI), for the detailed derivation)

$$\boldsymbol{x}(t) = \hat{\lambda}\left(\boldsymbol{I} - \boldsymbol{D}\hat{\boldsymbol{\theta}}\boldsymbol{B}\,\Xi\,(\boldsymbol{I} - \hat{\lambda})\right)^{-1}\boldsymbol{D}\hat{\boldsymbol{\theta}}\boldsymbol{B}\Gamma\boldsymbol{z}(t), \tag{7.62}$$

where $\Xi = [\xi_{sj}]$ is an $n_s \times n_y$ matrix with ξ_{sj} referring to the recovery yield of process scrap s in the production process of product j. The expression of the additional term includes an inverse matrix that originates from consideration of the full cycle where process waste is refined, redistributed, becoming process waste, and reentering the cycle.

Accompanying the flow of materials given by this equation is the flow of losses occurring in the collection of EoL products, refining scrap, and manufacturing final products. The amount of aggregate losses at $t \geq 1$ is,

$$\boldsymbol{\ell}(t) = \iota_n^\top(\boldsymbol{z}(t) - \boldsymbol{x}(t)), \tag{7.63}$$

which can be divided into its components (see [52], SI, for details):

$$\begin{aligned}
\boldsymbol{\ell}(t) &= \underbrace{(\iota_n' - \iota_S'\Gamma)\boldsymbol{z}}_{\text{Recovery loss}} \\
&\quad + \underbrace{\iota_R'(\boldsymbol{I} - \hat{\boldsymbol{\theta}})\boldsymbol{B}\left(\boldsymbol{I} + \Xi(\boldsymbol{I} - \hat{\lambda})\left(\boldsymbol{I} - \boldsymbol{D}\hat{\boldsymbol{\theta}}\boldsymbol{B}\,\Xi\,(\boldsymbol{I} - \hat{\lambda})\right)^{-1}\boldsymbol{D}\hat{\boldsymbol{\theta}}\boldsymbol{B}\right)\Gamma\boldsymbol{z}}_{\text{Refinery loss}} \\
&\quad + \underbrace{(\iota_n' - \xi)(\boldsymbol{I} - \hat{\lambda})\left(\boldsymbol{I} - \boldsymbol{D}\hat{\boldsymbol{\theta}}\boldsymbol{B}\,\Xi\,(\boldsymbol{I} - \hat{\lambda})\right)^{-1}\boldsymbol{D}\hat{\boldsymbol{\theta}}\boldsymbol{B}\Gamma\boldsymbol{z}}_{\text{Production loss}} \\
&= \boldsymbol{\ell}_1(t) + \boldsymbol{\ell}_2(t) + \boldsymbol{\ell}_3(t),
\end{aligned} \tag{7.64}$$

MaTrace can accommodate changing consumer behavior and market conditions by changing the elements of matrix $\boldsymbol{D}$. Changes in EoL management, such as better recovery yields or improved scrap sorting, can be considered by changing $\boldsymbol{\psi}$ and $\boldsymbol{\Gamma}$. Thermodynamic constraints in the allocation of scrap to refining processes are taken into account by matrix $\boldsymbol{B}$, the technical foundations of which can be provided by [53] describing the behavior of metals in refining processes based on thermodynamics (detailed in Appendix 1). Changes in product design and technological innovation can be accommodated by changing the relevant elements of the input coefficients matrix $\boldsymbol{A}$.

It is important to note that MaTrace, as formulated above, only considers one cohort, the virgin material input of the initial year, $\boldsymbol{x}(\boldsymbol{0})$, and traces the whereabouts of this single input over time. MaTrace can be extended to accommodate the case of multiple cohorts (see [52], the supporting information, for details).

7.2.2.2 Case Studies

The Fate of Car Steel

Reference [52] applied MaTrace to trace the car steel (mostly BF-BOF steel) produced in Japan for 100 years. The initial amount $\boldsymbol{x}_j(\boldsymbol{0})$ was set equal to 0 for products other than passenger cars. The path of exogenous variables that drive the model was obtained based on population projections, information from the World Steel Association about possible future changes in $\boldsymbol{\Gamma}$, and other sources. The use in products includes both domestic and export uses, with the latter assumed to follow exactly the same product life cycles and EoL treatments as their domestic counterparts.

Figure 7.7 shows the fate of car steel in an ELV over 100 years. The amount of products produced from car steel, $\boldsymbol{x}(\boldsymbol{t})$, decreases over time because a portion is lost during recovery cycles- and refining processes, which are repeated each time products reach the end of their lifetimes. The portion of recovered car steel recycled into a new car turned out to be fairly small, around 7–8%, confirming the limited extent of closed-loop recycling. The portion recycled was predominantly destined for use in machines, civil engineering, and buildings.

Figure 7.8 shows the transition among different products and losses of the location (or residence) of the stock of car steel during 100 years, which was obtained from Fig. 7.7 by integrating the flow of products, weighted by the ratio of material in stock and losses. Remarkable is the rapid decline, after 20 years, in the share of cars to less than 8%, with buildings, civil engineering, and machines occupying the rest. After 50 years, around 80% of the stock was still used in products, with buildings and civil engineering accounting for the vast majority, with the remaining 20% mostly in refinery losses. In 100 years, the share of use in products dropped to around 60%, with the rest mostly occurring as unrecovered EoL civil engineering and Others, and refinery losses.

7.2.2.3 MaTrace-Global

The above MaTrace study focused on Japan and thus disregarded the distribution of steel scrap to different regions via trade. Reference [54] extended the regional

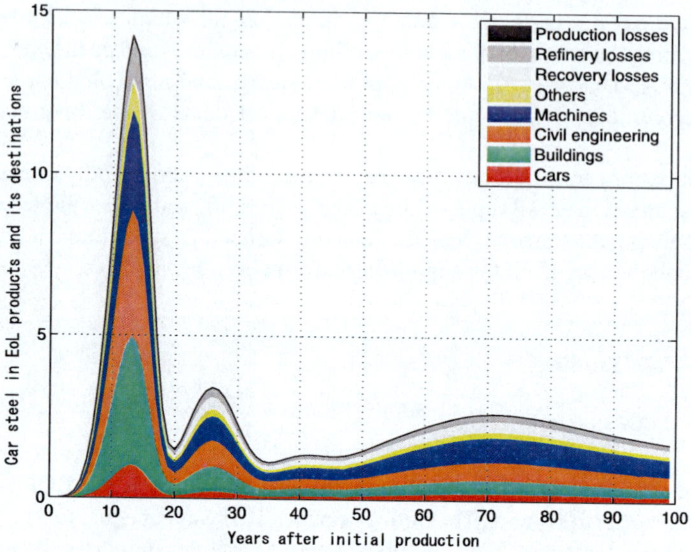

Fig. 7.7 Evolution of the amount of car steel in EoL products and its destination for new products and losses (as percentages of the initial production in year 0). Because the use in the categories "Containers" and "Others" happened to be small, they were aggregated into "Others". *Source* [52], Fig. 2

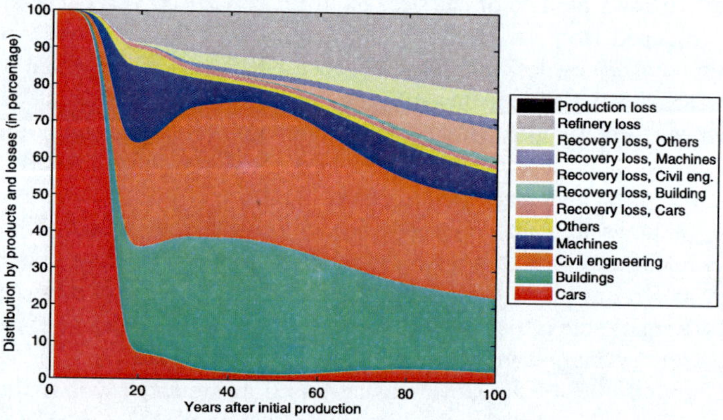

Fig. 7.8 Transition in the composition of the stock of car steel originally used for passenger cars in products and losses. Exports are assumed to follow the same product lives and are submitted to the same EoL and recycling processes as in Japan. Results of baseline scenario [52], Fig. 3

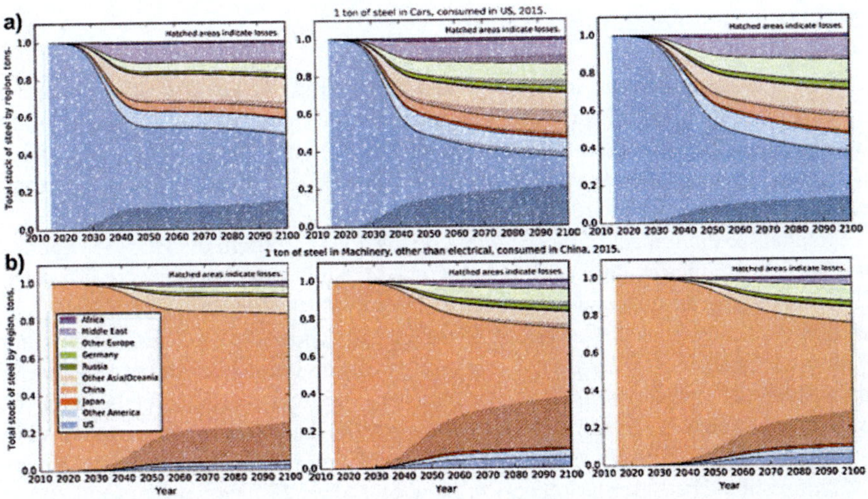

Fig. 7.9 Tracing of 10^3 kg of steel consumed in 2015 until 2100. Breakdown of the 10^3 kg of steel into 10 aggregated regions, where hatched areas indicate losses that accumulate within the different regions. **a** Steel in passenger vehicles consumed in the US in 2015. **b** Steel in machinery consumed in China in 2015. Left: Baseline scenario (the current loss rates, trade patterns, and EoL scrap through the EAF route), middle: Closed Loop scenario ((no cascading, EoL scrap through the BF-BOF route)), right: Closed Loop + High Recovery scenario + Longer product lives (+30%). *Source* [54], Fig. 4

scope of MaTrace to cover the world economy divided into 25 regions. The matrices representing the allocation of each process's outputs to the next process, $\boldsymbol{B}$ and $\boldsymbol{D}$, were spatially extended to consider international trading of scrap, secondary materials, and final products based on BACI (a refined version of the UN Comtrade database) and the EXIOBASE-v2 multi-regional supply and use table.

Figure 7.9 shows the results with regional resolution. With current trade patterns, for the US, only 40–50% of the steel in registered passenger cars will still be in the country by 2100, and the fraction of the metal that still provides useful service can be lower than 20%. Most of the losses will accumulate in the US, and the products that contain the remelted steel are distributed across world regions, where Africa, Other Europe, and China receive roughly equal shares of about 15%. In relative terms the Chinese recycling loop is less connected to the rest of the world than the US steel cycle, and therefore, between 75 and 85% of the steel consumed in a machine in China in 2015 will still be in the country by 2100 under the assumption of constant trade patterns. Most of the losses, between 20 and 30% by 2100, will also accumulate in China.

Tracking the Flow of Cobalt in the EU

Due to its high economic importance and supply risk, cobalt has been classified as a critical raw material for the EU and other economies. Securing its availability through fostering its efficient use and recycling is a part of the EU's strategy, which is affected by factors such as the amount of available end-of-life products and their collection-to-recycling rate.

To analyze the impact of these factors on the cobalt flows in the EU, [55] applied the MaTrace to predict the fate of cobalt embedded in finished products in use in the EU, considering the underlying life cycle phases within the technosphere. Seven high-end product categories of Co were studied: batteries (portable batteries and mobility batteries), catalysts (for hydroprocessing, hydroformylation, and the production of polyester (PET) precursors), dissipative uses (e.g., pigments), hard metals, magnets, superalloys, and other metallic uses. The original MaTrace was expanded to consider the hoarding of EoL products (the dead storage of a product that is no longer in use).

The results (the baseline scenario) show that after 25 years, around 8% of the initial stock of cobalt stays in use, 3% is being hoarded by users, 28% has been exported outside the EU, and 61% has been lost. The main contributors to the losses are the nonselective collection of EoL products and the export (outside the EU) of EoL products, recycled cobalt, and final products. Scenario analysis showed that higher collection-to-recycling rates and lower export could increase up to 50% the cobalt that stays in use in the EU.

A similar extension of MaTrace was used by [56] to trace the fate of aluminum in the EU. The results showed that after 25 years, 24% of the initial aluminium stays in use, 4% is hoarded, 10% has been exported and 61% has been physically lost, due mostly to nonselective collection.

Global Copper Flows

Reference [57] analyzed the global copper cycle with the world economy divided into nine regions. A distinguishing feature of this study is the consideration of the reuse option, where EoL products are reused instead of being transformed into scrap. The results showed that copper was lost at all life stages, though EoL collection was the main reason for losses. The total amount of copper in the use phase decreased rapidly, with 54% of copper lost after 35 years by 2050 under BAU. It was found that political measures (reuse, longer product life, efficiency of recovery and sorting) in the Consumer and Electronics sector have a remarkable impact on the resource efficiency of the copper cycle.

Remanufacturing

Reference [58] investigated the impacts of alternative recovery options of automotive engines, including remanufacturing, refurbishment, repair, direct reuse, and recycling, in terms of retaining the inherent value of materials and resources in products. The MaTrace model was extended to consider alternative recovery options.

Remanufacturing was found to improve the efficiency of resource use and reduce the physical loss of steel, Ni, and Cr within the system, and to avoid the cascading of high quality materials to the civil engineering and construction sectors.

7.2.2.4 *MaTrace-Alloy

In the original MaTrace [52], the cascading of steel scrap recycling occurs because steel scrap is mostly allocated to producing EAF crude steel, the major users of which are construction and civil engineering. While this modeling represents the actual situation of steel scrap recycling in Japan, the model does not explain the mechanism behind the cascading; instead, it is built into the model by the specific values of $\boldsymbol{B}$ and $\boldsymbol{D}$. To explore the mechanism behind the degradation of scrap quality through mixing metal species, the model has to consider multiple metals simultaneously. The original MaTrace focuses on a single metal/alloy.

To address this limitation of the original MaTrace, [59] developed an extended model, the MaTrace-alloy, that considers several metals combined into alloys, which allows one to trace different metals simultaneously through the waste streams, estimates the amounts of secondary alloys produced and their tramp metal contents, and quantifies losses and nonfunctional recycling of alloying metals.

The MaTrace-Alloy Model

Denote by $\boldsymbol{X}(t)$ an $n_P \times n_a$ matrix, with x_{ij} referring to the amount of alloy j, say, ferritic stainless steel, in product i, say a car, with $\boldsymbol{x} = \boldsymbol{X}\boldsymbol{\iota}_a$. The MaTrace-alloy represents the evolution of $\boldsymbol{X}(t)$ over time by

$$\boldsymbol{X}(t) = \underbrace{\Lambda \odot \boldsymbol{D}(t)\mathrm{diag}\Big(\underbrace{\Big(\boldsymbol{R}^\top \odot \boldsymbol{B}\underbrace{\boldsymbol{\Gamma}\underbrace{\underbrace{\sum_{r=0}^{t} \mathrm{diag}\Big(\boldsymbol{C}_{ap}(t-r)\big(\boldsymbol{\delta} \odot \boldsymbol{\psi}(r) \odot \boldsymbol{x}(t-r)\big)\Big)}_{\text{(a) Alloys in recovered EoL products}} \boldsymbol{C}_{ma}^\top(t-r)\Big)\boldsymbol{\iota}_m}_{\text{(b) Metals in recovered EoL products}}}_{\text{(c) Metal in scrap}}}_{\text{(d) Metals in secondary alloys}}\Big)}_{\text{(e) Metals in products made of EoL products}}, \tag{7.65}$$

where $\boldsymbol{C}_{ma}$ refers to the metal (as a chemical element) composition of alloys, $\boldsymbol{C}_{ap}$ to the alloy composition of products, $\boldsymbol{\Gamma}$ to the scrap transformation of alloys recovered from EoL products, $\boldsymbol{B}$ to the allocation of scrap to refinery processes, $\boldsymbol{R}$ to the yield of metals at the refinement of scrap into new alloys, $\boldsymbol{D}$ to the allocation of new alloys to products, Λ (an $n_P \times n_a$ matrix) to manufacturing yields, and $\boldsymbol{\delta}$ to the recovery yields of EoL products.

Particular to MaTrace-alloy is the occurrence of composition matrices $\boldsymbol{C}_{ma}(t)$ and $\boldsymbol{C}_{ap}(t)$, which are not exogenous constants but are endogenous variables. Starting with their initial values at $t = 0$ given by the initial design of alloys and products, $\boldsymbol{C}_{ma}(t)$ for $t > 0$ is determined based on the term (d) in (7.65), with $\boldsymbol{C}_{ap}(t)$ determined based on $\boldsymbol{D}$ adjusted by $\boldsymbol{\Lambda}$. Except for these points, MaTrace-alloy is identical to the original MaTrace (7.62) except for differences in dimensions to accommodate multiple metals and alloys. The matrix $\boldsymbol{R}^{\mathsf{T}} \odot \boldsymbol{B}$ corresponds to $\hat{\boldsymbol{\theta}}\boldsymbol{B}$, and $\boldsymbol{\Lambda}$ to $\hat{\boldsymbol{\lambda}}(\cdot)^{-1}$, while $\boldsymbol{D}$ and $\boldsymbol{\Gamma}$ have the same function in the two models. The term (b) in (7.65) corresponds to $z(t)$ in (7.62). The MaTrace-alloy collapses to the original MaTrace when $n_m = 1$ and $\boldsymbol{C}_{mp}(t)$ remains constant.

Tracking the Location of Cr and Ni Among 27 Steel Species

Reference [59] applied the MaTrace-alloy model to the large-scale input-output data on the steel flows for Japan developed by [29], which involves 27 steel species (of which 8 are carbon steels and 19 are nc-alloy steels) and 3 metals (Cr, Fe, and Ni). Five types of scrap were considered: austenitic stainless steel scrap, ferritic stainless steel scrap, Ni alloy scrap, other alloy steel scrap, and carbon steel scrap.

Figure 7.10 shows the transition in their alloy locations of Cr and Ni under alternative scrap sorting scenarios. Initially, around 50% of Cr and 80% of Ni were applied to austenitic stainless steel, with the rest of Cr divided into ferritic stainless steel and other noncarbon steel, while the rest of Ni was applied to noncarbon steel. Under the "maximum sorting" scenario (the middle panel), Cr and Ni remain in their initial applications, nc-alloy steels, whereas under the default sorting (the top panel), their increasing fractions are dissipated into carbon steels. Under the "minimum sorting" scenario (the bottom panel), their initial applications to nc-alloy steels are almost entirely replaced by those to carbon steels. For Cr, this results in sizable refinery (slag) losses because it is less noble than Fe (see Appendix 1 for further details). For Ni, it results in sizable functional losses because of its dissipation into carbon steels, which do not require its function.

7.2.2.5 MaTrace-Multi

Reference [60] extended the MaTrace model to trace the fate of a cohort of seven major elements in the global economy: Al, Cr, Fe, Ni, Cu, Zn, and Pb. The model was implemented using EXIOBASE3 [61] and data from [29] for chemical material composition of products calculated with WIO-MFA.

Their analysis covers wide-ranging topics, including measurement and sensitivity analysis of circularity and longevity. For two reasons, we focus on visualizing the unintentional mixing of metals over cycles of sorting and metallurgical processes by Sankey diagrams.[10] First, the unintentional mixing of metals and its consequences

[10] Sankey diagrams are named after Irish Captain Matthew Henry Phineas Riall Sankey, who used this type of diagram in 1898 to show the energy efficiency of a steam engine [62].

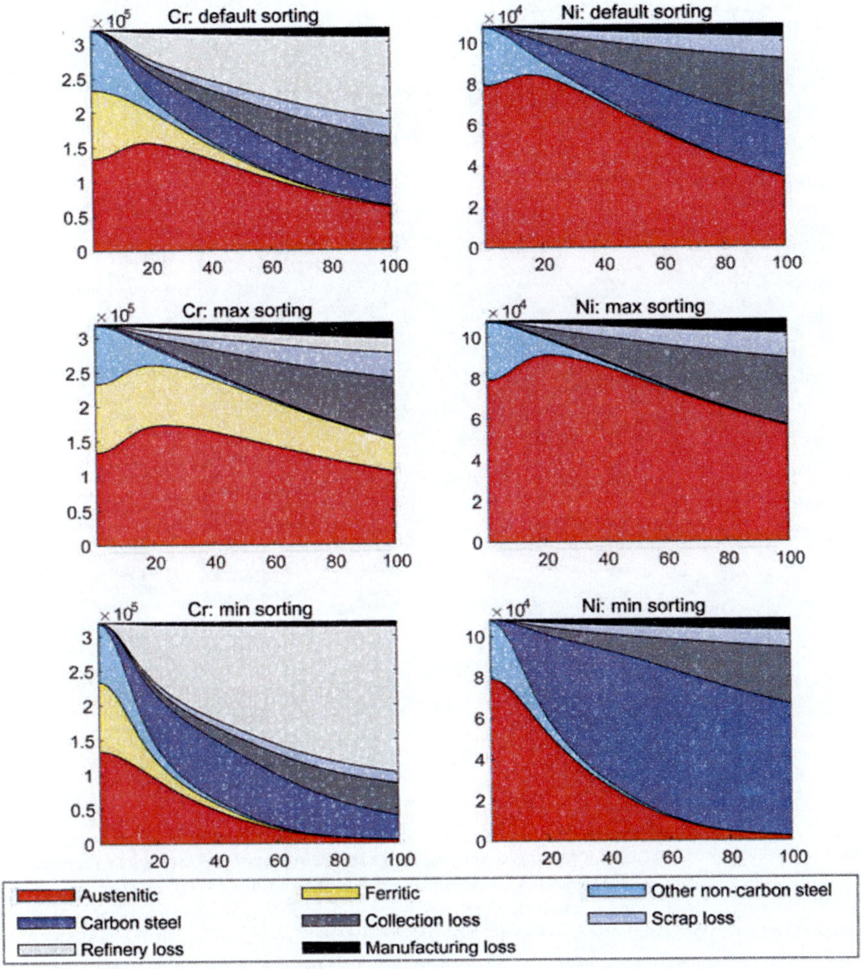

Fig. 7.10 Location of Cr and Ni among different alloys under alternative sorting schemes. The horizontal axis refers to the years after initial production. The vertical axis refers to the mass of each metal in 10^6 g. *Source* [59], Fig. 2

for the sustainable management of metals has been a major topic of this chapter. Secondly, despite its importance in MFA, we have yet to have a chance to show a real example of a Sankey diagram in this book.

A Sankey Diagram of the Global Flow of Seven Metals

Figure 7.11 shows, in the form of a Sankey diagram, the material flows for all seven metals in sorted scrap, allocated scrap, remelted secondary materials, sorting losses,

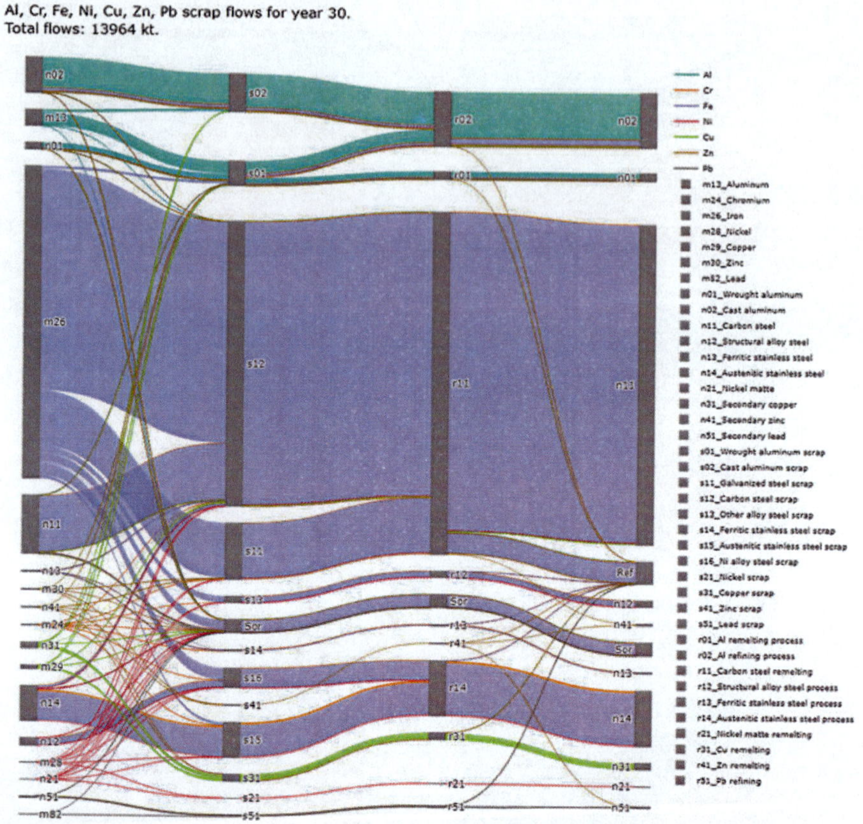

Fig. 7.11 Sankey diagram for scrap and remelting flows for the primary and secondary materials of Al, Cr, Fe, Ni, Cu, Zn, and Pb in year 30. Primary materials are indicated with index *m*, secondary materials with *n*, scraps with *s*, and remelting processes with *r*; Sor, sorting losses; Ref, remelting losses. The flow is divided into three sub-flows, each referring to a different stage: the first one from the left refers to the sorting of (primary (*m*) or secondary (*n*)) metals (in products) to scraps (*s*), the second one to the allocation of scraps to refining processes (*r*), and the last one to the transformation of scraps to secondary metals (*n*) and losses (Ref and Sor). *Source* [60], Fig. 3

and remelting losses 30 years after the initial period. In total, 14.0 Tg (Mt) is collected 30 years after the initial period, with iron making up 11.8 Tg, followed by aluminum with 1.54 Tg. All other metals combined only contributed 636 Gg (kt) of the collected end-of-life waste in that year.

In the sorting process (represented by the first sub-flow from the left), we observe that 30 years after the initial period many metals exit their first use-phase, implying that they are still collected as primary materials *m*, which applies particularly to iron (*m*26). For Al this is not the case: the flow of secondary cast aluminum waste (*n*02) is more significant than primary aluminum waste (*m*13). Furthermore, there are several wrong sorting cases, which leads to, for example, aluminum elements being mistakenly identified as steel scrap or pieces of iron entering the aluminum scrap flows.

In the second sub-flow representing the allocation of scrap to remelting/refining processes, one notices several occurrences of metal mixing. A sizable fraction of sorted wrought aluminum scrap ($s01$) enters the refining process of cast aluminum ($r02$). Galvanized steel scrap ($s11$) and carbon steel scrap ($s12$) are allocated to the carbon steel remelting process ($r11$). Austenitic stainless steel and Ni-alloy steel scrap are allocated to the same remelting process.

The third stage is where thermodynamics comes into play (see Appendix 1 for details). In the aluminum remelting and refining processes, only half the zinc content of scrap is removed. All other elements remain as contaminants or desired alloying elements in the secondary material. On the other hand, for more noble elements, such as Cu, Ni, and Pb, contamination with iron or aluminum is generally removed in the electrolysis. Some vaporized zinc is recollected in carbon steel remelting and becomes secondary zinc.

7.3 Summary and Remarks

As limitations that could hinder the use of MFA in company and governmental decision-making, [63] mentions, among others

1. the data are fragmented, inconsistent, and not harmonized [64],
2. the typically low technological detail can miss important physical constraints [65, 66].

As we have seen above, the first point on data can be, at least conceptually, addressed by resorting to IO tables, which are compiled in a fairly standardized form and are publicly available in many countries/regions. However, despite the conceptual closeness between IO and MFA, there are two major obstacles to using standard IO tables for MFA. First, standard IO tables are based on monetary units, with the flows expressed in monetary units, whereas the physical flow is needed for MFA. Secondly, the resolution in sector and region of IO tables used to be low, making it difficult to consider global flows at a detailed product/sector resolution. A possible solution to the first obstacle is to resort to the method of WIO-MFA, which enables one to transform a monetary IO into a physical flow table in terms of the masses of the materials of concern. As for measures to deal with the second obstacle (low sector/region resolution), we can mention the recent development of global MRIO databases with 100–160 sector resolution and 49–164 region resolution (see Sect. 4.3.4). Global MRIO databases with a high region and sector resolution, converted to physical flows, can provide a global database on material flows that is nonfragmented, consistent, and harmonized.

Within the context of metal recycling, the need for more technological details, as pointed out by [66] refers, among others, to the insufficient consideration of metallurgical thermodynamics governing the quality of refined/remelted metals. As we have seen above, the variants of MaTrace involving multiple metals, MaTrace-alloy [59] and MaTrace-multi [60], estimate the destination of metals in a remelting process to the gas (dust), liquid (molten metal), and solid (slag) phases based on metallurgical thermodynamics (see Appendix 1). The methods of MFA presented in this chapter indicate a possible direction to overcome the above two limitations in the current state of MFA.

We close this chapter by pointing to two limitations of the MaTrace models and possible future directions to cope with them. MaTrace focuses on tracing the fate of a single cohort of materials over the life cycles of products. It is important to note that the MaTrace model can easily be extended to trace the flows involving a large number of cohorts simultaneously; its focus on a single cohort is not its limiting feature.

The first limitation to be mentioned is the neglect of considering supply-demand balances, or the assumed absence of excess supply of scrap. Like standard IO models (and many dMFA models with a given $\boldsymbol{D}$ matrix, such as [67, 68]), the MaTrace models assume the absence of excess supply: the amounts of secondary alloys produced from recovered EoL products are 100% absorbed by the economy. Therefore, the issue of scrap surplus addressed by [39, 48–50] cannot be considered.

The second limitation (of MaTrace and most MFAs) is the absence of explicit consideration of the quantitative relationship between the flows captured by the model and the flows left uncaptured, such as energy and chemicals, which are vital for assessing environmental impacts associated with the flows of concern.[11]

These two limitations, which also apply to many models of dMFAs, can be coped with by integrating MaTrace with WIO and transforming WIO, which is static but considers supply-demand balances of all products and waste, into a dynamic one, dWIO, with the dynamics represented by MaTrace. Since WIO gives an IO-based representation of the full life cycle of products, including waste treatment and recycling (see Sect. 5.4), dWIO can be regarded as an IO-based integration of LCA and MFA. While [69] gives a conceptual sketch of dWIO, further elaborations are needed before it can be implemented with actual data.

Appendix 1: Metallurgical Thermodynamics and Refining Capabilities of Metal Remelting Processes

Unlike polymers such as plastics, textiles, and paper, which tend to lose their original function after a few life cycles, metals, as indestructible chemical elements, can be reused multiple times without deterioration or loss of functionality under normal conditions. However, in practical applications, metals are rarely used in isolation. They are often combined with other metal species either mechanically or as alloys. This simultaneous occurrence of different metal species in a product can lead to

[11] Reference [31] provides an elegant way of connecting the results of an MFA with life cycle energy requirements.

unintended mixing of metals during the recycling of EoL products. This unintended mixing, unlike deliberate alloying for specific functions, can result in the loss of original functionality and contamination of secondary materials with impurities. Addressing the issue of contamination and unintended mixing is crucial for the sustainable management of metal resources.

Metal scraps, such as iron and steel scrap and aluminum scrap, originate from diverse sources and typically undergo remelting processes before being recycled as secondary metals. These metal scraps are inputted into high-temperature melting furnaces, where they are transformed into gaseous (dust), liquid (molten metal), and solid (slag) outputs. To ensure the quality control of the output from the melting process, it is essential to understand how metal elements are distributed among the gas, liquid, and solid phases. For example, if unwanted elements that act as impurities or contaminants in the main metal output are distributed to the gas or solid phases, they can be easily removed. On the other hand, if the potential impurities or contaminants distribute to the liquid phase where the target metal is present, their removal becomes challenging unless the molten metal undergoes additional processes, such as electrolysis.

In light of these considerations, [26, 53, 70, 71] developed quantitative criteria to evaluate the refining capabilities of metal remelting processes (oxidation and evaporation) and investigated the equilibrium distribution ratios among the metal, slag, and gas phases, taking into account all relevant thermodynamic parameters.

The Gibbs Energy and the Distribution of Metal Solute Among Phases

Oxidization is a commonly used method for removing impurities in metallurgical processes. The direction of a chemical reaction can be effectively determined by examining the change in Gibbs energy.[12]

Suppose an element reacts with ambient oxygen and is distributed among different phases such as molten metal, oxide slag, and gas. In that case, the distribution can be quantified by calculating the equilibrium constant based on the change in the Gibbs energy. The equilibrium constant can then be used to determine the concentration

[12] The Gibbs energy, G, is defined as

$$G = H - TS \tag{7.66}$$

where H is enthalpy, and S is entropy. In a spontaneous change at constant pressure and temperature, the Gibbs energy decreases, $\Delta G < 0$. Every chemical reaction that is spontaneous under conditions of constant pressure and temperature are reactions that change in the direction of lower Gibbs energy ([72], p. 99). It is important to note that while the Gibbs energy shows the direction of change, it does not provide information about the speed of the reaction, which can vary greatly. Studying the rate of a chemical reaction is the subject of chemical kinetics.

of the element in each phase. We briefly discuss how the parameters controlling the equilibrium constant can be obtained.

For the evaporation of an impurity element M from a molten metal (metal solution)

$$\mathrm{M} = \mathrm{M}\ (\mathrm{gas}) \tag{7.67}$$

the equilibrium constant, K, is given by

$$K = \frac{p_M}{a_\mathrm{M}} = \frac{p_\mathrm{M}}{\gamma_\mathrm{M} x_\mathrm{M}} = p_\mathrm{M}^o \tag{7.68}$$

where p_M^o, p_M, a_M, γ_M, and x_M respectively refer to the vapor pressure of pure element M, the vapor pressure of M in the molten metal, the activity of M, the activity coefficient of M, and the mole fraction of M in the molten metal. The distribution ratio of M between the metal and gas phases is then defined by

$$L^{\mathrm{gas/metal}} = \frac{p_\mathrm{M}}{p_{\mathrm{solvent}}} = \frac{p_\mathrm{M}^o \gamma_\mathrm{M} x_\mathrm{M}}{p_{\mathrm{solvent}}} \tag{7.69}$$

where p_{solvent} is the partial vapor pressure of the solvent metal. A higher gas-metal distribution ratio indicates that the removal of M through evaporation is easier. For instance, when $L^{\mathrm{gas/metal}} = 100$, it means that the number of M atoms is 100 times larger than the number of solvent atoms in the gas phase. This suggests the possibility of a significant preferential removal of M into the gas phase through evaporation.

For the oxidation of M

$$m\mathrm{M} + \frac{n}{2}\mathrm{O_2} = \mathrm{M}_m\mathrm{O}_n \tag{7.70}$$

the corresponding expression for ΔG is

$$\Delta G = \Delta G^o + RT \ln\left(\frac{a_{\mathrm{M}_m\mathrm{O}_n}}{a_\mathrm{M}{}^m p_{\mathrm{O_2}}{}^{n/2}}\right) \tag{7.71}$$

where $p_{\mathrm{O_2}}$ is the oxygen partial pressure, ΔG^o the change in the standard Gibbs energy, R the gas constant, T the absolute temperature, and $a_{\mathrm{M}_m\mathrm{O}_n}$ the activity of oxide.[13] Since $\Delta G = 0$ at equilibrium, we have the following expression for the equilibrium constant K'

$$K' = \frac{a_{\mathrm{M}_m\mathrm{O}_n}}{a_\mathrm{M}{}^m p_{\mathrm{O_2}}{}^{n/2}} = \frac{\gamma_{\mathrm{M}_m\mathrm{O}_n} x_{\mathrm{M}_m\mathrm{O}_n}}{(\gamma_\mathrm{M} x_\mathrm{M})^m p_{\mathrm{O_2}}^{n/2}} = \exp\left(-\frac{\Delta G^o}{RT}\right) \tag{7.72}$$

---

[13] Please see standard textbooks on chemistry/physical chemistry for the derivation of this well-known formula.

where $\gamma_{M_mO_n}$ the activity coefficient of oxide, and $x_{M_mO_n}$ the mole fraction of oxide M_mO_n distributed in the slag phase by oxidization. Rearrangement of (7.72) gives the distribution ratio of M between slag and metal, $L^{\text{slag/metal}}$:

$$L^{\text{slag/metal}} = \frac{x_M}{x_{M_mO_n}} = \frac{\gamma_{M_mO_n}}{K'\gamma_M{}^m x_M{}^{m-1} p_{O_2}{}^{n/2}} \tag{7.73}$$

A lower metal-slag distribution ratio makes it easier for M to be removed into the slag phase through oxidation. The tendency of element distribution between the metal and slag can be distinguished by the boundary at $\log L^{\text{slag/metal}} = 0$.

Visualizing the Distribution of Elements Among Gas, Slag, and Metal Phases

Figure 7.12 gives a visual representation of the distribution tendencies of elements in the metal, slag, and gas phases of elements in the remelting of base metals such as steel, Cu, Pb, Zn, and Al. These tendencies were determined based on the above thermodynamic analysis and technical information on metallurgy processes. The chart, called an "Element Radar Chart" (or "metal pizza"), is a valuable tool for comparing the relative ease of impurity removal different base metals. This chart can be considered as an urban mining equivalent of the "metal wheel" [73, 74] discussed

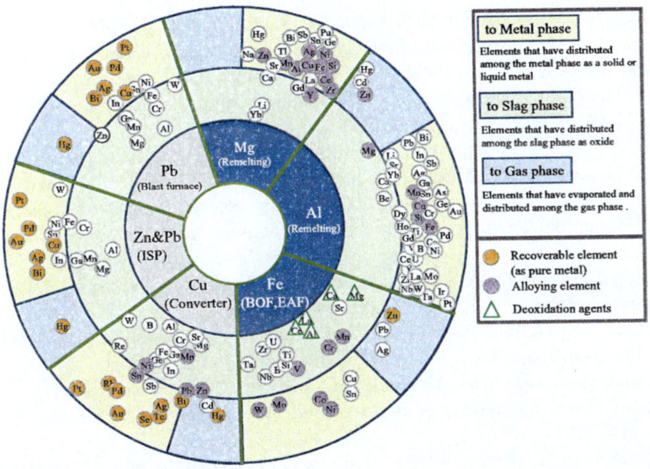

Fig. 7.12 Elemental radar chart representing the distribution tendencies in the metal, slag, and gas phases of the elements in the remelting of Al, Fe, Cu, Zn, Pb, and Mg. The radian direction indicates the element distribution tendency for oxidation, and the circular arc direction indicates that for volatilization. *Source* [70], Fig. 9

in Sect. 2.2.4.4, which illustrates the interconnections of metals in natural resource processing (see Table 2.10).

The chart reveals that removing impurities is notably more challenging for aluminum compared to iron/steel. In the case of iron/steel recycling, impurities such as Cu and Sn are present as tramp elements that are difficult to eliminate. On the other hand, for Al, most impurities, except for certain elements like Mg and Zn, are considered tramp elements that are hard to remove.

In contrast, the smelting of Cu and Pb incorporates an effective impurity removal process, electrolytic smelting, as a post-process, which acts as an excellent refiner for a number of impurities. The higher prices of Cu and Pb compared to Fe and Al, along with the co-production of highly valuable Au and Ag (see Sect. 2.2.4.4), justify the application of electrolytic smelting to Cu and Pb. However, such conditions do not apply to aluminum and iron remelting.

In aluminum and iron recycling, the disassembling processes and managing scrap quality play crucial roles in minimizing contamination. Compared to other base metals, the removal of impurity elements is particularly challenging for Mg, except for a very few elements like yttrium (Y). Ensuring a sustainable supply of Mg necessitates the careful design of advanced Mg alloys to prevent contamination or excessive accumulation of alloying elements, as well as the development of refining processes for EoL magnesium products [70].

References

1. Cullen, Jonathan M., and Julian M. Allwood. 2013. Mapping the global flow of aluminum: From liquid aluminum to end-use goods. *Environmental Science and Technology* 47 (7): 3057–3064.
2. Wiedmann, Thomas O., Heinz Schandl, Manfred Lenzen, Daniel Moran, Sangwon Suh, James West, and Keiichiro Kanemoto. 2013. The material footprint of nations. *Proceedings of the National Academy of Sciences* 201220362.
3. Hertwich, Edgar G., Thomas Gibon, Evert A. Bouman, Anders Arvesen, Sangwon Suh, Garvin A. Heath, Joseph D. Bergesen, Andrea Ramirez, Mabel I. Vega, and Lei Shi. 2015. Integrated life-cycle assessment of electricity-supply scenarios confirms global environmental benefit of low-carbon technologies. *Proceedings of the National Academy of Sciences* 112 (20): 6277–6282.
4. Brunner, Paul H., and Helmut Rechberger. 2016. *Handbook of material flow analysis: For environmental, resource, and waste engineers*. CRC Press.
5. Laner, David, and Helmut Rechberger. 2016. Material flow analysis. In *Special types of life cycle assessment*, Chap. 7, ed. Matthias Finkbeiner, 293–332. Springer.
6. Cullen, Jonathan M., Julian M. Allwood, and Margarita D. Bambach. 2012. Mapping the global flow of steel: From steelmaking to end-use goods. *Environmental Science and Technology* 46 (24): 13048–13055.
7. Kemeny, Laurie, and John Snell. 1976. *Finite markov chains*. Number 3. Springer.
8. Yamada, Hiroyuki, Ichiro Daigo, Yasunari Matsuno, Yoshihiro Adachi, and Yasushi Kondo. 2006. Application of Markov chain model to calculate the average number of times of use of a material in society: An allocation methodology for open-loop recycling Part 1: Methodology Development. *International Journal of Life Cycle Assessment* 11 (5): 354–360.

9. Duchin, Faye, and Stephen H. Levine. 2010. Embodied resource flows and product flows: Combining the absorbing markov chain with the input-output model. *Journal of Industrial Ecology* 14 (4): 586–597.
10. Matsuno, Yasunari, Ichiro Daigo, and Yoshihiro Adachi. 2007. Application of Markov chain model to calculate the average number of times of use of a material in society: An allocation methodology for open-loop recycling - Part 2: Case study for steel. *International Journal of Life Cycle Assessment* 12 (1): 34–39.
11. Streeck, Jan, Stefan Pauliuk, Hanspeter Wieland, and Dominik Wiedenhofer. 2023. A review of methods to trace material flows into final products in dynamic material flow analysis: From industry shipments in physical units to monetary input-output tables, Part 1. *Journal of Industrial Ecology* 1–21.
12. Chen, Wei Qiang, and T.E. Graedel. 2015. Improved alternatives for estimating in-use material stocks. *Environmental Science and Technology* 49 (5): 3048–3055.
13. Krausmann, Fridolin, Dominik Wiedenhofer, Christian Lauk, Willi Haas, Hiroki Tanikawa, Tomer Fishman, Alessio Miatto, Heinz Schandl, and Helmut Haberl. 2017. Global socioeconomic material stocks rise 23-fold over the 20th century and require half of annual resource use. *Proceedings of the National Academy of Sciences* 114 (8): 1880–1885.
14. Lanau, Maud, Gang Liu, Ulrich Kral, Dominik Wiedenhofer, Elisabeth Keijzer, Yu. Chang, and Christina Ehlert. 2019. Taking stock of built environment stock studies: Progress and prospects. *Environmental Science and Technology* 53 (15): 8499–8515.
15. Miller, Ronald E., and Peter D. Blair. 2022. *Input-Output Analysis Foundations and Extensions*. Cambridge University Press.
16. Nakamura, Shinichiro, and Kenichi Nakajima. 2005. Waste input-output material flow analysis of metals in the Japanese economy. *Materials Transactions* 46 (12): 2550–2553.
17. Nakamura, Shinichiro, Kenichi Nakajima, Yasushi Kondo, and Tetsuya Nagasaka. 2007. The waste input-output approach to materials flow analysis: Concepts and application to base metals. *Journal of Industrial Ecology* 11 (4): 50–63.
18. Nakamura, Shinichiro, and Yasushi Kondo. 2009. *Waste input-output analysis*. Dordrecht: Springer Science & Business Media.
19. Yokoi, Ryosuke, Jun Nakatani, and Yuichi Moriguchi. 2018. An extended model for tracking accumulation pathways of materials using input-output tables: Application to copper flows in Japan. *Sustainability (Switzerland)* 10 (3).
20. Nakamura, Shinichiro, Kenichi Nakajima, Yoshie Yoshizawa, Kazuyo Matsubae-Yokoyama, and Tetsuya Nagasaka. 2009. Analyzing polyvinyl chloride in japan with the waste input-output material flow analysis model. *Journal of Industrial Ecology* 13: 706–717.
21. Nakamura, Shinichiro, Yasushi Kondo, Kazuyo Matsubae, Kenichi Nakajima, and Tetsuya Nagasaka. 2011. UPIOM: A new tool of mfa and its application to the flow of iron and steel associated with car production. *Environmental Science and Technology* 45 (3): 1114–1120.
22. Simpson, David, and Jinkichi Tsukui. 1965. The fundamental structure of input-output tables, an international comparison. *Review of Economics and Statistics* 47: 434–446.
23. Jiang, Daqian, Wei-Qiang. Chen, Wei Liu, and Marian Chertow. 2017. Inter-sectoral Bisphenol A (BPA) flows in the 2012 Chinese economy. *Environmental Science & Technology* 51 (15): 8654–8662.
24. Castro, Maria B.G., Johannes A.M. Remmerswaal, and Markus A. Reuter. 2003. Life cycle impact assessment of the average passenger vehicle in the Netherlands. *The International Journal of Life Cycle Assessment* 8 (5): 297–304.
25. Castro, M.B., J.A.M. Remmerswaal, J.C. Brezet, Schaik A. van, and M.A. Reuter. 2005. A simulation model of the comminution-liberation of recycling streams Relationships between product design and the liberation of materials during recycling. *International Journal of Mineral Processing* 75 (3–4): 255–281.
26. Nakajima, Kenichi, Osamu Takeda, Takahiro Miki, Kazuyo Matsubae, Shinichiro Nakamura, and Tetsuya Nagasaka. 2010. Thermodynamic analysis of contamination by alloying elements in aluminum recycling. *Environmental Science and Technology* 44 (14): 5594–5600.

27. Houpert, C., V. Lanterl, M. Guttmann, and M. Jallon. 1997. Influence of tramp elements in the production of high quality steels using the scrap/electric arc furnace route. *La Revue de Métallurgie-CIT Novembre* 1369–1384.
28. Ichiro Daigo, Leo Fujimura, Hideo Hayashi, Eiji Yamasue, Satoshi Ohta, Tran Duc Huy, and Yoshikazu Goto. Quantifying the total amounts of tramp elements associated with carbon steel production in japan. *ISIJ International*, 57(2):388–393, 2017.
29. Ohno, Hajime, Kazuyo Matsubae, Kenichi Nakajima, Shinichiro Nakamura, and Tetsuya Nagasaka. 2014. Unintentional flow of alloying elements in steel during recycling of end-of-life vehicles. *Journal of Industrial Ecology* 18 (2): 242–253.
30. Ohno, Hajime, Kazuyo Matsubae, Kenichi Nakajima, Yasushi Kondo, Shinichiro Nakamura, and Tetsuya Nagasaka. 2015. Toward the efficient recycling of alloying elements from end of life vehicle steel scrap. *Resources, Conservation and Recycling* 100: 11–20.
31. Wang, Tong, Peter Berrill, Julie B. Zimmerman, and Edgar G. Hertwich. 2021. Copper recycling flow model for the United States Economy: Impact of scrap quality on potential energy benefit. *Environmental Science & Technology*.
32. Schneider, Falk, Sophie Parsons, Sally Clift, Andrea Stolte, and Marcelle C. McManus. 2018. Collected marine litter - A growing waste challenge. *Marine Pollution Bulletin*, 128 (December 2017): 162–174.
33. Arp, Hans Peter H., Dana Kühnel, Christoph Rummel, Matthew Macleod, Annegret Potthoff, Sophia Reichelt, Elisa Rojo-Nieto, Mechthild Schmitt-Jansen, Johanna Sonnenberg, Erik Toorman, and Annika Jahnke. 2021. Weathering plastics as a planetary boundary threat: Exposure, fate, and hazards. *Environmental Science and Technology* (1).
34. Bank, Michael S., Peter W. Swarzenski, Carlos M. Duarte, Matthias C. Rillig, Albert A. Koelmans, Marc Metian, Stephanie Wright, Jennifer F. Provencher, Monica Sanden, Adrian Jordaan, Martin Wagner, Martin Thiel, and Yong Sik Ok. 2021. Global plastic pollution observation system to aid policy. *Environmental Science & Technology*.
35. Nakatani, Jun, Tamon Maruyama, and Yuichi Moriguchi. 2020. Revealing the intersectoral material flow of plastic containers and packaging in Japan. *PNAS* 117 (33): 19844–19853.
36. Müller, Daniel. 2006. Stock dynamics for forecasting material flows-Case study for housing in The Netherlands. *Ecological Economics* 59: 142–156.
37. Mueller, Esther, Lorenz M. Hilty, Rolf Widmer, Mathias Schluep, and Martin Faulstich. 2014. Modeling metal stocks and flows: A review of dynamic material flow analysis methods. *Environmental Science & Technology* 48 (4): 2102–2113.
38. Lauinger, Dirk, Romain G. Billy, Felipe Vásquez, and Daniel B. Müller. 2021. A general framework for stock dynamics of populations and built and natural environments. *Journal of Industrial Ecology* 25 (5): 1136–1146.
39. Løvik, Amund N., Roja Modaresi, and Daniel B. Müller. 2014. Long-term strategies for increased recycling of automotive aluminum and its alloying elements. *Environmental Science & Technology* 48 (8): 4257–4265.
40. Pauliuk, Stefan, Tao Wang, and Daniel Beat Müller. 2012. Moving toward the circular economy: The role of stocks in the Chinese steel cycle. *Environmental Science & Technology* 46 (1): 148–154.
41. Müller, Daniel B., Jing Cao, Elif Kongar, Matthew Altonji, Paul-Harvey Weiner, and a TE Graedel. 2007. Service lifetimes of mineral end uses. *Minerals Resources External Research Program* (USGS Award nr. 06HQGR0174): 31.
42. Weymar, Elisabeth, and Matthias Finkbeiner. 2016. Statistical analysis of empirical lifetime mileage data for automotive LCA. *International Journal of Life Cycle Assessment* 21 (2): 215–223.
43. Zeng, Xianlai, Saleem H. Ali, Jinping Tian, and Jinhui Li. 2020. Mapping anthropogenic mineral generation in China and its implications for a circular economy. *Nature Communications* 11 (1): 1–9.
44. Kagawa, Shigemi, Keisuke Nansai, Yasushi Kondo, Klaus Hubacek, Sangwon Suh, Jan Minx, Yuki Kudoh, Tomohiro Tasaki, and Shinichiro Nakamura. 2011. Role of motor vehicle lifetime extension in climate change policy. *Environmental Science and Technology* 45 (4): 1184–1191.

45. Kagawa, Shigemi, Shinichiro Nakamura, Yasushi Kondo, Kazuyo Matsubae, and Tetsuya Nagasaka. 2014. Forecasting replacement demand of durable goods and the induced secondary material flows: A case study of automobiles. *Journal of Industrial Ecology* 19 (1): 10–19.
46. Kagawa, Shigemi, Tomohiro Tasaki, and Yuichi Moriguchi. 2006. The environmental and economic consequences of product lifetime extension: Empirical analysis for automobile use. *Ecological Economics* 58 (1): 108–118.
47. Watari, Takuma, Keisuke Nansai, Damien Giurco, Kenichi Nakajima, Benjamin McLellan, and Christoph Helbig. 2020. Global metal use targets in line with climate goals. *Environmental Science and Technology* 54 (19): 12476–12483.
48. Hatayama, Hiroki, Ichiro Daigo, Yasunari Matsuno, and Yoshihiro Adachi. 2012. Evolution of aluminum recycling initiated by the introduction of next-generation vehicles and scrap sorting technology. *Resources, Conservation and Recycling* 66: 8–14.
49. Modaresi, Roja, and Daniel B. Müller. 2012. The role of automobiles for the future of aluminium recycling. *Environmental Science and Technology* 16: 8587–8594.
50. Billy, Romain G., and Daniel B. Müller. 2023. Aluminium use in passenger cars poses systemic challenges for recycling and GHG emissions. *Resources, Conservation and Recycling* 190 (September 2022): 106827.
51. Nakamura, Shinichiro, Yasushi Kondo, Kazuyo Matsubae, Kenichi Nakajima, Tomohiro Tasaki, and Tetsuya Nagasaka. 2012. Quality- and dilution losses in the recycling of ferrous materials from end-of-life passenger cars: Input-output analysis under explicit consideration of scrap quality. *Environmental Science and Technology* 46 (17): 9266–9273.
52. Nakamura, Shinichiro, Yasushi Kondo, Shigemi Kagawa, Kazuyo Matsubae, Kenichi Nakajima, and Tetsuya Nagasaka. 2014. MaTrace: Tracing the fate of materials over time and across products in open-loop recycling. *Environmental Science and Technology* 48 (13): 7207–7214.
53. Nakajima, Kenichi, Osamu Takeda, Takahiro Miki, and Tetsuya Nagasaka. 2009. Evaluation method of metal resource recyclability based on thermodynamic analysis. *Materials Transactions* 50 (3): 453–460.
54. Pauliuk, Stefan, Yasushi Kondo, Shinichiro Nakamura, and Kenichi Nakajima. 2017. Regional distribution and losses of end-of-life steel throughout multiple product life cycles-Insights from the global multiregional MaTrace model. *Resources, Conservation and Recycling* 116: 84–93.
55. León, María Fernanda Godoy, Gian Andrea Blengini, and Jo Dewulf. 2020. Cobalt in end-of-life products in the EU, where does it end up? - The MaTrace approach. *Resources, Conservation and Recycling* 158 (September 2019): 104842.
56. Jácome, Gabriela Jarrín, María Fernanda Godoy León, Rodrigo A.F. Alvarenga, and Jo Dewulf. 2021. Tracking the fate of aluminium in the EU using the MaTrace model. *Resources* 10 (7).
57. Klose, Stefanie, and Stefan Pauliuk. 2021. Quantifying longevity and circularity of copper for different. *Journal of Industrial Ecology* 1–15.
58. Zhang, Zhengyang, Kazuyo Matsubae, and Kenichi Nakajima. 2021. Impact of remanufacturing on the reduction of metal losses through the life cycles of vehicle engines. *Resources, Conservation and Recycling* 170: 105614.
59. Nakamura, Shinichiro, Yasushi Kondo, Kenichi Nakajima, Hajime Ohno, and Stefan Pauliuk. 2017. Quantifying recycling and losses of Cr and Ni in steel throughout multiple life cycles using MaTrace-alloy. *Environmental Science and Technology* 51 (17): 9469–9476.
60. Helbig, Christoph, Yasushi Kondo, and Shinichiro Nakamura. 2022. Simultaneously tracing the fate of seven metals at a global level with MaTrace-multi. *Journal of Industrial Ecology* 1–14.
61. Stadler, Konstantin, Kjartan Steen-Olsen, and Richard Wood. 2014. The "Rest of the World" - Estimating the economic structure of missing regions in global multi-regional input-output tables. *Economic Systems Research* 26 (3): 303–326.
62. Schmidt, Mario. 2008. The sankey diagram in energy and material flow management part I: History. *Journal of Industrial Ecology* 12: 82–94.
63. Baars, Joris, Mohammad Ali Rajaeifar, and Oliver Heidrich. 2022. Quo vadis MFA? Integrated material flow analysis to support material efficiency. *Journal of Industrial Ecology* 26 (4): 1487–1503.

64. Müller, Daniel, Maren Lundhaug, and Mark Simoni. 2020. Challenges, systems and data (Synthesis report D2. 2). Technical Report 730330.
65. Gaustad, Gabrielle, Elsa Olivetti, and Randolph Kirchain. 2011. Toward sustainable material usage: Evaluating the importance of market motivated agency in modeling material flows. *Environmental science & technology*.
66. Schaik, Antoinette van, and Markus A. Reuter. 2010. Dynamic modelling of E-waste recycling system performance based on product design. *Minerals Engineering* 23 (3): 192–210.
67. Davis, Jenny, R. Geyer, J. Ley, J. He, R. Clift, A. Kwan, M. Sansom, and T. Jackson. 2007. Time-dependent material flow analysis of iron and steel in the UK Part 1: Production and consumption trends 1970–2000. *Resources, Conservation and Recycling* 51: 101–117.
68. Davis, Jenny, R. Geyer, J. Ley, J. He, R. Clift, A. Kwan, M. Sansom, and T. Jackson. 2007. Time-dependent material flow analysis of iron and steel in the UK. Part 2. Scrap generation and recycling. *Resources, Conservation and Recycling* 51 (1): 118–140.
69. Nakamura, Shinichiro, and Yasushi Kondo. 2018. Toward an integrated model of the circular economy: Dynamic waste input-output. *Resources, Conservation and Recycling* 139: 326–332.
70. Hiraki, Takehito, Osamu Takeda, Kenichi Nakajima, Kazuyo Matsubae, Shinichiro Nakamura, and Tetsuya Nagasaka. 2011. Thermodynamic criteria for the removal of impurities from end-of-life magnesium alloys by evaporation and flux treatment. *Science and Technology of Advanced Materials* 12: 1–10.
71. Lu, Xin, T. Hiraki, K. Nakajima, O. Takeda, K. Matsuabe, H.-M. Zhu, S. Nakamura, and T. Nagasaka. 2012. Thermodynamic analysis of separation of alloying elements in recycling of end-of-life titanium products. *Separation and Purification Technology* 89: 135–141.
72. Atkins, Peter, and Julio de Paula. 2009. *Elements of Physical Chemistry*. Oxford University Press.
73. Castro, M.B.G., J.A.M. Remmerswaal, M.A. Reuter, and U.J.M. Boin. 2004. A thermodynamic approach to the compatibility of materials combinations for recycling. *Resources, Conservation and Recycling* 43: 1–19.
74. Verhoef, E., P. Gerard, Markus Dijkema, and P. Reuter. 2004. Process knowledge, system dynamics, and metal ecology. *Journal of Industrial Ecology* 8 (1): 23–43.

Appendix A
Basics of Matrix Algebra

We review some of the basic operations and fundamental properties of elementary matrix algebra that are relevant to input-output analysis.[1] A matrix $\boldsymbol{C}$ of size $m \times n$ is the $m \times n$ rectangular array of scalars c_{ij} given by

$$\boldsymbol{C} = \begin{pmatrix} c_{11} & c_{12} & \dots & c_{1n} \\ c_{21} & c_{22} & \dots & c_{2n} \\ \vdots & \vdots & & \vdots \\ c_{n1} & c_{n2} & \dots & c_{nn} \end{pmatrix} \tag{A.1}$$

which will be simply identified as $\boldsymbol{C} = [c_{ij}]$, with c_{ij} referring to its ith row jth column element. If $m = n$, $\boldsymbol{C}$ is called a square matrix of order m. In this book, a matrix is distinguished from a scalar using bold font.

$$\boldsymbol{Z} = \begin{pmatrix} x_{11} & x_{12} \\ x_{21} & x_{22} \end{pmatrix} \tag{A.2}$$

An $m \times 1$ matrix, denoted by $\boldsymbol{c}$,

$$\boldsymbol{c} = \begin{pmatrix} c_1 \\ c_2 \\ \vdots \\ c_m \end{pmatrix} \tag{A.3}$$

is called a column vector. On the other hand, an $n \times 1$ matrix $\boldsymbol{d} = [d_i]$

$$\boldsymbol{d} = \begin{pmatrix} d_1 & d_2 & \cdots & d_n \end{pmatrix} \tag{A.4}$$

---

[1] See James E. Gentle. 2007. *Matrix algebra*. Springer, for further details.

S. Nakamura, *A Practical Guide to Industrial Ecology by Input-Output Analysis*,
https://doi.org/10.1007/978-3-031-43684-0

is called a row vector. We will use bold case capital letters to denote matrices and bold case lower letters to denote vectors. A matrix is called square when it has the same number of columns and rows, and rectangular when it is not.

A square matrix with all its off-diagonal elements zero is called a diagonalized matrix. A diagonalized matrix of order n with $\boldsymbol{c}$ in (A.3) as its diagonal elements is denoted $\hat{\boldsymbol{c}}$ or $\mathrm{diag}(\boldsymbol{c})$

$$\hat{\boldsymbol{c}} = \begin{pmatrix} c_1 & 0 & 0 & \cdots & 0 \\ 0 & c_2 & 0 & \cdots & 0 \\ \vdots & \vdots & \ddots & \vdots & \vdots \\ 0 & \cdots & \cdots & \cdots & c_n \end{pmatrix} \tag{A.5}$$

A diagonal matrix with all its diagonal elements equal to unity is called an identity matrix and is denoted by $\boldsymbol{I}$.

Alternatively, for $\boldsymbol{c} = [c_i]$, we denote by $\mathrm{diag}[c_i]$ the diagonal matrix of $\boldsymbol{c}$.

A.1 Basic Operations of Matrices

A.1.1 Transpose

For a rectangular matrix $\boldsymbol{A} = [a_{ij}]$ of order $m \times n$, the matrix obtained from $\boldsymbol{A}$ with its i, j elements interchanged with j, i elements, is called the transpose of A, and is denoted by $\boldsymbol{A}^\top$, that is, $\boldsymbol{A}^\top$ is an $n \times m$ matrix with its i, j element given by a_{ji}.

A.1.2 Addition

Let there be two matrices $\boldsymbol{Q} = [q_{ij}]$ and $\boldsymbol{R} = [r_{ij}]$ of orders $m \times n$ and $o \times p$. The sum of two matrices $\boldsymbol{Q}$ and $\boldsymbol{R}$ is defined if they have the same number of rows and the same number of columns; in this case

$$\boldsymbol{Q} + \boldsymbol{R} = [q_{ij} + r_{ij}] \tag{A.6}$$

A.1.3 Multiplication

The multiplication of $\boldsymbol{Q}$ by a scalar α gives an $m \times n$ matrix:

$$\alpha \boldsymbol{Q} = [\alpha q_{ij}] \tag{A.7}$$

The pre-multiplication of $\boldsymbol{R}$ by $\boldsymbol{Q}$ is defined only if the number of columns of $\boldsymbol{Q}$ is equal to the number of rows of $\boldsymbol{R}$. Thus, if $n = o$, then $\boldsymbol{T} = \boldsymbol{Q}\boldsymbol{R}$ will be an $m \times p$ matrix $\boldsymbol{T} = [t_{ij}]$ with

$$t_{ij} = \sum_{k=1}^{n} q_{ik} r_{kj}. \tag{A.8}$$

Note that the right-hand side involves the ith row elements of $\boldsymbol{Q}$, $\boldsymbol{Q}_{i\cdot}$ and the jth column elements of R, $\boldsymbol{R}_{\cdot j}$. Accordingly, (A.8) can also be represented by

$$t_{ij} = \boldsymbol{Q}_{i\cdot}\boldsymbol{R}_{\cdot j}. \tag{A.9}$$

In the special case where R is a diagonal matrix, this becomes

$$t_{ij} = q_{ij} r_{jj} \tag{A.10}$$

On the other hand, the post-multiplication of $\boldsymbol{R}$ by $\boldsymbol{Q}$ is defined only if the number of rows of $\boldsymbol{Q}$, m, is equal to the number of columns of $\boldsymbol{R}$, p. Thus, if $m = p$, then $\boldsymbol{U} = \boldsymbol{R}\boldsymbol{Q}$ will be an $o \times n$ matrix $\boldsymbol{U} = [u_{ij}]$ with

$$u_{ij} = \sum_{k=1}^{m} r_{ik} q_{kj}. \tag{A.11}$$

For the transpose of the product of matrices, we have

$$(\boldsymbol{Q}\boldsymbol{R})^{\top} = \boldsymbol{R}^{\top}\boldsymbol{Q}^{\top} \tag{A.12}$$

A.1.4 A Vector with All Elements Being One

Denote by $\boldsymbol{\tau}_n$ an $n \times 1$ vector with all elements being one

$$\boldsymbol{\tau}_n = \begin{pmatrix} 1 \\ 1 \\ \vdots \\ 1 \end{pmatrix} \tag{A.13}$$

$\boldsymbol{\tau}_n$ is useful to obtain the row (column) sum of a matrix. From the rule of multiplication of matrices, it follows

$$\tau_m^\top Q = \sum_{i=1}^{m} q_{ij}$$
$$Q\tau_n = \sum_{j=1}^{n} q_{ij} \tag{A.14}$$

A.2 Determinants, Minors, and Cofactors

Consider the following square matrix of $n = 2$

$$A = \begin{pmatrix} a_{11} & a_{12} \\ a_{21} & a_{22} \end{pmatrix} \tag{A.15}$$

The determinant of A, $|A|$, is defined as

$$|A| = a_{11}a_{22} - a_{12}a_{12} \tag{A.16}$$

The determinant of a scalar is the value of its sole element: $|a_{11}| = a_{11}$. Accordingly, we have

$$\begin{aligned} |A| &= a_{11}(-1)^{1+1}|a_{22}| + a_{12}(-1)^{1+2}|a_{21}| \\ &= a_{22}(-1)^{2+2}|a_{11}| + a_{21}(-1)^{2+1}|a_{12}| \end{aligned} \tag{A.17}$$

The determinants $|a_{22}|$, $|a_{21}|$, $|a_{11}|$, and $|a_{12}|$ are respectively called the minors of a_{11}, a_{12}, a_{22}, and a_{21}. The minor of a_{ij}, $|M_{ij}|$, refers the determinant of the sub-matrix M_{ij} obtained from A by deleting the ith row and jth column: $M_{11} = a_{22}$ and $|M_{11}| = |a_{22}|$ in the current simple case. Accordingly, the determinant of a square matrix is computed by adding up its minors $|M_{ij}|$ signed by -1^{i+j}, or expanded as:

$$|A| = \sum_i a_{ij}(-1)^{i+j}|M_{ij}| = \sum_i a_{ij}c_{ij} \tag{A.18}$$

where $c_{ij} = a_{ij}(-1)^{i+j}|M_{ij}|$ is called the cofactor of a_{ij}.

For a square matrix of $n = 3$

$$A = \begin{pmatrix} a_{11} & a_{12} & a_{13} \\ a_{21} & a_{22} & a_{23} \\ a_{31} & a_{32} & a_{33} \end{pmatrix} \tag{A.19}$$

we have

$$
\begin{aligned}
|A| &= a_{11}(-1)^{1+1}|M_{11}| + a_{12}(-1)^{1+2}|M_{12}| + a_{13}(-1)^{1+3}|M_{13}| \\
&= a_{11}c_{11} + a_{12}c_{12} + a_{13}c_{13} \\
&= a_{11}\begin{vmatrix} a_{12} & a_{13} \\ a_{32} & a_{33} \end{vmatrix} - a_{12}\begin{vmatrix} a_{21} & a_{31} \\ a_{23} & a_{33} \end{vmatrix} - a_{13}\begin{vmatrix} a_{21} & a_{12} \\ a_{31} & a_{32} \end{vmatrix}
\end{aligned} \tag{A.20}
$$

Note that

$$
\sum_j a_{ij}c_{hj} = 0 \quad \text{for any } i \neq h, \tag{A.21}
$$

which follows from the fact that c_{hj}, $h \neq i$, refer to cofactors that come from all rows of A excluding the h ($h \neq i$)th, and so contain the ith row: (A.21) refers to the expansion of a determinant with the same two rows, which is zero. In the case of $n = 2$, we have

$$
a_{11}c_{21} + a_{12}c_{22} = -a_{11}a_{12} + a_{12}a_{11} = 0 \tag{A.22}
$$

A.3 Inverse Matrices

A square matrix B is called the inverse of a square matrix A, if their product gives an identity matrix

$$
AB = I \tag{A.23}
$$

and is denoted by A^{-1}. Denoting its cofactor by $C = [c_{ij}]$, we have

$$
A^{-1} = \frac{C^{\top}}{|A|} \tag{A.24}
$$

which follows from the fact that the diagonal elements of AA^{-1} are equal to (A.18), that is, $|A|$, while the off-diagonal elements are equal to (A.21), and hence zero. For the case of $n = 2$, we have

$$
A^{-1} = \frac{\begin{pmatrix} a_{22} & -a_{21} \\ -a_{12} & a_{22} \end{pmatrix}}{a_{11}a_{22} - a_{12}a_{21}} \tag{A.25}
$$

From (A.24) it follows that A^{-1} does exist only if $|A| \neq 0$. This property is called the square matrix A is nonsingular. A is called a singular matrix when $|A| = 0$.

What are the conditions under which $|A| \neq 0$ holds? A square matrix A is nonsingular if and only if all the columns (rows) of A are mutually independent. A square matrix A of order m is linearly independent if for an $m \times 1$ vector b:

$$\boldsymbol{A}\boldsymbol{b} = \boldsymbol{0} \Rightarrow \boldsymbol{b} = \boldsymbol{0} \tag{A.26}$$

The maximum number of linearly independent columns is called the rank of the matrix, and is denoted by $r(\boldsymbol{A})$. If $\boldsymbol{A}$ is nonsingular, $r(\boldsymbol{A}) = m$. When the rank of a square matrix is equal to its order, the matrix is said to have a full rank.

A.3.1 Inversion of a Product and a Transpose

The inverse of a product is the product of the inverse taken in reverse order

$$(\boldsymbol{A}\boldsymbol{B})^{-1} = \boldsymbol{B}^{-1}\boldsymbol{A}^{-1} \tag{A.27}$$

provided $\boldsymbol{B}^{-1}$ and $\boldsymbol{A}^{-1}$ exist.

A repetitive application of the above rule gives

$$(\boldsymbol{A}\boldsymbol{B}\boldsymbol{C})^{-1} = (\boldsymbol{A}(\boldsymbol{B}\boldsymbol{C}))^{-1} = (\boldsymbol{B}\boldsymbol{C})^{-1}\boldsymbol{A}^{-1} = \boldsymbol{C}^{-1}\boldsymbol{B}^{-1}\boldsymbol{A}^{-1} \tag{A.28}$$

The inverse of a transpose is the transpose of the inverse

$$(\boldsymbol{Q}^{\top})^{-1} = (\boldsymbol{Q}^{-1})^{\top} \tag{A.29}$$

A.3.2 Inversion of Partitioned Matrices

Consider partitioning a square matrix $\boldsymbol{P}$ of order n into four submatrices as

$$\boldsymbol{P} = \begin{pmatrix} \boldsymbol{A} & \boldsymbol{B} \\ \boldsymbol{C} & \boldsymbol{D} \end{pmatrix} \tag{A.30}$$

where $\boldsymbol{A}$ and $\boldsymbol{D}$ are invertible, and $\boldsymbol{B}$ and $\boldsymbol{C}$ are of conformable dimension. We then have[2]

$$\boldsymbol{P}^{-1} = \begin{pmatrix} \boldsymbol{A}^{-1} + \boldsymbol{A}^{-1}\boldsymbol{B}\boldsymbol{X}\boldsymbol{C}\boldsymbol{A}^{-1} & -\boldsymbol{A}^{-1}\boldsymbol{B}\boldsymbol{X} \\ -\boldsymbol{X}\boldsymbol{C}\boldsymbol{A}^{-1} & \boldsymbol{X} \end{pmatrix} \tag{A.31}$$

where

$$\boldsymbol{X} = (\boldsymbol{D} - \boldsymbol{C}\boldsymbol{A}^{-1}\boldsymbol{B})^{-1} \tag{A.32}$$

---

[2] Henderson, Harold V., and Shayle R. Searle. 1981. On deriving the inverse of a sum of matrices. *SIAM Review* 23 (1): 53–60.

A.3.3 The Kronecker Product

If $\boldsymbol{A}$ is an $m \times n$ matrix and $\boldsymbol{B}$ is an $p \times q$, then the Kronecker product of $\boldsymbol{A}$ and $\boldsymbol{B}$, denoted by $\boldsymbol{A} \otimes \boldsymbol{B}$, is the $mp \times nq$ matrix

$$\boldsymbol{A} \otimes \boldsymbol{B} = \begin{pmatrix} a_{11}\boldsymbol{B} & a_{12}\boldsymbol{B} & \cdots & a_{1n}\boldsymbol{B} \\ a_{21}\boldsymbol{B} & a_{22}\boldsymbol{B} & \cdots & a_{2n}\boldsymbol{B} \\ \vdots & \vdots & \vdots & \vdots \\ a_{m1}\boldsymbol{B} & a_{m2}\boldsymbol{B} & \cdots & a_{mn}\boldsymbol{B} \end{pmatrix} \tag{A.33}$$

Unlike ordinary matrix multiplication, the Kronecker product $\boldsymbol{A} \otimes \boldsymbol{B}$ is defined regardless of the sizes of $\boldsymbol{A}$ and $\boldsymbol{B}$.

For $\boldsymbol{B} = \iota_2$, we obtain

$$\boldsymbol{A} \otimes \boldsymbol{B} = \begin{pmatrix} a_{11} & a_{12} & \cdots & a_{1n} \\ a_{11} & a_{12} & \cdots & a_{1n} \\ a_{21} & a_{22} & \cdots & a_{2n} \\ a_{21} & a_{22} & \cdots & a_{2n} \\ \vdots & \vdots & \vdots & \vdots \\ a_{m1} & a_{m2} & \cdots & a_{mn} \\ a_{m1} & a_{m2} & \cdots & a_{mn} \end{pmatrix} \tag{A.34}$$

Appendix B
Capital Goods, Capital Stock, and Capital Services

This section serves two purposes. Firstly, it presents the capital flow matrix for Japan (Table B.1), which is discussed in Sect. 4.4.2. Secondly, it provides a concise introduction to the measurement of capital stock, capital services, and the price of capital services in economics.

B.1 Capital Goods in the Japanese Capital Flow Table 2015

The following is an excerpt from the Compilation Manual of Japanese IO tables dealing with the definition of fixed capital formation (2015 Input-Output Table Basic Guidelines, MIC, 2017).

The components of fixed capital formation Fixed capital formation consists of the acquisition of fixed assets, such as construction, machinery, equipment, defense equipment, and intellectual property (including research and development and software). It includes associated direct costs, such as installation costs, freight margins, and brokerage fees. Fixed capital formation refers to assets produced from production processes and does not include non produced assets. Therefore, land is not included in fixed capital formation, but land development and improvement costs are included. In the case of asset retirement or disposal, the necessary costs of restoration are also included.

The range of capital goods The range of capital goods, defined as fixed assets, refers to those used repeatedly or continuously in production for more than one year. However, items such as hand tools that are cheaply and stably purchased are regarded as current transactions and are not included in fixed capital formation.

Maintenance and repair Regular asset maintenance and repairs are not considered capital formation. However, when extending the useful life of an asset, major repairs and alterations contingent on the asset are, in principle, accounted for in

S. Nakamura, *A Practical Guide to Industrial Ecology by Input-Output Analysis*,
https://doi.org/10.1007/978-3-031-43684-0

Table B.1 Capital goods in the Japanese capital flow table 2015

1	Fruit	62	Electronic application equipment
2	Other beverage crops	63	Electrical measuring instrument
3	Other dairy products	64	Electric lighting fixtures
4	Other livestock	65	Other electrical equipment
5	Rope/Net	66	Wired telecommunications equipment
6	Woven garment	67	Mobile phone
7	Knitted garment	68	Wireless telecommunications equipment (Excluding mobile phones)
8	Bedding	69	Radio and television receivers
9	Carpets and floor coverings	70	Other telecommunications equipment
10	Wood products not elsewhere classified	71	Video equipment/digital camera
11	Wooden furniture	72	Electroacoustic equipment
12	Metal furniture	73	Personal computer
13	Other furniture and equipment	74	Computer body (Excluding personal computers)
14	Nuclear fuel	75	Electronic computer accessory
15	Construction metal products	76	Passenger car
16	Gas/oil equipment/heating/cooking equipment	77	Trucks, buses and other automobiles
17	Metal containers and can manufacturing sheet metal products	78	Two-wheeled vehicle
18	Metal products not elsewhere classified	79	Steel ship
19	Boiler	80	Other vessels
20	Turbine	81	Marine internal combustion engine
21	Engines	82	Rail car
22	Pump/compressor	83	Aircraft
23	Material handling machine	84	Bicycle
24	Refrigerator/temperature and humidity adjustment device	85	Industrial transport vehicle
25	Power transmission	86	Transportation equipment not classified elsewhere
26	General-purpose machines not classified elsewhere	87	Exercise equipment
27	Agricultural machinery	88	Clock
28	Construction and mining equipment	89	Instrument
29	Textile machinery	90	Other manufactured products
30	Food machinery and equipment	91	Residential building (Wooden construction)
31	Wood working machinery	92	Residential construction (Nonwooden construction)
32	Pulp equipment/paper manufacturing machinery	93	Nonresidential building (Wooden construction)

(continued)

Table B.1 (continued)

33	Printing/bookbinding/paper converting machinery	94	Nonresidential construction (Nonwooden construction)
34	Packaging and packing machinery	95	Construction repair
35	Chemical machinery	96	Road-related public works
36	Casting equipment	97	Rivers, sewers, and other public works
37	Plastic processing machinery	98	Agriculture and forestry-related public works
38	Metal machine tools	99	Railway track construction
39	Metalworking machinery	100	Power facility construction
40	Machine tools	101	Telecommunications facility construction
41	Semiconductor manufacturing equipment	102	Other civil engineering construction
42	Mold	103	Wholesale
43	Vacuum equipment/vacuum equipment	104	Retail
44	Robot	105	Real estate brokerage and management business
45	Other production machines	106	Rail freight
46	Copier	107	Road freight transport (Excluding private transport)
47	Other office machines	108	Coastal/inland water freight transportation
48	Vending machine	109	Harbor transportation
49	Entertainment equipment	110	Domestic air freight Forwarding
50	Other service equipment	111	Freight forwarding
51	Measuring equipment	112	Warehouse
52	Medical equipment	113	Software industry
53	Optics/lens	114	Production of video/audio/text information (Excluding newspapers/publishing)
54	Weapon	115	Natural science research institution (National/public)
55	Generator equipment	116	Humanities/social science research institutions (National/public)
56	Electric motor	117	Natural science research institute (Nonprofit)
57	Transformer	118	Humanities and social sciences research institutes (Nonprofit)
58	Switching control device/distribution board	119	Natural science research institute
59	Other industrial electrical equipment	120	Humanities and social sciences research institutes
60	Consumer air conditioner	121	In-house R&D
61	Consumer electrical appliances (Excluding air conditioners)	122	Other business services

capital formation. Additionally, replacement work for railway and tramway tracks, power transmission and distribution facilities, signaling facilities, cable facilities for the telecommunications industry, and power transmission and distribution facilities for the electric power industry are recorded as capital formation.

Long-term production Assets with long-term production (long-term products) are held in inventory until the point at which the user is deemed to take ownership. For self-account (production of capital for own use), since the user has acquired the ownership, the amount of progress is recorded as capital formation, even if it is a work-in-progress. However, in the case of work-in-progress construction, the amount of construction progress is recorded in capital formation even if there is no transfer of ownership. For livestock used for draft, breeding, dairy, competition, wool, and other capital services, the increase in growth is included in capital formation, even if the animals are not adults. However, the increased growth of pre-sale livestock owned by producers specializing in breeding is recorded in inventory. Own accounts of plants providing capital services, such as fruit trees, mulberries, and tea trees, account for increased growth in capital formation.

Machine, construction, civil engineering, and shipbuilding The purchase of capital goods in production activities by any sector is recorded in the domestic gross fixed capital formation, except for the following four cases.

Machine built-in This refers to the case where a capital good becomes a part of a new and different machine.

Construction detour Capital goods, such as elevators and boilers, become a part of a building. Construction sectors use these capital goods as intermediate inputs.

Civil engineering bypass Civil engineering work is necessary to install capital goods, such as bridges and floodgates. Civil engineering sectors use these goods as intermediate inputs.

Shipbuilding detour In shipbuilding, capital goods such as boilers and communication equipment become a part of ships.

B.2 Capital Services: Quantity and Price

This section deals with the concept and measurement of the price and quantity of capital services used in economics.[3] While the topics to be discussed in this section may divert from the rest of this book, I found it necessary to discuss them due to the growing interest in the IE community in utilizing the KLEMS database (see Sect. 4.4.3.6). The KLEMS database is based on the concept of economics and refers

[3] This section mostly refers to Jorgenson, Dale W. 1974. The economic theory of replacement and depreciation. In *Econometrics and economic theory*, ed. W. Sellekaerts, 189–221. London: Palgrave Macmillan.

to five groups of inputs: Capital (K), Labor (L), Energy (E), Materials (M), and Services (S). Each of these groups is an aggregate of its components.[4]

B.2.1 The Basics of Depreciation and Replacement

We explore the depreciation and replacement of capital goods. The efficiency of a capital good, relative to its normal or designed productive capacity, declines over time due to aging and other factors. The efficiency of a capital good (the ratio of effective to the normal or designed productive capacity of the capital good) after t periods of use, say $\eta_i(t)$, will decline over time due to its aging (wear and tear, malfunctioning, material failure, etc.): $\eta_i(t) \geq \eta_i(t+r),\ r > 0$, with

$$\lim_{t \to \infty} \eta_i(t) = 0. \tag{B.1}$$

where $\eta_i(0) = 1$ for normalization: the efficiency is relative because of this normalization. It's worth noting that land is an exception to this decline in efficiency. The efficiency of land will not change over time.[5]

Henceforth in this section, we consider capital goods with declining efficiency over time. The sequence of coefficients $\eta_0, \eta_1, \ldots,$ is termed the efficiency distribution. The stock of capital good i at the end of t, $K_i(t)$, is the sum of past investments, $I_i(t), I_i(t-1), \ldots,$ each weighted by efficiency

$$K_i(t) = \sum_{r=0}^{\infty} \eta_i(t) I_i(t-r) \tag{B.2}$$

Note that we diverted from the notation rule of this book by choosing to denote capital stock by K, following the long tradition in economics, presumably dating back to "Das Kapital" of Karl Marx published in 1867. Implicit in this formulation is the assumption that capital goods remain homogeneous over time except for the decline in efficiency: the investment in capital good i is not distinguished by the time of investment: the vintage structure (the age composition of capital stock) does not matter: the stock of old capital goods with low efficiency would work as well as a small number of high efficiency one if they outperform the latter by their number.

Assume that for a given stock of capital good i existing at the beginning of t, a fraction proportional to the decline in its efficiency during t is decommissioned

---

[4] One of the earliest studies on "KLEMS production functions" is Nakamura, Shinichiro. 1984. *An inter-industry translog model of prices and technical change for the West German economy*. Springer.

[5] Christensen, L.R., and D.W. Jorgenson. 1969. The measurement of US real capital input, 1929–1967. *Review of Income and Wealth* 15 (4): 293–320.

by the end of t and needs to be replaced in $t + 1$. This fraction is referred to as the mortality rate, denoted by $m_i(t)$, and is calculated as follows

$$m_i(t) = -(\eta_i(t) - \eta_i(t-1)),\ t = 1, 2, \ldots \tag{B.3}$$

The sequence of coefficients $m_i(1), m_i(2), \ldots$ is known as the mortality distribution.

Since a capital good has a finite life, the sum of the mortality distribution over an infinite period is equal to 1

$$\sum_{t=1}^{\infty} m_i(t) = 1. \tag{B.4}$$

For a given mortality distribution, the replacement requirements of capital good i in t, denoted as $R_i(t)$, are determined by the fraction of stock decommissioned during t

$$R_i(t) = \sum_{r=1}^{\infty} m_i(r) I_i(t-r) \tag{B.5}$$

The geometric distribution is one of the most widely used efficiency distributions for measuring capital stock. It is defined as

$$\eta_i(t) = (1 - \delta_i)^t,\ t = 0, 1, \ldots \tag{B.6}$$

where δ_i is a constant. In this section, we assume (B.6), unless stated otherwise. The corresponding mortality distribution is also geometric, as expressed by

$$m_i(t) = \eta_i(t-1) - \eta_i(t) = (1-\delta_i)^{t-1} - (1-\delta_i)^t = \delta_i(1-\delta_i)^{t-1} \tag{B.7}$$

which implies that the replacement requirements at t, as given by (B.5), can be written as

$$R_i(t) = \sum_{r=1}^{\infty} \delta_i (1-\delta_i)^{r-1} I_i(t-r) \tag{B.8}$$

The capital stock at the end of t, denoted as $K_i(t)$, is calculated using (B.2) as

$$K_i(t) = \sum_{r=0}^{\infty} (1-\delta_i)^r I_i(t-r) \tag{B.9}$$

By taking the first difference of this equation, we obtain

$$K_i(t) - K_i(t-1) = I_i(t) - \sum_{r=1}^{\infty}(1-\delta_i)^r I_i(t-r) = I_i(t) - \delta_i K_i(t-1) \quad \text{(B.10)}$$

which further leads to

$$K_i(t) = I_i(t) + (1-\delta_i)K_i(t-1) \quad \text{(B.11)}$$

B.2.2 Capital Services: Price and Quantity

This section mostly refers to Christensen and Jorgenson (1969).[6]

The flow of capital services i in current prices is given by the product of the (imputed) rental price $p_i^K(t)$ and the amount of capital service utilized over time, $p_i^K(t)K_i(t)$. In order to measure real capital input, it is necessary to separate this product into its price and quantity components. For a capital asset with an active rental market, the price of capital services can be observed directly as the rental price for the use of the capital asset. However, in the majority of cases relevant to IE, capital assets do not have an active rental market, making this method inapplicable.

Christensen and Jorgenson (1969) proposed an alternative estimation method based on the assumed correspondence between asset prices and service or rental prices, which arises from the equality between an asset's value and the discounted value of its services. For a capital asset i, denote by $q_i^A(t)$ the asset price, $r(t)$ the rate of return, and δ_i the rate of replacement. The user cost (the imputed rental price) of capital of i at time t, $p_i^K(t)$, is then given by (for the sake of simplicity, the terms referring to taxes are omitted) (Christensen and Jorgenson (1969), p. 302)

$$p_i^K(t) = q_i^A(t)r(t) + q_i^A(t)\delta_i - \left(q_i^A(t) - q_i^A(t-1)\right) \quad \text{(B.12)}$$

The same rate of return is assumed to apply to all durable assets: $r(t)$ does not depend on i. It is important to note that in this expression, the rate of return $r(t)$ is the only variable that is not directly observable. The capital stock $K_i(t)$ is obtained from (B.11), and ${q_i}^A(t)$ and δ_i are directly observable. Therefore, the estimation of $r(t)$ is necessary, which we will discuss next.

Let there be m types of durable assets. We assume that the sum of the product of p_i^K and K_i over all assets used in a sector is equal to the amount of capital compensation (or corporate property income in the case of a corporate sector)

$$\sum_{i=1}^{m} p_i^K(t)K_i(t) = \text{capital compensation} \quad \text{(B.13)}$$

[6] Christensen, L.R., and D.W. Jorgenson. 1969. The measurement of US real capital input, 1929–1967. *Review of Income and Wealth* 15 (4), 293–320.

where p_i^K given by (B.12). It is important to note that "capital compensation" is directly observable, leaving only $r(t)$ as the unknown variable in this equation. Therefore, the equation can be solved to determine $r(t)$.

It is important to note that the value of capital services to the owner, also known as capital compensation or capital cost, is defined as the value of output less labor cost and intermediate consumption (Statistics Canada `https://www150.statcan.gc.ca`) or gross value added minus labor compensation.[7] The value of capital includes the corporate surplus and is not equivalent to the asset's depreciation for a particular year.

The next step involves separating the amount of capital compensation or the value of capital input into price and quantity components, denoted as $p^K(t)$ and $K(t)$, respectively. To achieve this, an accounting identity is employed, stating that the value of all capital services or property compensation in each sector is equal to the sum of the values of individual capital services (Christensen and Jorgenson (1969), p. 314)

$$p^K(t)K(t) = \sum_{i=1}^{m} p_i^K(t)K_i(t) \tag{B.14}$$

where $p^K(t)$ and $K(t)$ represent the Divisia index numbers of capital service price and quantity, respectively. The Divisia index $K(t)$ in its discrete form is given by

$$K(t)/K(t-1) = \prod_{i=1}^{m} (K_i(t)/K_i(t-1))^{(w_i(t)+w_i(t-1))/2} \tag{B.15}$$

where $w_i(t)$ is the share of asset i in the total capital cost at t

$$w_i(t) = p_i^K K_i(t) / \sum_{i=1}^{m} p_i^K(t)K_i(t) \tag{B.16}$$

which is available because the denominator is directly observable (capital compensation), $p_i^K(t)$ is obtained by the above mentioned method, and $K_i(t)$ by (B.11). By specifying a reference value $K(0)$, (B.15) gives a time series for $K(t)$.

The above formula assumes that the amount of capital service utilized at t, $K_i(t)$, is proportional to the amount of capital stock. If needed, adjustments can be made to accommodate changes in the level of utilization of capital stock, for example, based on the amount of power use. Once $K(t)$ is obtained, the aggregate price index of capital services $p^K(t)$ can be obtained from (B.14) by dividing the amount of capital compensation by $K(t)$.

---

[7] O'Mahony, M., and M.P. Timmer. 2009. Output, input and productivity measures at the industry level: The EU KLEMS database. *Economic Journal* 119(538).

B.2.3 The Divisia Index of Capital Services

This section mostly refers to Diewert, Erwin W. 1976. Exact and superlative index numbers. *Journal of Econometrics* 4 (2): 115–145. The readers unfamiliar with the economics literature may be puzzled by the sudden (out of the blue) occurrence of the functional form (B.15), called the Divisia index. To be exact, it is a discrete approximation to the Divisia index, which is continuous, by Törnqvist (and hence called the Törnqvist index). Notice that K is an aggregate of its components $K_i, i = 1, \ldots, m$, which cannot be obtained without a function to aggregate the components. One needs an aggregator function f to obtain K from $K_k, k = 1, \ldots, m$,

$$K = f(K_1, K_2, \ldots, K_m) \tag{B.17}$$

The choice of this functional form is not trivial because it represents the contribution of each capital asset i to production via aggregate capital K. For instance, if one chooses to add up all the elements

$$K = \sum_i K_i \tag{B.18}$$

it implies that the m assets are perfect substitutes for each other: a given amount of K can be attained by a single asset only without using any of the remaining $m - 1$ assets. If this were true, a photovoltaic (PV) plant could be built using "Electric cable" only without even using "Solar module." Accordingly, using a simple sum as the aggregator function, f, is inappropriate.

Since the functional form of f is generally unknown, one widely adopted procedure in economics is to approximate it by a second-order expansion of the logarithms of f, known as the translog function

$$\ln K = \sum_i \alpha_i \ln K_i + 1/2 \sum_{i,j} \beta_{ij} \ln K_i \ln K_j \tag{B.19}$$

Here, $\beta_{ij} = \beta_{ji}$ and the following restrictions

$$\sum_i \alpha_i = 1, \sum_i \beta_{ij} = 0 \tag{B.20}$$

These parameters restrictions ensure that (B.19) is homogeneous of degree one, meaning that a proportional increase (decrease) in all inputs leads to a proportional increase (decreases) in the output with the same proportion. For further details on the translog function within the context of input-output analysis, refer to Nakamura, S., and Y. Kondo. 2009. *Waste input-output analysis*. Springer, Chap. 4.

Due to its second-order nature, this translog function can accommodate a wide range of substitutability among the K_i's. Furthermore, Diewert (1976) demonstrated that (B.19) is "exact" to (B.15), meaning that the following relationship holds

$$\ln[K(t)/K(t-1)] = \sum_i 1/2\,[w_i(t) + w_i(t-1)]\ln[K_i(t)/K_i(t-1)] \quad \text{(B.21)}$$

which is the logarithmic form of (B.15). If one is solely interested in obtaining the time series of an aggregate, such as $K(t)$, $t = 0, 1, \ldots$, the knowledge of the individual parameters α's and β's is not necessary. The aggregate consistent with (B.19) can be obtained by simply calculating the discrete form of the Divisia index (B.15). The fact that the information on individual parameters αs and βs is not required to obtain the aggregate K results from the marginal (first order cost-minimization) conditions in economics

$$w_i = \frac{p_i K_i}{\sum_j p_j K_j} = \frac{\partial \ln f}{\partial \ln K_i} = \alpha_i + \sum_j \beta_{ij} \ln(K_j) \quad \text{(B.22)}$$

The nominal share of each asset, which is directly observable, reflects the information on the unknown parameters. Unless one is specifically interested in the values of individual parameters, they are not needed to obtain the aggregate K. For further details, refer to Diewert (1976).

For many questions of interest to IE, the aggregation of individual items of capital stock (or any input items), as shown above, may not be highly relevant. Instead of focusing on the capital aggregates, K, the information about individual capital stock items, K_k, becomes more important for questions relevant to IE, where details matter. This stands in sharp contrast to Computable General Equilibrium (CGE) models, which are mostly formulated in terms of highly aggregated variables (refer to Zhou, L., and Z. Chen. 2021. Are CGE models reliable for disaster impact analyses? *Economic Systems Research* 33 (1): 20–46, for a recent review of CGE models).

Appendix C
WIO-MFA: A Numerical Example

For illustration, we consider a simple numerical example using the input coefficients matrix given in Table C.1. In this example, we have four product sectors and two materials denoted as α and β. Three of the four product sectors produce physical products (a, b, and c), while the fourth sector is a service/energy sector. Product c is made of products a and b, while none of c is used in producing a and b: c is a final product, while a and b are intermediate products. In Table C.1, the columns corresponding to the sectors producing materials α and β have been removed because they are not used in the calculation of $\boldsymbol{C_{MP}}$.

It is important to note that products a, b, and c are measured in monetary units, whereas materials α and β are measured in physical units: Producing one monetary unit of product b requires 0.15 monetary units of product a, 0.02 monetary unis of product b, 0.2 monetary units of Service/energy, three physical units of material α, and two physical units of material β. We assume that the mass filter $\boldsymbol{\Phi}$ and the yield matrix $\boldsymbol{\Gamma}$ are respectively given by Tables C.2 and C.3. While one can put any number to the column and row elements of Service/energy, we have put zero.

S. Nakamura, *A Practical Guide to Industrial Ecology by Input-Output Analysis*,
https://doi.org/10.1007/978-3-031-43684-0

Table C.1 The A matrix: a numerical example

	Product *a*	Product *b*	Product *c*	Service/energy
Product *a*	0.02	0.15	0.2	0
Product *b*	0.2	0.02	0.3	0
Product *c*	0	0	0.01	0
Service/energy	0.2	0.2	0.25	0.3
Material α	2	3	0	0
Material β	4	2	0	0

Table C.2 The mass filter $\boldsymbol{\Phi}$

	Product *a*	Product *b*	Product *c*	Service/energy
Product *a*	1	1	1	0
Product *b*	1	1	1	0
Product *c*	1	1	1	0
Service/energy	0	0	0	0
Material α	1	1	1	0
Material β	1	1	1	0

Table C.3 The yield matrix $\boldsymbol{\Gamma}$

	Product *a*	Product *b*	Product *c*	Service/energy
Product *a*	0.95	0.9	0.9	0
Product *b*	0.9	0.95	0.9	0
Product *c*	0.95	0.95	0.95	0
Service/energy	0	0	0	0
Material α	0.8	0.7	0.7	0
Material β	0.75	0.8	0.8	0

From (7.32), the matrix $\boldsymbol{A}$ adjusted for mass and yield is given by

$$\widetilde{\boldsymbol{A}} = \boldsymbol{\Gamma} \odot \boldsymbol{\Phi} \odot \boldsymbol{A} = \begin{pmatrix} 0.019 & 0.135 & 0.18 & 0 \\ 0.18 & 0.019 & 0.27 & 0 \\ 0 & 0 & 0.0095 & 0 \\ 0 & 0 & 0 & 0 \\ \hline 1.6 & 2.1 & 0 & 0 \\ 3 & 1.6 & 0 & 0 \end{pmatrix} = \begin{pmatrix} \widetilde{\boldsymbol{A}}_{PP} \\ \widetilde{\boldsymbol{A}}^{*}_{MP} \end{pmatrix} \tag{C.1}$$

From (7.34) we then obtain the material composition matrix $\boldsymbol{C}_{MP}$ as

$$
\begin{aligned}
\boldsymbol{C_{MP}} &= \tilde{\boldsymbol{A}}^{*}_{\boldsymbol{MP}}(\boldsymbol{I} - \tilde{\boldsymbol{A}}_{\boldsymbol{PP}})^{-1} \\
&= \begin{pmatrix} 1.6 & 2.1 & 0 & 0 \\ 3.0 & 1.6 & 0 & 0 \end{pmatrix} (\boldsymbol{I} - \tilde{\boldsymbol{A}}_{\boldsymbol{PP}})^{-1} \\
&= \begin{pmatrix} 2.076 & 2.426 & 1.038 & 0 \\ 3.444 & 2.105 & 1.199 & 0 \end{pmatrix}
\end{aligned}
\tag{C.2}
$$

It follows that one monetary unit of product c consists of 1.038 physical units of material α and 1.199 physical units of material β. If these materials are the only ones used in the economy, then the mass of one monetary unit of product c will be 2.237 physical units.

Appendix D
The Weibull Distribution

This section refers to McCool, John. 2012. *Using the Weibull distribution.* Wiley. For a random variable $x > 0$, the Weibull cumulative distribution function F is given by

$$F(x) = 1 - \exp\left(-\left(\frac{x}{\eta}\right)^{\beta}\right) \tag{D.1}$$

where $\beta > 0$ is the shape parameter (the Weibull slope) and $\eta > 0$ is the scale parameter. The probability that the life of a device or subject will exceed a given value, say y, is

$$Prob[x > y] = \exp\left(-\left(\frac{y}{\eta}\right)^{\beta}\right) \tag{D.2}$$

while the probability that the life of a device or subject will not exceed y is

$$Prob[x < y] = 1 - \exp\left(-\left(\frac{y}{\eta}\right)^{\beta}\right) \tag{D.3}$$

The probability density function of the Weibull distribution, f, is

$$f(x) = \frac{dF(x)}{dx} = \frac{\beta}{\eta}\left(\frac{t}{\eta}\right)^{\beta-1} \exp\left(-\left(\frac{t}{\eta}\right)^{\beta}\right) \tag{D.4}$$

S. Nakamura, *A Practical Guide to Industrial Ecology by Input-Output Analysis*,
https://doi.org/10.1007/978-3-031-43684-0

The expected value of x is expressed as

$$E(x) = \int_0^{\infty} x dF(x) = \eta\Gamma\left(1 + \frac{1}{\beta}\right) \tag{D.5}$$

where Γ is the Gamma function

$$\Gamma(z) = \int_0^{\infty} t^{z-1} e^{-t} dt \tag{D.6}$$

Index

S. Nakamura, *A Practical Guide to Industrial Ecology by Input-Output Analysis*,
https://doi.org/10.1007/978-3-031-43684-0

O

P

R

S

X

Y

Z